£35.00
1553

Richard Breger.

Corporate Communications Management

Corporate Communications Management

The Renaissance Communicator in Information-Age Organizations

Diane Gayeski

Focal Press
Boston London

Focal Press is an imprint of Butterworth–Heinemann.

Library of Congress Cataloging-in-Publication Data

Gayeski, Diane M. (Diane Mary), 1953–
Corporate communications management : the Renaissance Communicator in information-age organizations / Diane Gayeski.
p. cm.
Includes bibliographical references.
ISBN 0-240-80139-3
1. Communication in management.
2. Communication in organizations. I. Title.
HD30.3.G38 1993
658.4′5—dc20 93-6592
CIP

British Library Cataloguing-in-Publication Data

A catalogue record for this book is available from the British Library.

Butterworth–Heinemann
80 Montvale Avenue
Stoneham, MA 02180

10 9 8 7 6 5 4 3 2 1

Printed in the Unites States of America

CONTENTS

Preface

Like most books, *Corporate Communications Management* is the product of several decades of thinking and doing. In my own case, I've moved from producing media to designing instructional materials and managing corporate communication projects to my current fascination with and work toward moving corporate communication up a notch in terms of its practices and image. I have described the process of writing this text to friends and colleagues as "backing up my mental hard disk," and it does feel good to get these ideas out of my head and onto paper.

Hundreds of colleagues, clients, current and former students, and other writers have played important roles in the development of the approaches presented. My husband and business partner, David V. Williams, has been an especially important influence; his discipline as a psychologist, his creativity as a Welshman, and his expressiveness as an extrovert have provided me with innumerable new ideas and challenged assumptions. He and our son, Evan, have been patient cohabitants, as well as enthusiastic supporters of "Mom's National Book Week."

If this book provides some unexpected and controversial views of professional communication, it is in large part due to the fact that David can talk and I can type. This combination of inclinations and talents has afforded us the opportunity to work, through our firm, OmniCom Associates, with clients from around the world on projects that have spanned new information technologies, the design and development of training programs, video and interactive media production, advertising and public relations campaigns, and communications management and strategy development. Through these projects and the support and encouragement of our clients, my skills have been polished and I have been able to see glimpses of the shortcomings, as well as the possibilities, that exist in our field. I also acknowledge and appreciate the loyal support of our project manager and colleague, Rob Gearhart, who has been a versatile member of our team and a good listener over the years, during many project meetings and brainstorming sessions.

Many clients and their projects are presented as models for the way communication and training *should* be practiced today. Some of these include: Ed Nathan, manager of Lederle International's field sales training department; Jay Zimmerman, Joan Mesic, Mary Dunn, David Lavoie, and Reid Brownscombe in the training, public affairs, and policies & procedures departments at the Bank of Montreal; Phyllis Wittner, manager of organizational development at New York State Electric and Gas; C.L. Price, manager of employee development and quality performance at Dow Chemical; Ron Brown, manager of the audiovisual department at Amway; and Roger Williams, president of Espar Products.

Another important influence has been David Berlo, one of the founders of the academic field of communications, and currently, in his own words, "a teacher in private practice." Berlo's 1960 text on communication; his more recent ideas about information overload, leadership, and standards for communication; and the examples of professional practice and scholarship that he offered to me have been invaluable sources of inspiration.

Writing this book gave me the opportunity to talk with many individuals who are making important contributions to the practice of formal communication within organizations. Scores of them are mentioned in the book's scenarios, and there are too many of them to list here although I want to again acknowledge their contributions. Several people, however, were especially significant in terms of the novelty of their approaches and the candor of their comments. Among them are: Ralph Meyer, president of Guthrie Healthcare Group; Carl Johnson, vice president of communications at New York State Electric and Gas; Roger D'Aprix of William Mercer, Inc.; Ron Martin, executive vice president of communications at American Express; Stephen Bean, assistant vice president of information products at NCR; Les Landes of Landes Communications; Mitch Curran, "the PR queen" at Ben & Jerry's Homemade; and Larry McMahon, vice president of human resources at Federal Express.

Last, but not least, Ithaca College has provided me with an excellent undergraduate preparation in communications, talented and curious students, freedom to develop new curricula, opportunities to observe communication cultures and practices in operation, and a 1-year sabbatical to write this book.

Never before has communication been such a vital process within organizations. Further, communication practitioners have never been so challenged, by sponsors of public relations, training, and information systems projects to demonstrate how these interventions play a strategic part in achieving their organizations' objectives and in reflecting their values. It is my hope that you will be inspired by the visions and guided by the models and concepts presented, because exciting possibilities for professional growth and organizational effectiveness await us.

1 What Is Corporate Communications?

Communication is a topic often discussed but rarely understood in contemporary corporate circles. Most people realize that good communication skills are essential in today's business world and that new technologies are enabling more rapid, sophisticated, and varied forms of information exchange. But what exactly do we mean by the word *communication*—especially as it's applied to organizations? Talking? Holding meetings? Publishing newsletters and annual reports? Developing and using media? Managing press relations? Training new employees? On-line exchange of data among computers? Or the new telephone system?

Currently, there's no right answer, and the closest we can come to a response is "all of the above." As this book explains, the practice of corporate communication is rapidly developing and changing as new organizational practices and technologies emerge. Although there are no official definitions, procedures, or practices for corporate communication, there are a number of concepts and models that can help professional communicators to understand the different contexts and manage the activities in this dynamic field during the last few decades or what has been called the *information age.*

DEFINING THE FIELD

Although everybody in an organization communicates, some individuals are directly responsible for formal communication policies and practices; these individuals are sometimes referred to as *professional communicators* or *corporate communication practitioners.* But what does such a person actually do in today's business environment? Yesterday's concepts of a communicator as a speech maker, a writer, or an audiovisual technician cannot adequately encompass the wide range of technologies and applications that are currently involved in corporate communication.

Consider the myriad of formal communication systems and practices that exist in contemporary enterprises. For example, who proposes and manages the establishment of satellite teleconferencing, who chooses desktop publishing systems, who acts as official company spokespersons to the press, who develops training courses, who decides which persons get electronic mail (e-mail) accounts, who facilitates meetings on total quality management, who designs brochures, who orchestrates the annual sales meeting, who conducts employee attitude surveys, and who produces the print and video components of the annual report? These are only a few of the more visible and commonplace responsibilities of professional communicators within organizations.

The communications field is larger and more varied than the traditional view of it. Most people think of corporate communications as primarily an in-house writing function; however, in today's information-age organizations, the technologies and techniques of internal and external communication are complex and their deployment is crucial to an organization's ultimate success. Professional communicators may design messages and campaigns, develop public relations techniques, manage information technologies, assist with organizational development and change, or produce media programs.

If you pick up any business magazine, you'll read stories (such as the one from *USAir Magazine* shown in Figure 1.1) about how enterprises are using new interactive computer technologies to train their employees, how they provide support to and solicit feedback from their customers, how their executives try to influence workers to buy into their new visions, and how they use information systems to collect and to process data quickly. Communication is no longer a sideline; in fact, formal communication systems are an organization's most sensitive and powerful tools for attracting and informing employees, customers, and investors. Professional communicators do more than just write and produce media; they are, or should be, analysts, managers, and members of their organizations' strategic planning teams. Some of the major themes we'll consider throughout this book are

- Who manages corporate communication projects and systems?
- What are the theories, techniques, and technologies behind them?
- How can communication professionals become increasingly valuable members of their organizations?

We need a definition of our field to guide us. *Corporate communication* is the professional practice of developing and implementing communication rules and tools in order to enhance the dissemination, comprehension, acceptance, and application of information in ways that will help to achieve an organization's goals. We'll also explore the ways in which communication professionals engage in their practice and the ways in which their roles are evolving.

Before we can more fully answer the question of "what is corporate communication?", we need to consider the terms *corporate* and *communication.* Although many people refer to our field as corporate communications or business communication, practitioners don't necessarily work only in for-profit firms. Hospitals, universities, government agencies, charitable foundations, and religious organizations are all served by professional communicators, either as employees or as independent agencies or consultants.

The importance of communication functions to organizations is pointed out by this definition: an *organization* is a group of people who are engaged in some joint activity. Since people can't work together unless they communicate, communication is the primary activity of an organization. It's what all organizations do. Organizations process information to reduce uncertainty and to coordinate actions in order to achieve their goals. The more that tasks vary, that employees are interdependent, and that the environment is uncertain, the more that frequent and effective communication is necessary.

FIGURE 1.1
Interviewing, training, testing, teamwork, customer relations, and the use of high-tech simulators are crucial aspects of running a modern organization, and each is an important component of the practice of formal communication. [From Schofield, S.E. (1992, April). Commentary: Highly trained professionals. *USAir Magazine,* 9. Reprinted with permission from *USAir Magazine.*]

COMMENTARY

Highly Trained Professionals

You've probably heard airlines describe their employees, especially the pilots and flight attendants responsible for your safety and security, as being "highly trained professionals" or "experienced and dedicated." Airlines like to tout that their employees are the best in the business.

In our case, I know it's true. Our pilots and flight attendants are highly trained, experienced, and dedicated professionals. Let me explain.

It starts with USAir's high standards for hiring. For pilots, USAir requires 2,000 hours of flight time, which means they already have years of flying experience. After initial screening, pilot applicants undergo a stringent medical exam and take psychological and aptitude tests. The next hurdle is a test in a flight simulator to evaluate the applicant's flying skills.

USAir requires that flight-attendant applicants demonstrate a high degree of professionalism and responsibility, and a friendly and caring personality. Typically, USAir interviews about 100 applicants for every three flight attendants hired.

As for training, both new pilots and pilots changing aircraft types undergo the same training. All pilot-training programs are approved by the Federal Aviation Administration (FAA), and the USAir programs are well in excess of FAA requirements. Initial training includes 10 days of ground school and an average of 10 days of flight-simulator training.

In my opinion, the most important advancement in pilot training was the development of the flight simulator. Flight simulators, which realistically simulate flight, allow pilots to train for the expected and the unexpected. Many levels of emergency training can be performed in the simulator—much more than in the actual aircraft. Every kind of abnormal situation, including wind shear, lightning, and aborted takeoffs, can be simulated. With this year's acquisition of a Boeing 757 simulator, USAir will have 14 flight simulators that are literally flown around-the-clock to train our pilots.

Before pilots receive an aircraft type rating, they must demonstrate to an FAA examiner or an FAA-designated examiner that they can perform various types of takeoffs and landings, inflight maneuvers, and emergency procedures.

For recurrent training, captains return to their training base twice a year; first officers come in once a year. Pilots are also checked once a year during actual flight by a check pilot. There is an annual physical examination. Home-study materials, which include written tests, are regularly distributed to all USAir pilots.

There has been increased attention to "cockpit resource management," or improving communication and teamwork among flight-crew members. USAir has integrated this focus into all levels of pilot training, including videotaping simulator sessions so pilots can evaluate how well they worked together. Communication between pilots and flight attendants is also an integral part of training.

Initial flight-attendant training is an extensive five-week course with classroom training and hands-on training in aircraft mock-ups that simulate operating emergency exits. Safety training includes simulated emergencies such as preparing the cabin for evacuation, evacuating the cabin, and fighting a live fire. All newly-hired flight attendants are CPR and first-aid certified. In annual recurrent training, flight attendants must pass proficiency drills and written exams, and are taught new safety methods and regulations.

Flight-attendant training incorporates review and simulation of actual incidents. Use of videotapes of USAir flight attendants talking about what happened, telling how their training and procedures helped them respond, and explaining exactly what being in an emergency situation is like has been well received.

USAir's flight attendants also participate in the two-day Putting People First seminars that foster teamwork among employee groups and promote a greater awareness of how to meet customers' needs.

As for experience, on average, our pilots have been flying with us for 10 years, and our flight attendants have been responsible for your safety and security for eight years.

USAir begins with you refers to our commitment to provide our customers quality service. Inside the family, "you" means our highly trained, experienced, and dedicated professionals. I appreciate their contribution.

Seth E. Schofield
President and
Chief Executive Officer

Each month Seth E. Schofield discusses issues of concern to USAir and its customers. If you have comments or would like to suggest a topic, please write to Commentary, USAir Magazine, H850, 2345 Crystal Drive, Arlington, VA 22227.

APRIL 1992 USAIR MAGAZINE 9

Communication is the use of symbol systems to share meaning and information. However, when we say corporate communication, we mean more than merely talking or writing memos at work. We generally mean formal, official systems or practices for communication. For example, corporate communication deals with official company announcements, public meetings, policies for communication, training programs, and information processing and distribution systems. Following is a list of some of the reasons that organizations communicate:

- to motivate employees
- to stimulate sales
- to garner public support
- to draw investors
- to teach employees new skills
- to aid in internal decision making
- to coordinate internal activities
- to lobby for favorable legislation
- to attract qualified employees
- to comply with regulations
- to promote their general industry
- to respond to mass media inquiries
- to fulfill their roles as good community citizens
- to maintain a supportive internal social system
- to elicit suggestions for improvement

Ackoff (1957), a management theorist, has identified three major types of communication within organizations:

- *environmental*—messages that describe the actions or events in the environment that change the probabilities of choice
- *motivational*—messages that assert goals and change the value of the outcomes
- *instructional*—messages that indicate particular approaches and change the efficiencies of the courses of action of an organization

Organizations use a variety of tools to provide environmental, motivational, and instructional messages to internal and external audiences. Figure 1.2 shows an example of one increasingly popular new technology—interactive media—which aids in presenting a corporation's capabilities.

FIGURE 1.2 Organizations, such as Reliance Electric, use tools like interactive video for internal employee training and external customer information. (Courtesy of AVID Communications.)

A variety of sources are telling us that effective communication is central to an organization's success in today's social and economic environment. Starting with the influential book, *The Coming of Post-Industrial Society* (Bell, 1973), and the book, *The Third Wave* (Toffler, 1980), terms such as *information age, corporate culture,* and *high tech–high touch* have crept into our language. Indeed, some theorists like Weick (1969) say that organizations don't exist, they evolve, and the mechanism for that evolution is communication. These notions, however au courant, are not new.

Chester Barnard, the president of New Jersey Bell Telephone in the 1930s, wrote an influential book, *The Functions of the Executive,* in which he argued that the primary function of an executive is to establish and maintain a system of communication. Indeed, one vice president for public relations who was interviewed characterized his CEO (chief executive officer) as the principal communication expert in his organization. Machlup (1962), an economist, first measured the percentage of the U.S. economy associated with the production and distribution of information; even in 1958, it accounted for 29% of the gross national product and 31% of the labor force. Needless to say, this segment of the economy has continued to grow exponentially.

Whether it's communicating with the customer, gathering up-to-the-second data on financial transactions, motivating the sales force, or training new recruits, information transfer is the key to staying profitable, even alive. Goldhaber, a well-known organizational communication specialist, claims that more than 10% of U.S. businesses fail each year primarily because of communication problems. In one of his studies, he found that employees generally receive insufficient or untimely messages about their jobs and organizations, and that there are usually no effective mechanisms for two-way dialogue between management and line workers (Goldhaber, 1983). As organizations are faced with stiffer competition (both domestically and from abroad), stricter regulations, and narrower profit margins, often the only area that can significantly affect the bottom line is communication.

Changes in our economy and social structures are creating the need for newer and more effective communication systems. It's not just that new communication tools, such as e-mail, interactive multimedia, and satellite teleconferencing, are available; it's that new policies, systems, and tools are mandatory. As you can see from Table 1.1, the significant differences between industrial-age and information-age organizations dictate new approaches to formal and informal communication practices.

Poor communication can lead to disasters like the Three Mile Island (TMI) nuclear power plant accident. The Presidential Commission's investigation report found that the Nuclear Regulatory Commission (NRC), Metropolitan Edison Company, Babcock and Wilcox (designer and supplier of the system), and General Public Utilities Corporation (the holding company) were all lacking in information and in effective communication practices. Here are some of the Presidential Commission's findings and biting criticisms (Meltzer, 1981):

- The response to the emergency was dominated by total confusion.
- A lack of communication existed at all levels.
- Recommendations were made by individuals who did not have accurate and up-to-date information.
- Little evidence of the impact of modern information technology appeared in the control room.
- Information was not presented in a clear form.
- There had not been enough exchange of information within the NRC.
- There was a serious lack of communication about critical safety matters within the companies involved in building and operating TMI.

TABLE 1.1
Contrasting Industrial-Age and Information-Age Organizations

Industrial Age	Information Age
Management by a few at the top	Self-managed employees
Stable economy and products	Highly unstable economy, products, and services
Competitiveness based on price and product features	Competitiveness based on information and service packaged with the product
Production machinery is most valuable asset	Information is most valuable asset
Homogeneous work force	Diverse work force
Workers expected to obey	Workers want explanations and a voice in decisions
Highly specialized workers	Broadly educated workers
Assembly-line production of routine products	Collaborative production of unique products
Work force trained once at school or on the job	Continual learning mandatory
Centralized	Dispersed in time and space
Human communication (talking) was sufficient	Mediated communication systems are required
Costs of communication failures or inefficiency relatively low	Costs of communication failures or inefficiency extremely high

Similar communication problems were reportedly a major cause of the space shuttle Challenger's fatal explosion. Poor communication can get an organization into a lot of trouble and, quite literally, lead to disaster. Due to the rapid changes in technologies and techniques for communication, not only must professionals in this area be fluent with the basics of current systems, but they must also be able to envision and implement new strategies that will fit into their organizations' cultures and that will comply with their goals. Too often, new technologies are introduced without the knowledge of how they will affect the human communication system. Or, opportunities are passed up because employees aren't armed with the most effective means for gathering and disseminating data.

Technology is changing the way we communicate and is fuzzing the lines that define departmental territories. Mrs. Fields Cookies, for example, equips each of the more than 600 cookie stores with an inexpensive personal computer linked to headquarters in Utah. These systems plan production, maintain inventory, enable communication with top managers, monitor progress toward goals, deliver employee training, and organize accounting. They include procedures for developing daily sales projections and monitoring hourly sales activities; for example, the system might direct a manager to stand outside a store and give away samples when sales are below expectations. Guidance systems for common problems, such as dealing with a jammed cash register, are also included in troubleshooting modules. In this dispersed and highly successful organization, there are only about 130 people at headquarters—this efficiency being due, in large part, to the effective application of communication technologies (Handy, 1989; Alter, 1992).

A large chemical company is investigating the possibility of providing a laptop computer to each sales representative to help the reps automate their work. The initial purpose of this technology would be to contain a data base of current and potential customers and to provide the means to track leads and sales; however, a competent professional communicator would look at this opportunity more broadly.

- From an employee communications standpoint, how could the laptops provide instant newsletters to sales reps around the world by downloading information through modems?
- From a human resources development standpoint, how can the laptops be used to train reps in the field by means of computer-based training?
- From a marketing standpoint, how can the computers be used to create customized slide shows for sales calls?
- From a telecommunications standpoint, how can the laptops be used for e-mail, to reduce phone and mail charges?
- From an organizational communication standpoint, how will this new technology be accepted by the sales force, and might it reduce face-to-face meetings?
- From a media production department standpoint, will these computers reduce the need for hard-copy charts and slides for marketing and training?
- From a line management standpoint, how can the timely and comprehensive data collected by the computers be better used to manage and to evaluate sales reps?

As organizations grapple with the complexities of automation and communication technologies, the most serious problem is that few people seem to have the big picture. Indeed, in a company as large as an international chemical distributor, many people would be involved in deliberating the questions just posed. The problem is that those questions seldom get raised. The sales force automation project is often assigned to one department or individual with little input from other areas that will be affected by it. Usually, that department or individual approaches the problem from a technical perspective and doesn't have a handle on the managerial and social implications of technology. Or, often, a potentially good solution gets caught in turf wars between departments who each want to have a stake.

I have seen a number of companies who would love to use computers for training but can't afford to purchase them for their field offices; however, these field offices already have computers in some departments, but claim that they were bought by a different department and can't be used for training. In other cases, technologies are put into place without any regard for future or concurrent uses; for example, expensive videodisc systems are purchased without a thought as to how they can be updated when the information changes. Sometimes companies have communication systems installed without considering the physical and social environment. For instance, hundreds of video playback units were purchased for each branch of a bank system to be used to deliver new product information to the tellers and platform representatives. However, most branch banks have no training or conference area and the units, in many cases, ended up being placed in the small staff lounges used for lunch breaks. One can imagine what kind of programming is actually being played on these machines, given the employees' mind-sets while in this space.

Unfortunately, people have been led to believe that humankind falls into two categories, the *techie* and the *people person*. The techie is good with cables and DOS commands, and the people person is charming and can make good jokes during an after-dinner speech. However, as the media continues to intervene in the communication processes in organizations, an even greater sensitivity to the human side of the matter is necessary. For example, Zuboff (1988) cites a number of instances in which computer-based automation has drastically changed the nature of work, management, and communication within several companies. Many negative outcomes in terms of morale and actual productivity were not even imagined by the decision makers; thus, communicators can be more than icing on the cake of business. Their skills and knowledge can mean the difference between bottom-line success and failure.

STORYTELLERS, SPEECH MAKERS, AND ELECTRONIC JESTERS

Society has always needed the professional communicator. Storytellers, priests, medicine men, teachers, and jesters all have had similar roles—to maintain and enrich the culture by preserving its history and disseminating its news in powerful ways that could be understood and felt by all members of the society. For example, through chants and songs, people could memorize important facts about their tribe's history; through rites and rituals, people could define their own culture and learn its mores; through religious ceremonies, individuals of all ages and classes could find a common identity and were taught objectives for good living; and through the buffoonery of court jesters, people could learn important lessons while laughing.

There was no school for jesters or wise men, and no books. In fact, a field of study in communications did not even exist until the 1920s, when the first speech training for executives was begun (notably, with Dale Carnegie whose influence persists today), and later in the 1950s, when the first departments of communications within universities were established.

Organizations and their communication systems have undergone rapid transformations in the last century. Until the early 1900s, there were virtually no large organizations; most people worked on their family farms or within small production shops. As the industrial age dawned and mass-production techniques spawned the growth of much larger companies, the only models of management and communication systems that existed were the military and the church, both of which were extremely patriarchal and formal systems that were built on the assumptions of strict hierarchical rank and obedience.

It shouldn't be surprising, then, that communication systems in these early organizations were primarily mechanisms for giving orders and keeping financial records. Most managers also owned their firms and could obtain and impart information by just walking around. The first organizations to need more sophisticated and mediated forms of communication were the railroads, which had stockholders, dispersed offices, and a need for explicit rules and clear communication to avoid accidents. The railroads began what may have been two of the first types of media-based internal communication, circular letters or memos, which were sent to various posts to make announcements, and little books of rules (Yates, 1989).

As all kinds of organizations began to realize the value and the complexity of communication, research and, eventually, programs of study were developed. Since the late 1940s, scholars in various traditional disciplines have been working in the area of communication. In 1949, the National Society for the Study of Communication was formed (now the International Communication Association). In the 1940s and 1950s, the primary focus of organizational communication was on effective managerial communication; how to influence, inspire, sell, and lead. World War II spawned an increase in research on propaganda, persuasion, and efficient training techniques that were later applied to advertising, marketing, education, and training.

In the 1960s and 1970s, courses and academic majors in speech, journalism, and mass media developed. Although many individuals with this kind of preparation ended up practicing their professions within organizations, there were relatively few, if any, specific courses in organizational communication. The field we now are defining as corporate communications emerges from the following areas:

- *organizational communications*—the study of human communication in organizations
- *educational technology*—the application of media and design systems to instruction
- *organizational development*—the improvement of organizations through planned change and the professional development of individual members
- *telecommunications*—the use of telephone and computer systems
- *technical writing*—the application of writing and layout techniques for documentation and instruction
- *information science*—the use of cataloging and retrieval systems to archive and access information
- *public relations*—the use of communication techniques to build a positive public image
- *mass media*—the use of audiovisual technologies to create, store, and disseminate messages

The professional associations in the field give us a clue as to the current size of the activities that exist within the broad area of corporate communications. For example, the IABC (International Association of Business Communicators) has more than 11,500 members worldwide involved in public relations and employee communication; the ASTD (American Society for Training and Development) boasts a membership of more than 55,000 professionals involved in human resources development and the provision of employee training programs; the ITVA (International Television Association) consists of almost 10,000 corporate video professionals; the ICA (International Communication Association) serves scholars and professors in communication; this association must not be confused with the other ICA (International Communications Association) whose members manage large corporate telecommunications services, such as telephone and data systems.

While outsiders may think that business communication is a step-child of broadcast media or education in terms of budgets, consider that U.S. organizations spent $43 billion on training in 1990 alone (Gordon, 1991), and many in-house corporate video departments manage budgets of more than $1 million. IABC's 1989 survey found that their typical member earns about $47,000 and manages a budget of $250,000. This many people and this much money obviously make up a large, and continually increasing, portion of our economy, and for good reason.

As businesses get larger and are more dispersed, it's more difficult for top management to communicate with the individuals on the line. Forty years ago, most company presidents were in more or less daily contact with their staffs, although some popular books today recommend management by walking around, which is often not physically possible. Someone else or something else has to be the mediator of the messages; that mediator may be a teleconferencing system, a company newsletter, or the director of corporate communications.

Today, organizations need to create and to maintain their corporate cultures. We might look over our shoulders at the past to give us some perspective on how this can be done. David V. Williams, a psychologist and partner in OmniCom Associates, created an interesting appellation for today's corporate communications specialist, the *electronic jester*. Before you become offended by any negative connotations of this term, let's consider for a moment the role of court jesters in history.

These people traveled from court to court, mixing with everyone from peasants to kings and popes. They were welcomed at almost any event because they provided entertainment while poking fun at current events or personalities. Through multimedia communication, they were able to present their messages so that everyone could understand and appreciate them. In addition, they served as an important conduit of information from the top of the organization (kings and popes) to the bottom of the organization (the peasants) and back again. By poking fun at the noblemen in front of the rest of the court, they made the high and mighty seem a little more human. Similarly, important and sensitive information could be presented to the royalty in a humorous fashion, and things could be said by jesters tha could never be suggested in another way. These jesters and buffoons were a lot more than weird or pitiful individuals. "They triumphed among princes while learned poets, eloquent orators, and subtle philosophers languished in obscurity" (Welsford, 1966, p 15). We still need such go-betweens today.

Many communication professionals say that they love their jobs because they're the one person in their company who's on a first-name basis with everybody, from the CEO to the janitor. That person, like the jester of old, may communicate important and sensitive messages through a combination of tricks—perhaps a funny video program where the CEO shows he's really human, a computer bulletin board that allows customers to anonymously leave messages and questions for each other, a suggestion box system to increase employee input, or the use of facilitation techniques to mediate a dispute between warring departments.

Hallmark Cards in Kansas City, Missouri, has added an unusual new position to its ranks—the creative paradox. The person who holds this job, Gordon MacKenzie, is one of the few creativity consultants in the United States whose job is to "subvert corporate stultification from the inside." He encourages employees to grow and to risk by holding workshops and brainstorming sessions that bring out the creativity he believes is inborn in everyone. Wearing a T-shirt and jeans, he can often be seen skipping down the halls of Hallmark on his way to a meeting. Says MacKenzie. "Large organizations are like giant hairballs. Every decision adds another hair. There is existence but no life in a hairball. You have to expend creative energy to avoid getting all tangled up" (Gerstner, 1991).

The nature of corporate communication has changed as dramatically as the channels we now use. While early corporate communication began as issuing orders, more humanistic approaches to management fostered the development of internal newsletters, union meetings, and committees. In today's business environments, the need for rapid exchange of information, participatory decision making, and ongoing training have not only made it mandatory that systems for two-way exchange of messages exist, but also put communication at the center of effective management practices.

THE ROLE OF PROFESSIONAL COMMUNICATORS

The range of activities, organizations, styles of work, and backgrounds of corporate communicators is immensely varied. Some people in the field have no specific educational background but are talented artists, producers, or writers who make effective creative contributions to publications, speeches, presentation graphics, or audiovisual programs. Others hold degrees in a subspeciality of the field, such as public relations, media production, technical writing, computer science, training and development, or speech. These individuals are primarily engaged in designing and managing projects and systems, such as developing communication campaigns, planning marketing strategies, installing and maintaining communication technologies, designing training programs, or producing documentation and performance support systems. Finally, a significant percentage of practitioners hold advanced degrees in such areas as instructional design, organizational communication, or information science. They may also be involved in managing communication departments, conducting analyses of internal and external communication, developing communication strategies, and researching and implementing new technologies.

Obviously, many corporate communications professionals work in *corporate communications* departments. Typically, these groups design and produce newsletters and annual reports; handle employee communications, such as orientation programs and information on benefits and corporate decisions; and sometimes write speeches and coordinate large meetings. However, formal communication activities in most organizations are much more extensive, and technologies and practices are rapidly blurring the distinctions between traditional roles and departments.

For example, an ad in *The New York Times* in September 1988 described the ideal candidate for a large organization's vice president of human resources as a goal-driven strategic thinker who was to report directly to the vice president (VP) and the COO, and to provide expertise in the areas of manager selection and development, training, communications, and professional selection and placement. The company sought a person who was able to negotiate win-win resolution of conflicts and who had exposure to formalized interview and organizational development training.

Communication professionals can be found in the departments of documentation, public affairs, employee communications, information technology, policies and procedures, library and information systems, corporate media, training and development, or advertising and marketing. As information exchange has become more important and specialized during the past few decades, these areas have developed into separate islands within many organizations (Figure 1.3). However, as can be seen later, this trend of dispersion is, and should be, reversed in many cases.

Most people think of communication specialists as being employed by big business, or perhaps by public relations agencies or media production houses. Although the majority of corporate communications practitioners do work within such organizations, they also work in almost any kind of organization that has a need to develop and manage communication systems and programs for its employees or its customers. Many small companies or not-for-profit agencies don't believe that they have enough work to do to hire even one person whose sole responsibility is organizational communication. Most people don't think much of communication at all. The common notion is that everybody does it, so there's nothing special to be known. However, most organizations that don't think they're big enough to support separate departments or even individual staff positions in employee communications, training, marketing, and public relations could easily support one person or a department with a broad range of responsibilities relating to communication. The small and growing enterprise really needs to pay attention to these issues in order to survive.

Typically, communication roles are more specialized in large organizations where different departments share the responsibility for the various types of communication systems and projects that take place. For instance, in most large businesses, people who design and provide training don't generally get involved in the annual report or in marketing materials. Jobs exist in which an individual can just write, produce media, design instructional programs, conduct audience research, or manage technology systems.

Smaller organizations offer opportunities to do a little of each. A communication generalist for a small not-for-profit agency, for example, could handle volunteer recruitment, training for board members and volunteers, fund raising, and press relations. Someone with similar broad skills in a small sales-oriented firm could oversee the communication components of several areas, such as advertising and trade shows, telephone sales and support, logo and brochure design, and sales training. A media-oriented individual might work at a small college to design and produce teaching materials, create posters for extracurricular events, and design and coordinate catalogs, admissions materials, and recruiting programs.

Many small organizations don't hire communication specialists because they

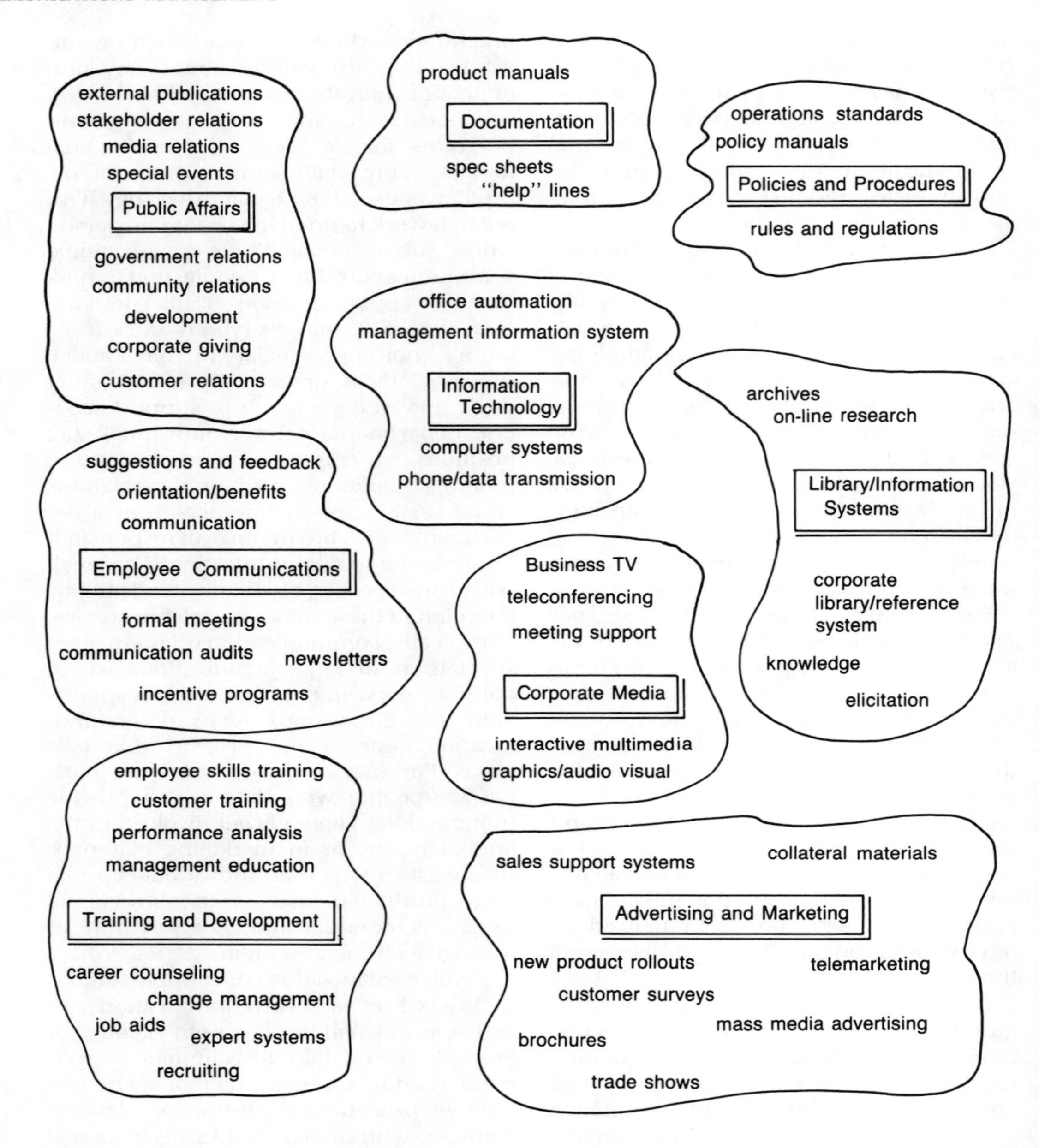

FIGURE 1.3
The islands of corporate communication.

don't know what to call them or where to put them in the organization. There are neat slots for receptionists, accountants, sales managers, and CEOs, but not for communication people. Some organizations try to hire them, but don't know how to develop a job description how to find such people.

A medium-sized manufacturing company saw the need for a media person to create videos for sales and training. So, the company advertised the job as being a camera operator position, envisioned hiring a TV/Radio major straight out of college, and set the salary quite low. Management was confused when, after several weeks of running the ad, they had no acceptable candidates. After talking with them, I determined that they didn't need someone merely to point and shoot a camcorder. What they wanted was someone who could design and produce programs, and more important, decide what kinds of programs to do. Even more broadly, they needed a person to create marketing and training content and strategies. They realized that their company had grown rapidly, and that little or nothing had been done to develop job descriptions, to write down company operating procedures, or to document equipment operation. Even before any videos could be done, the company had to decide who they were training and what the content would be! I convinced them to hire an instructional designer who spent the major part of her first 2 years doing human resources development work, creating policy manuals, job descriptions, and equipment documentation while managing to produce several short marketing videos.

Of course, there's an important role for effective generalists in large, complex organizations as well. While these companies will probably continue to need staffs of specialists, the ranks of upper management aren't complete without someone to pay attention to the communication implications and opportunities throughout the organization. According to surveys conducted by the IABC, an increasing number of individuals hold the title Vice President for Communications, and this trend will continue as the profession matures.

Current and Emerging Jobs

As organizations realize the power of information, idea-sharing, and culture-building, more and different jobs will be created for communicators. In particular, the human resources departments of companies will be challenged to describe these jobs and to recruit or train individuals to fill them. CEOs will need to create visions of their organizations that include the ways in which new and old communication technologies help them to meet their goals. Universities may be called on to define the conceptual territory of organizational communication and to prepare young professionals to command it. Here are some of the job titles prevalent in today's workplace:

- manager of corporate communications
- employee information specialist
- newsletter editor
- training developer
- recruiter/employment interviewer
- meeting facilitator
- media producer
- director of information systems
- audiovisual supervisor
- public information officer
- director of advertising
- graphic artist
- presentation support specialist
- counselor
- manager of total quality program
- workshop leader
- manager of publications
- technical writer
- electronic performance support developer
- performance technologist
- information systems manager
- instructional technologist

> Our future is very bright and full of opportunities—but we'll have to recognize those opportunities and we must do our homework. We'll start being invited to participate in those meetings on Mahogany Row, instead of merely videotaping them. What's the catalyst responsible for this? You guessed it: Technology. The fact that the technical folks can change their reputations and take their place in the mainstream of business operations through technology is a paradox. . . . What makes us so valuable is our ability to design, develop, and deliver products that solve communications objectives. What kinds of meetings should take place in the videoconferencing rooms? What kind of information would be distributed by the new VCR network? How can interactive video help in the operations of my employer? This is the bottom line of technology.
> (Schwartz, 1988, p. 94)

Many practitioners created their own positions by convincing management of the benefits of the activities in which they are capable of engaging. Although these people had to do much more than just answer an ad, they did often get to write their own job descriptions. Many people also start in one narrowly defined role and gradually expand their roles and increase their salaries by seeking new responsibilities and areas of opportunity for the company. Often, new jobs or expanded roles are established when new facilities, such as the teleconferencing system shown in Figure 1.4, are installed.

FIGURE 1.4
Meetings still occur around the conference table. The only difference is that participants may sit at tables separated by thousands of miles, such as this one at AT&T, thanks to teleconferencing and other such technologies. (Courtesy of Lake Systems, Inc.)

In 1991, Charles Black became vice president of corporate communications at General Public Utilities Corporation, a large holding company that owns a number of utility companies throughout the United States. The route to his current position could certainly not be considered direct, although it is not unlike the routes taken by many of his peers. Black began working for GPU in 1965, taking a variety of jobs in human resources until he finally became the training and development director and then the personnel services director of Jersey Central Power and Light (JCP&L) (one of GPU's subsidiaries). After two 4-year stints in those positions, he was asked to become the community and public information director of JCP&L, a post that he also held for 4 years before being given the title Vice President of Corporate Communications of the parent company.

He comments that human resources used to be considered a dead-end job, but because he managed to keep on the leading edge and demonstrate that he was willing to take on greater challenges, new opportunities soon opened up for him. Training in his company was once just skills training; later, he moved it into management development and, eventually, change facilitation. His work with upper management and his willingness to take on community projects on behalf of the company made his talents and energies visible. His people skills and experiences in training transfer well to his new position, which includes customer education and teleconferencing. His advice to young professionals is to be a value-added person to your organization.

Perhaps the best way to obtain a better understanding of the field is to hear the way several professionals describe their jobs. At the time this book was being written, my former students, clients, and colleagues characterized their primary activities as follows.

Professional Profiles

Frank Oliver is the vice president of development and communications at the Guthrie Healthcare System. He directs and manages the fund-raising plan and is responsible for community relations and internal and external publications, such as the annual report, internal newsletters, and press releases. He reports directly to the president of the corporation and is a member of the institutional management team. Frank also advises managers on the development of the organization's marketing strategies. He holds a bachelor's degree in communications and a master's degree in public management.

Theresa Lyczko is the marketing media specialist for the Healthy Heart Program in both Cayuga and Tompkins Counties in New York State. She is responsible for establishing and maintaining a communication network that promotes cardiovascular health and services. In doing so, she identifies and produces promotion devices, such as workshops, presentations, video programs, public service announcements, and community events. She also works with local and regional health experts and providers, as well as the state health department and the Centers for Disease Control. For example, one of her recent projects was to write a proposal for a mini-grant to fund the production of a 29-second rap video aimed at preteens. This video was shown on regional cable channels during Tobacco-Free Awareness Week. Theresa holds a master's degree in communications.

Craig Douglass is the creative director for Frontier Media Group, a production house that designs and produces video and interactive media for marketing and training applications. He is responsible for the development of creative treatments and the production of video, audio, and computer graphics. To do this, he manages a diverse team of scriptwriters, graphic artists, production crews, and computer programmers to combine a variety of media disciplines into a cohesive program that is engaging, easy to use, and satisfies their client's objectives. Craig has a bachelor's degree in corporate communications and previously worked for several years as a media producer at a college.

Holly Walkland is the manager of diversity at Eastman Kodak Company. After completing her bachelor's degree in educational communications, she held a variety of positions in training and communications, starting out as a video producer at the Raymond Corporation, which manufactures forklift trucks. She joined Kodak several years later as a developer of interactive training programs when that technology was in its infancy. From her work on a task force investigating satellite TV, she became the first manager of KBTV, Kodak's Business TV network. She began work on her MBA and was then rotated through a few positions within Kodak's communications functions, including public relations, to give her a more expanded view of the business. After she completed her MBA she began the transition from her position as manager of training and communication support services for Kodak's Customer Equipment Services Division, where she supervised a staff of writers, media producers, systems specialists, and illustrators, to her present position.

Marc Rosenberg is a project manager in AT&T's Corporate Training Support Group. His responsibilities include providing support for the planning and future direction of AT&T's worldwide training and education functions and developing integrated performance management systems. As an internal consultant, Marc helps other AT&T training and human resources professionals and line managers learn more about new approaches to instruction and performance improvement. He also manages research and development projects that demonstrate the use of new techniques and technologies, such as computer-based interactive performance support systems. Marc holds a PhD and has had experience as a course developer, instructor, and training evaluator.

Cathleen Wolf runs her own employee communication consulting firm, CW Communication, Inc., which specializes in employee attitude surveys, communication audits, and action plans. She works with large organizations in such businesses as computers, agriculture, and aerospace to identify gaps in internal communication systems and to recommend ways to correct them. For example, when I asked her to describe her work, she was busy analyzing a room full of data from 50 focus groups and 706 questionnaires. She developed her own data analysis methods that help her to follow an organization's progress as improvements in their communication practices are made. Cathleen has owned her own business since 1988, and before that worked in several corporate communications departments.

Ghenno Senbetta is a manager in the Change Management Services Division of Andersen Consulting, which is a part of one of the world's largest accounting firms, Arthur Andersen. His responsibilities include consulting with outside clients in the areas of instructional technology, computer-based training, and strategic planning for integrated performance support services. Prior to joining Andersen Consulting, Ghenno worked at Andersen's Center for Professional Education as a senior CBT specialist and at Advanced Systems, Inc., as developer and manager of computer-based training and interactive video simulations. He holds a PhD in instructional technology.

Nick Cerro is a technical communications specialist for the Syracuse, New York, office of Blasland & Bouck, a firm of professional engineers and geoscientists. Nick writes and produces print and graphic materials that become part of major project proposals, presentations, scientific papers, internal publications, and government reports. For example, he writes the sections of proposals that discuss his firm's experience, personnel, and qualifications. He also writes and produces slides and other visual support for client meetings and conference presentations. In doing so, he is a key player in helping his firm win bids and gain professional recognition. Recently, he was involved in developing a document development and style manual, and in writing segments of a model proposal. Each of these will be used to make Blasland & Bouck's print materials more consistent and easier to produce. Nick has a bachelor's degree in English and began this job after he completed a master's degree in corporate communications.

Sean Quinn is an instructional designer for Ford Motor Credit, Inc., a subsidiary of Ford Motor Company. He analyzes training needs and designs; develops training materials, including video and interactive video; and evaluates the success of training programs. He also writes RFPs (requests for proposals), which are sent to potential vendors who produce some of their training materials, and, subsequently, he manages those projects. Sean previously worked in media production and youth counseling, holds a master's degree in communications, and is working toward a PhD in educational technology.

Ed Nathan is manager of international sales training at Lederle, a multinational pharmaceutical company. He manages the development of manuals and computer-based training courses for their sales reps in more than 100 countries, covering such areas as basic anatomy and physiology, the applications of particular drugs, and general sales skills. He also develops and conducts seminars for sales and marketing managers around the world (mainly in Europe), on topics such as negotiation, product management, and marketing. Ed also edits Lederle's bimonthly training newsletter, *HORIZONS,* which the company uses to recognize their affiliate training efforts and to promote their own training programs. Ed holds an MBA and was previously a sales representative for Lederle.

Brian Manning is the head of the information systems department at Hughes Research Labs. His group is responsible for the computing and telecommunications systems and services at the lab, which include mini-computers, local area networks (LANs) composed of IBM-compatible and Macintosh personal computers, the telephone system, and both voice mail and e-mail. His group of eight people buys, develops, and maintains the hardware and software information systems that support office automation (e.g., word processing and budgeting), scheduling, an employee building locator, and scientific computing. Since many functions, such as creating charts and graphs for presentations and for laying out proposals, are done by end-users on their desktop computers rather than by a separate communication department, the systems that Brian and his department support are integral to the internal and external communication activities of the lab. He also researches and recommends new technologies (e.g., teleconferencing). Brian holds a bachelor's degree in speech and a master's degree in communications with a concentration in new technologies.

THE REDEFINED COMMUNICATION PROFESSIONAL

Communication people are learning more about training, human resources people are becoming more involved in communication issues, and information systems professionals are becoming proficient in telephone and satellite transmission of data. All these professionals are learning how to use new tools and techniques to disseminate information within and outside the organization. They're meeting each other in boardrooms and talking about corporate culture, recruiting programs, motivational strategies, and electronic communication systems.

In many ways, we need to rewire communication systems within organizations. Strategies and systems must be coordinated and brought into alignment with an organization's goals and culture and with today's participatory, lean organizations (Gayeski, 1992). Many tools are available; in fact, communication technology has far surpassed our ability to effectively use it. However, we must learn to use these technologies—otherwise, they use us. The way in which systems are designed—so that machines work for and with people—is the key to enhanced communication that, as we will see, can transform organizations.

The Renaissance Communicator

A new breed of professional is emerging—the Renaissance communicator. This communicator's role will probably mean the difference between success and failure for many information-age organizations.

> The term "Renaissance man" calls to mind a person of broad intellectual interests encompassing a full spectrum of available knowledge and wisdom. From the fourteenth through the seventeenth century, Europe experienced the Renaissance, a time of great intellectual and artistic revival during which the creative spirit was given free rein. Truly, the Renaissance represented a time of renewal for the human spirit unique in history. . . . The ethical practice of communication in business can be the least costly common denominator in regaining lost markets, re-inspiring the workforce, and rebuilding the public's dwindling faith in American industry. We believe the renaissance in American corporate leadership must be solidly based on the principles of good communication. . . . American corporate leadership is clearly in a state of crisis. Communication is, we believe, the pathway out of this crisis. (Ruch & Goodman, 1983, pp. 21–22)

This ideal, however, has eluded business and academe for several decades. For the last century, our society has been creating increasingly more specialized jobs, academic departments, and organizational units, which fragment knowledge and individuals' responsibilities. However, the assembly line approach of dividing jobs into the smallest possible units and having them repetitively cranked out by human machines is backfiring. Contemporary organizations are discovering that this kind of division of labor leads to dull jobs, bored and unmotivated people, and the general tendency of employees to think and act like the machine for which the job was created. Therefore, many organizations are trying to enrich jobs by giving employees broader responsibilities and more opportunities to learn new skills and, thus, to have a larger impact on the organization.

More specifically, in communications we have tended to narrowly cast people into roles as trainers, marketing specialists, human resources development practitioners, writers, media producers, or speech makers. Although in the 1960s and 1970s one might have gone to college to major in English or Communications, taking courses in speech, drama, literature, writing, media production, and journalism, today each of those areas is a major in itself.

Not only has this academic structure been motivated by increasing specialization in theory and research, but it is a reflection of the industry into which colleges place their graduates. As we've seen, corporations have developed separate departments to do their advertising, information systems management, electronic communication, employee newsletters, recruiting, annual reports, documentation, media production, speech writing, press relations, and training. The relative explosion of knowledge in each of these areas, as well as rapidly changing technologies for the production of print and audiovisual media, has made it seem virtually impossible for anyone to stay on top of it all. Yet, the result of this fragmentation has been the following:

- a lack of coherence in (or even blatantly conflicting) organizational messages and communication styles
- duplication of efforts and technologies
- data overload
- poor application of technologies

Such fragmentation leads to massive bleeding of corporate resources by the production and consumption of information and instruction. People begin to spend more and more time attending meetings, participating in training courses, and reading memos and manuals. The very information that is designed to improve their performance winds up, by its sheer volume and lack of consistency, creating confusion and anxiety. Getting a better grip on communication, which undoubtedly consumes the most time of organization members, may become a significant challenge.

The time has come to "draw maps and build bridges" to connect the islands that make up the "archipelago of information" in modern organizations (McKenney & McFarlan, 1982). Training, public affairs, advertising, marketing, policies and procedures, employee communication, and information systems are all part of the basic communication infrastructure in every business; therefore, creating cohesive strategies and control mechanisms for them is essential.

The challenge may seem daunting. Can any one person really master adult learning theory, instructional design, employment interviewing laws, data base management, video editing, electronic typesetting, speech writing, satellite teleconferencing, conflict resolution techniques, computer graphics, public relations tools, and newsletter editing? Increasingly, the answer is yes. The *Renaissance communicator* is an individual with a broad base of skills and concepts in human and mediated communication who can apply the theory and technology of modern communication systems to assist individuals and organizations in meeting their goals (Gayeski, 1991). The domain of the Renaissance communicator includes the following:

- *internal communications*—in-house publications, meetings, displays, management information systems, e-mail
- *external communications*—advertising, public affairs, annual reports, customer communications, community relations
- *performance improvement*—feedback and incentive systems, expert systems, job aids, information retrieval systems, work redesign, human and computer interfaces

New Tools, New Roles

How do we deal with the inherent conflict between the explosion of specifics and the need for generalists? The answer is not simple but it does revolve around one basic principle; that is, new communication tools and a more comprehensive view of the profession enable individuals to successfully engage in a broader range of corporate communication activities. We'll look at specific tools and techniques for research, training, information dissemination, and media production in later chapters, but for now it is good to keep in mind that the technical aspects of information gathering, packaging, and dissemination are becoming easier to accomplish, while at the same time, the conceptual aspects are becoming more challenging. This is largely due to advancing information technologies and rapid changes in the business environment.

I'm not a graphic artist by a long shot (it's hard to tell whether my 5-year-old or I drew a picture), but I can create dazzling 3-D bar charts to display some data on the acceptance of videodiscs in about 5 minutes using my PC and a business graphics package. I display those charts during my workshops directly from my laptop computer, which is hooked into a video projector. I'm no expert in the legal aspects of the trucking industry, but we search out instant information for our client who sells to this market by keying in a few terms using our on-line information system that scans current magazines, journals, and congressional reports. I can then edit the information to create newsletters for the client's distributors on our desktop publishing system. From my home-based office overlooking a garden and pond in upstate New York, I can help a client in Arizona who needs some information on using voice synthesis within a computer-based training program for the Prosecutor's Office in Phoenix. I talked with him by phone, sent a message by CompuServe to our programmer in Japan who created the software our client is using, got the programmer's reply the next morning, imported it into my word processor, created a memo to our client, and then faxed it to him. The client's initial phone call to me was at 6 P.M.; he had his answer waiting at his fax machine when he walked into work the next morning. This morning, we had a phone conference with a team of trainers in Canada with whom we are working to create a set of guidelines for the print and electronic materials that their department creates. Communication technologies can multiply our capabilities and eliminate constraints of time and space.

As the preceding case illustrates, new technologies are creating jobs for us and changing the ways in which we perform them. These systems can leverage our capabilities and eliminate constraints of time and space. Never in history have organizations had such a need to communicate effectively and consistently. We have laws about what we may not say to job candidates and what we must say to workers who are exposed to potential risks on the job. We have created high-tech jobs that require extensive training of employees. We are also experiencing tough business environments in which we must compete for customers and investors and the good will of our communities. However, we have never before had so many theories, models, systems, and tools to accomplish these jobs.

If you read through the press release in Figure 1.5 you'll see numerous examples of the roles of information and communication in contemporary organizations. To

FIGURE 1.5
This press release, produced for U.S. Healthcare by Weightman Public Relations, contains many examples of communication techniques and technologies. (Courtesy of Weightman Public Relations.)

FOR IMMEDIATE RELEASE

FOR MORE INFORMATION CONTACT:
Steve Rosen or Donna Bleznak
Weightman Public Relations
(215) 561-6100

U.S. HEALTHCARE OPENS FUTURISTIC CUSTOMER SERVICE CENTER

BLUE BELL, PA, October 18, 1990 -- U.S. Healthcare, a recognized leader in the managed care industry, enters a new era today when it officially opens its Customer Service Center -- an innovative facility designed to provide quality service for years to come. Located at 1425 Union Meeting Road in Blue Bell, PA, the new building will house one thousand employees and has the capacity to ultimately accommodate a staff of over two thousand. Today's grand opening ceremony will be attended by dozens of corporate leaders as well as state and local officials.

Conceived and built to accommodate major corporate expansion as well as innovative methods of providing quality service, the Center will enhance service by combining computerized technology and automated systems, innovative design and forward-thinking employee training programs. As an operator of health maintenance organizations, U.S. Healthcare currently serves more than 1,100,000 members in six Northeastern states.

"Quality and customer service is the name of the game in an industry like ours, and the new Customer Service Center will no doubt help us better serve our members," said Marcy Abramson, Vice President, Operations, at U.S. Healthcare.

Highlights of today's ceremony include participation of tennis legend Arthur Ashe who will receive the first U.S. Healthcare Humanitarian Award for his outstanding philanthropic efforts. Ron Zemke, a well-known author, consultant and leader in the field of quality service and productivity improvement will officially dedicate the Center.

In addition, State Representative George E. Saurman will join with U.S. Healthcare officials in sealing a package of supplies to be sent to the United States troops in the Middle-East, and discuss corporate responsibility during turbulent times. And, to top off the day's events, attendees will be invited to tour the Center and view the advanced customer service technology incorporated in the building.

(continued)

FIGURE 1.5
(*Continued*)

In order to realize the dream of superior customer service in the new Center, innovative technology and progressive management techniques were merged to create an atmosphere where quality service and efficiency are synonymous.

For starters, U.S. Healthcare renamed, reorganized and retooled many of its departments. These include TRIAD, a new department that incorporates three departments into one unit, offering a single source for customer service, response to inquiries and problem resolution. The new building is also home to a state-of-the-art computer system called "Target" which performs high-tech research and helps provide automated solutions to customer service needs.

A physical environment has also been created within the Customer Service Center that works to encourage highly responsive employees. From ergonomic furniture and workstations, to "Vital Signs" -- a publicly displayed billboard that communicates intradepartmental service achievements, every element has been carefully chosen to facilitate and reward the highest level of customer service. In addition, bright colors, unique artwork and a "no doors" environment have been created to build employee morale and an atmosphere conducive to optimal customer service.

The new Customer Service Center is a big step forward for U.S. Healthcare. But, as Abramson points out, "As in any area related to business growth and competition, standards of achievement must be upgraded continually, and so-called "limits" must be exceeded, and that is U.S. Healthcare's mission."

begin with, the press release itself is the creation of U.S. Healthcare's public relations agency, designed to gain publicity for the opening of its new customer service center. The opening ceremony was a carefully executed media event in itself; speakers included tennis pro Arthur Ashe (who was given an award), a well-known author on productivity improvement, and a representative of the state government who joined U.S. Healthcare officials in sending supplies to the troops engaged in the Desert Storm War in the Middle East. Now read through the descriptions of the new facility itself; it uses state-of-the art computer systems to provide research and recommendations on how to solve customers' problems, a public electronic billboard to display employee achievements, and interactive employee training systems. By engaging in these activities, U.S. Healthcare has positioned itself as a forward-looking organization, dedicated to high levels of customer service, efficiency, employee satisfaction, and public service. (See Figure 1.5)

CONTEMPORARY ISSUES

A number of important conceptual and practical problems face professional communicators today. Through professional practice, the tracking of professional and scholarly literature, and the many interviews and conversations that I initiated in writing this book, I compiled a list of these issues, each of which I discuss in more detail in the following chapters.

Defining the Field

What is corporate communications anyway? The word *communication* has no standard definition, and depending on whom you're talking to, it can mean talking, data transmission, public relations, media, or counseling. Interestingly, the French call the emerging communication industry *l'informatique.* There are currently many related sciences and disciplines, including bibliometrics, linguistics, semantics, semiotics, cognitive science, computer science, communication arts and sciences, library science, management science, speech science, communication theory, and telecommunications research, all of which are concerned with information. Before becoming influential within organizations, communicators must agree on a description of their capabilities, duties, and methods, and have their work recognized as an established profession.

Professionalizing Communication

David Berlo, communications scholar and teacher, has remarked that communication must be elevated to major product status within organizations. It must be recognized as an essential asset, and one that must be brought into fiscal control by relating it to the bottom line. Currently, communication is often considered a frill—something nice to do because it makes people feel good—and professional communicators are often ill-prepared to handle the challenges that face them. However, as we have seen, it is far from a peripheral concern. It is what organizations do. Many professional communicators wish that they were able to participate in formulating corporate strategy and to play a proactive rather than reactive role; yet, in many organizations, communication directors are still not members of the executive management team. This needs to change.

Managing Information

Most people realize that we are now in an information economy, where data and concepts are often more valuable than objects. Most work involves developing and exchanging ideas and symbols (communication) rather than developing and exchanging actual objects (production and distribution). However, no one has decided what being in this information age truly means.

Toffler (1990) maintains that information is power, as important as natural resources and money have been in the past. In other words, those individuals and organizations who have access to accurate and rapid information will succeed. Fast cultures will overpower the slower cultures. Of course, not everyone agrees with this analysis; for example, Roszak (1988) claims that buzzwords like "the information economy" and "the information society" are the "mumbo jumbo of a widespread public cult" (p. x). Regardless of what one calls this era in which we live, it will, to a large extent, be up to professional communicators to define and manage the role of information in our increasingly symbolic world.

Helping Organizations Grow and Change

Today, many organizations are grappling with the need to change the management and communication structures that date back to the industrial age. As organizations become more complex, expand geographically, and endeavor to increase the speed at which they do business, the methods and policies of communication must change as well. Whether it's helping to launch a new product, explaining the total quality program, dealing with the public's wariness of the organization's toxic waste disposal practices, training line workers to manage their own teams, or installing an e-mail system to communicate with the new office in Eastern Europe, professionals in communication are at the center of the process. For example, Corning, Inc., uses large group, highly interactive meetings to introduce new organizational design ideas to its employees and to solicit their input. (See Figure 1.6.)

FIGURE 1.6 Large-scale meetings, such as this one at Corning Incorporated, are powerful means of affecting organizational planning and change. (Courtesy of Corning Incorporated.)

Balancing Information Overload and Underload

There has never been more information available to more people. In fact, most of us are drowning in it. Bennis (1976) observed that the people who don't merely receive information, but who filter it into a meaningful pattern, are the ones who succeed in organizations. Are communicators contributing to this overload and actually bogging down their audiences in data rather than increasing productivity? How do we know how much is enough? Rapid changes in our society demand access to information and instruction in order to remain competitive, and many people believe (whether it's true or not) that more information increases job satisfaction. However, a crucial issue facing corporate communicators is how to provide complete and current information without overwhelming their audiences.

Developing Theories, Policies, and Practices

There are still no established theories, policies, or practices in corporate communication. The field has evolved from other disciplines such as journalism, speech, psychology, and educational technology, and most professional communicators merely feel their way through, developing their own standards. "No single criticism has been echoed with such vigor as the fragmentation of the field, and no prescription for a cure has been forwarded so frequently as the advice to elaborate coherent theory. And, no injunction has broached so little remedy" (Fulk & Boyd, 1991, p. 407). Several professional associations, such as the IABC and the Public Relations Society of America (PRSA), have established accreditation programs and published codes of ethics. However, as organizations and communication technologies change, new sets of conceptual frameworks, skills, and benchmarks for communication programs need to be developed.

Managing Technologies

Overall, there is widespread awareness of new communication technologies. Even the least technically sophisticated manager probably owns a VCR and a camcorder and can produce somewhat professional-looking print materials on a personal computer. Whereas access to communication media used to be limited by its expense and complexity, today almost everybody has the tools to become a mini-TV channel or a mini-newspaper. It is extremely difficult for professional communicators to keep on top of new technologies such as interactive multimedia, teleconferencing, and desktop video. Moreover, it is a challenge to maintain control over increasingly decentralized communication now that every department can produce its own newsletter or video memo.

Social and Ethical Concerns

Those who control access to information and dissemination channels have tremendous power. This power can be used in a democratic, constructive manner, or it can be wielded to exploit and to disadvantage individuals or classes of people. Issues such as access to information, honesty, and privacy are escalating in importance within organizations, especially as the general public recognizes how both information and images can be manipulated.

Promoting Flexibility and Diversity

Organizations and the individuals within them are becoming more diverse. The old image of the manager as the man in the grey flannel suit has just about been erased. Organizations may operate from an electronic cottage, a skyscraper, or several hundred international offices. Their employees are male and female, young and old, full- and part-time, average or exceptional in mental and physical abilities, and have a variety of ethnic, racial, and religious backgrounds. This increasing diversity makes communication more difficult, but even more necessary. Communication professionals need to learn not only to cope with diversity, but also to foster the creativity and vigor that it can yield within organizations.

Demonstrating the Value of Communication

When asked about results of a communication program, most professional communicators back off and say that the effects really can't be measured. Although organizations spend hundreds of thousands or even millions of dollars on internal communication and training (not to mention advertising and public relations), they're not sure of what they're getting for their money. Therefore, in tight economic times, often the first thing to be cut from the budget is communications. In order to be taken seriously, communicators must find a way to specify their objectives and rigorously measure their results. Communication programs should not be gifts that are given in order to make people feel good without any expectation of payback. They are an investment that should have a measurable return.

A ROAD MAP

It seems appropriate to offer a road map of the remainder of this book. This introductory chapter has outlined what the field of corporate communications is, where it came from, and the challenges we face as professional communicators. From here, Chapter 2 will move on to an examination of the major theories and concepts we need to know in order to talk about and think about communication within organizations. In Chapter 3, we'll look at where we want to go—new-age or information-age organizations—and how communication policies and practices can help companies change and thrive.

Once we know where we want to go, we'll need to examine how to get there. Chapter 4, on analysis and design, explains how to find communication problems, uncover opportunities, and plan programs. Chapter 5, on interventions, describes the various tools and techniques, from meetings through multimedia, used to reach an organization's communication goals.

However, communication is much more than a collection of individual interventions. The professional communicator needs to know how to manage communication systems. Chapter 6, on managing corporate communications, provides guidelines for establishing communication functions and departments within organizations and discusses the issue of communicators' roles. In today's society, there are many challenging issues facing corporate communicators, and Chapter 7 discusses the future directions for the profession and the individuals involved in it.

The goal of this book is to provide the concepts, tools, and models needed by the Renaissance communicator to play an influential role in an organization's success. You'll find the insights and candid observations of many successful practitioners and noted scholars. As you read, keep this question in mind: what would an ideal organizational communication system look like, and how can I create and manage it?

REFERENCES/ SUGGESTED READINGS

Ackoff, R. (1957). Towards a behavioral theory of communication. *Management Science, 4,* 218–234.

Alter, S. (1992). *Information systems: A management perspective.* Reading, MA: Addison-Wesley.

Barnard, C. (1938). *The functions of the executive.* Cambridge, MA: Harvard University Press.

Bell, D. (1973). *The coming of post-industrial society.* New York: Basic Books.

Bennis, W. (1976). *The unconscious conspiracy: Why leaders can't lead.* New York: AMACOM.

Fulk, J., & Boyd, B. (1991). Emerging theories of communication in organizations. *Journal of Management, 17* (2), 407–446.

Gayeski, D. (1991, April). What's my next job? *Performance & Instruction,* 37–40.

Gayeski, D. (1992, March). Rewiring corporate communication. *Communication World,* 23–25.

Gerstner, J. (1991, August). Hanging loose in a bureaucracy. *IABC Communication World,* 31–33.

Goldhaber, G.M. (1983). *Organizational communication,* 4th ed. Dubuque, IA: Wm. C. Brown.

Gordon, J. (1991, October). Training budgets: Recession takes a bite. *Training,* 37–45.

Handy, C. (1989). *The age of unreason.* Boston: Harvard Business School Press.

Machlup, F. (1962). *The production and distribution of knowledge in the United States.* Princeton, NJ: Princeton University Press.

Machlup, F., & Mansfield, U. (1983). *The study of information: Interdisciplinary messages.* New York: John Wiley & Sons.

McKenney, J.L., & McFarlan, F.W. (1982, September/October). The information archipelago—Maps and bridges. *Harvard Business Review,* 109–119.

Meltzer, M. (1981). *Information: The ultimate management resource.* New York: AMACOM.

Nadler, L., & Nadler, Z. (1989). *Developing human resources,* 3rd ed. San Francisco: Jossey-Bass.

Roszak, T. (1988). *The cult of information.* New York: Random House.

Ruch, R.R., & Goodman, R. (1983). *Image at the top: Crisis and renaissance in corporate leadership.* New York: The Free Press.

Schofield, S.E. (1992, April). Commentary: Highly trained professionals. *USAir Magazine,* 9.

Schwartz, M. (1988, September/October). A presentation manager by any other name. . . . *Presentation Products Magazine,* 94.

Toffler, A. (1980). *The third wave.* New York: Bantam Books.

Toffler, A. (1990). *PowerShift: Knowledge, wealth, and violence at the edge of the 21st century.* New York: Bantam Books.

Weick, K. (1969). *The social psychology of organizing.* Reading, MA: Addison-Wesley.

Welsford, E. (1966). *The fool, his social and literary history.* Gloucester, MA: Peter Smith.

Yates, J. (1989). *Control through communication.* Baltimore: The Johns Hopkins University Press.

Zuboff, S. (1988). *In the age of the smart machine: The future of work and power.* New York: Basic Books.

2 The Concept System

Before we start practicing corporate communication, we have to be informed as to where the current ideas and systems started and what to call the concepts and actions that make up the profession. The concept system presented in this book is broader than those you might find in other books on organizational communication, media, or training. We will take the perspective of the Renaissance communicator, who holistically directs the communication process within an organization. The theories with which we must be fluent are

- communication
- learning
- management/organizational development

Why have a theory you might ask? Why bother studying definitions? Most communicators are biased toward action and creative brainstorming, and not toward scientific analysis and standardization. Unfortunately, the result is that in our adolescent profession, we have no agreed-upon definitions, methods, or theories. Lacking them, we have not been taken seriously by other professions. Even within our field, we find it difficult to communicate with our colleagues. It's difficult because we don't know what we're talking about: words, such as "communication" and "information," have no set referents.

Theory is not just abstract or academic. It makes us more efficient in explaining and in controlling our environment. It is not in opposition to practice, rather it is the necessary flip side to practice. A *theory* is a system of ideas used to describe or predict a phenomenon. *Practice* is using theory (whether formal or informal) to design, produce, and predict the outcome of phenomena in order to attain some goal.

Theories can be descriptive (how things exist) or prescriptive (how things should be done). In the following overview of theories, you'll get a sense of how scholars and practitioners in communication, management, and learning have attempted to discuss, predict, and recommend practices and outcomes in the communications field. You'll be able to see how ideas developed and to better discern the underlying operating assumptions of many organizations and their managers.

THEORIES OF COMMUNICATION

Communication theory attempts to describe and to predict the processes and outcomes of information transmission among individuals or groups either directly or through various media. Although theories of persuasion date back to the early Greek civilization, communication as a system was not really conceptualized until the 1950s.

There are three different levels from which we can view communication: (1) from a technical perspective (e.g., How accurately is information transmitted?); (2) from a semantic perspective (e.g., How effective are symbols in conveying the correct meaning?); and (3) from an effectiveness perspective (e.g., How effectively does the received meaning affect an individual's conduct in the desired way?). For example, as a communication manager, you might decide to set up a series of meetings to inform employees about a new flextime program. You would need to ensure that the policy is communicated accurately; that is, that the facts are correct each time the message is relayed down the chain of command, and that the data is correctly typed and transposed. You need to be concerned about the words and images used; for example, do they convey what the vice president for human resources has in her mind? Finally, you need to consider the impact of the message; for instance, what actions did you envision happening as a result of this information, and are they occurring as planned?

The root of the word *communication* is the Latin *communicare,* which means "to share." In common speech, we use the word in many ways. We may say that after several counseling sessions, a couple is really communicating. A manager might tell us to communicate his new expectations about sales quotas to the field representatives. When the phone lines go down, we say we've had an interruption in the communication system. We even talk about communicating a disease!

A common dictionary definition of communication is "information transmission," but referring to communication as just the exchange of information is generally inadequate because there has to be a useful context. People can't transmit information directly by passing a concept from their brain to someone else's. Rather, people have to use words, symbols, images, models, or sounds to try to encode their meanings from both the available information and their individual expertise.

A number of contemporary communication scholars have constructed their own definitions of communication:

> shared meaning created among two or more people through verbal and nonverbal transaction. (Daniels & Spiker, 1991, p. 29)

> the process that enables people to co-orient their behaviors . . . [and] empowers people to establish functional interpersonal relationships that allow them to work together toward goal attainment. (Kreps, 1986, p. 5)

> The process of constructing shared realities. (Shockley-Zalabak, 1988, p. 29)

> the act, by one or more persons, of sending and receiving messages that are distorted by noise, occur within a context, have some effects, and provide some opportunity for feedback. (DeVito, 1982, p. 4)

We can see from these definitions that most people believe the following about communication:

- It is any form of sharing meaning among people.
- It has a purpose.
- It must have both a sender and a receiver.
- It operates within a complex context.

As human beings, we constantly communicate; it is one of our species' inborn traits. Many people, in fact, argue that we cannot *not* communicate. For example, if an organization refuses to divulge any information about the possibility of a plant closing, it is, in fact, communicating something. Perhaps it is conveying management's distrust of unions or, even more likely, that there will be a plant closing. As individuals, people communicate through gestures, styles of dress, and word choices. As organizations, we communicate through the design of facilities, logos, the accessibility of executives, and the amount and the kind of information made available through various kinds of media.

Within communication systems, there are two components, message and meaning, which sometimes get confused. This is what makes the process difficult. *Messages* are symbols we construct in order to create or share meanings. They are the outward acts that a receiver can see and respond to. *Meanings* are mental images or feelings that we create as we perceive and try to make sense of the world.

When people receive a message, they decode or interpret it and try to create meanings for themselves. Although a message may remain constant, meanings will always vary due to individual responses, values, and backgrounds, and due to the inadequacy of certain media to capture someone else's meanings, whether it be the spoken word, a picture, or even a video clip. When I say the word "chair," what comes to your mind? Do you think you're seeing the same picture that I am? As a matter of fact, I happen to be thinking of my colleague who is chairperson of my undergraduate department in corporate communication.

Although communication is natural, we aren't innately good at it. As people become more aware of the process, they engage not only in communication, but also in metacommunication. *Metacommunication* is communication about communication—it relates to the communication process as well as the relationship between communicators. Metacommunication occurs when people reflect on the communication process itself; for instance, "I'm having a hard time getting through to you," or "we're not on the same wavelength anymore."

Organizational communication, a specific branch of the discipline of communication, focuses on the context of organizations and various aspects of group and managerial communication. It is the exchange of information internally (up, down, and across organizational lines), as well as externally (to stakeholders and to the public).

Although the kinds of organizations that may first come to mind are businesses, there are many kinds of organizations—clubs, religious groups, professional societies, or even an ad hoc group of people coming together for a weekend to build a community playground. An *organization* is two or more people who interdependently pursue a common goal within a given context and set of relationships. In order to achieve their objectives, they communicate for

- *production*—coordination of activities to meet goals
- *innovation*—stimulation of new ideas
- *maintenance*—to build and foster interpersonal relationships

Every organization has a structure—its own rules and networks. In the field of organizational communication, scholars focus not only on the transmission of messages, but also on the way the organization is designed in order to form structures for information flow. They also focus on the underlying corporate culture, which consists of styles and values.

Information Theory

Although communication has a number of purposes in society, a primary function of communication within organizations is to share information in order to move toward meeting organizational goals. As Chapter 1 points out, a major issue within organizations today is information overload. What exactly is information? Most of us would say it's data or ideas that we can use to make decisions.

If I tell you that the sun rose this morning, would you consider that message information? You'd probably say it's a fact, but you don't consider it information because it's not something new; in other words, you already knew or assumed that fact. If I tell you that the price on hog futures in Argentina rose 12% yesterday, is that information? It is not something that you knew, but it is also something that you probably can't do anything about or interpret because it is probably out of your sphere of action or expertise.

Information is another one of those words that people commonly use, but for which they have no standard definition. Since information is the basic commodity that professional communicators manage, it's essential that we seek more precision in its use. In information theory, *information* is the measure of uncertainty, or entropy, in a situation. The greater the uncertainty, the more the information. If a message is completely predictable, it contains no information.

Miller (1967), a psychologist, has used concepts from statistics to explain the notion of information. We can look at the content of a message and how it varies from what we knew or what we could have predicted before we were presented with it. The amount of variance equals the amount of information. If we are ignorant, we receive much information from a message; if the variance is small, we receive little information from a given communication.

If a communication system is effective, there should be a systematic relationship between the communication input (the message) and its output (a receiver's interpretation or restatement of that message). In statistical terms, there should be a high correlation between the input and the output. When we evaluate a communication system, we look at how much output variance is attributable to input and how much is due to flaws in the system (Miller, 1967).

Information is measured in *bits* (binary digits). One bit reduces uncertainty by half; it is all we need to make a decision between two equally likely alternatives. If two cars are outside in the parking lot and I need to know which is yours, I only need one bit of information to clarify the situation. Two bits of information are needed to choose among four alternatives, three for eight, four bits for 16, and so on.

All communication systems have a *channel capacity,* which determines how much information can be pumped through the system and still be transmitted accurately. We may think about this in technical terms, such as how many phone conversations can occur on a single wire, but human beings have limited channel capacities, as well. A human's channel capacity is the greatest amount of information that we can give a person and still have him be able to match his responses exactly to the stimuli (Miller, 1967). For example, I can probably tell you my telephone number quickly, and you can repeat it within a few seconds. However, if I tell you my bank account number, which is much longer, you'll probably have greater difficulty in repeating it.

At the semantic level, the communication of information reduces uncertainty. However, within most organizations, we want to do more than reduce uncertainty or share news. At the effectiveness level, we concentrate on the impact of the information. A person or system has some goal, which is generally to change someone's behavior. Therefore, professional communicators need to reduce data but increase information by providing only data that is not predictable and that relates to a specific goal.

Models of Communication

Starting in the 1950s, engineers and scholars have attempted to create models or descriptive theories about communication. The first of these grew out of the new sciences involved in the engineering of transmission systems, such as the telegraph and telephone. Engineers were interested in creating the most efficient and accurate means of sharing information through the use of a particular medium. The models that grew from this school are classified as linear models because they emphasize the one-way flow of information from one point to another.

Shannon and Weaver (1949) are recognized for their seminal work in developing a model of the communication system. Shannon was an engineer at Bell Telephone who developed the definition of information as the amount of entropy (randomness or unexpectedness) in a system. Shannon and Weaver's classic book, *A Mathematical Theory of Communication* (1949), with pages full of Greek letters and formulas, reads more like a physics text than a communication book. For these communication theorists, information is the degree of uncertainty surrounding a proposition, a measure of the rate of information acquisition, and the statistical unexpectedness of an item of information selected from a given set. They do not discuss the concepts of meaning or relevance within their model. (See Figure 2.1.)

Shannon and Weaver (1949) defined the communication system as an information source or transmitter, a channel, and a destination or receiver. Within such a communication system there exists noise. Noise, in communication theory, means more than a sound. *Noise* is anything that inhibits the precise reception and interpretation of a message—anything that impedes communication. Therefore, noise in a communication system can be static on the phone line, the fact that it's hot in the room and the audience is falling asleep at a meeting, or a narrator's outlandish hairdo that detracts from her credibility and the audience's concentration.

Another significant model of the communication process was created by Berlo (1960), one of the founders of the academic discipline in communication. His SMCR model of the ingredients in communication (see Figure 2.2) include the source, the message, the channel, and the receiver. The source and the receiver are persons with communication skills, attitudes, and

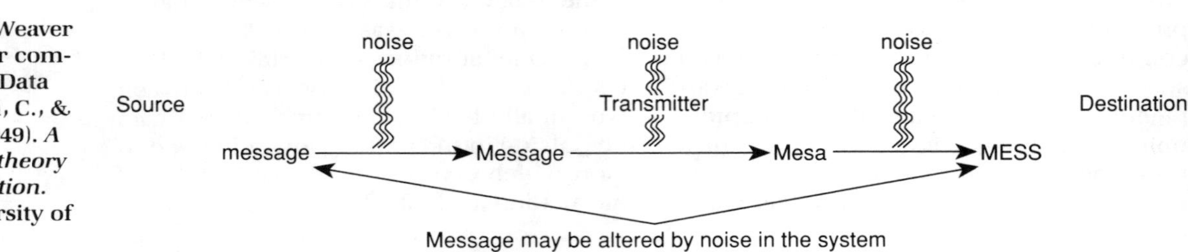

FIGURE 2.1 Shannon and Weaver model of linear communications. [Data from Shannon, C., & Weaver, W. (1949). *A mathematical theory of communication.* Urbana: University of Illinois Press.]

knowledge, operating within a social system and culture. The degree to which there is commonality of these factors between the source and the receiver determines how effective or easy the communication of a message will be. The message consists of content (the data), treatment (how it will be structured), and the code (how it will be encoded so that it can pass from the mind of the sender to some sensory perception of the receiver). Any message has a channel through which it is sent and received; this includes the five senses.

The previous two models are typical of the early linear models. They stress the expression and encoding of messages, the transmittal of messages, and describe a one-way flow of information. Contemporary criticism of this school of communication points out that communication is not easily broken down into simple messages. Critics address meaning and two-way communication. More current models of communication build on the linear models to embody the concept of two-way interactional or transactional communication. They emphasize communication as a process of message exchange rather than message transmission.

Schramm (1954) created one of the first two-way models of communication (see Figure 2.3). In it, he indicated that two people within a communication system were both senders (encoders) and receivers (decoders); therefore, emphasizing that communication was a two-way process.

Many theories and models of communication have been developed during the past decades, influenced by the human relations movement that was popular in the 1960s and 1970s. These models deemphasize telling or transmission and stress the importance of the receiver, participation, and listening. In particular, these theories feature the concept of *feedback*, which is a receiver's verbal and/or nonverbal cues and responses to messages. As we know, feedback is an important element in the communication process. We judge how well we're doing by our audience's expression (bored, enthusiastic, or puzzled) and by the outcomes of our messages. Unless we receive feedback, we don't really know if we were successful or not.

The notions of feedback and interaction now move us away from the linear or one-way model of communication to a two-way system or, as we now call it, the *communication loop*. In fact, although relatively few people have studied communication models, it's not uncommon to hear supervisors or clients talk about closing the communication loop or getting feedback.

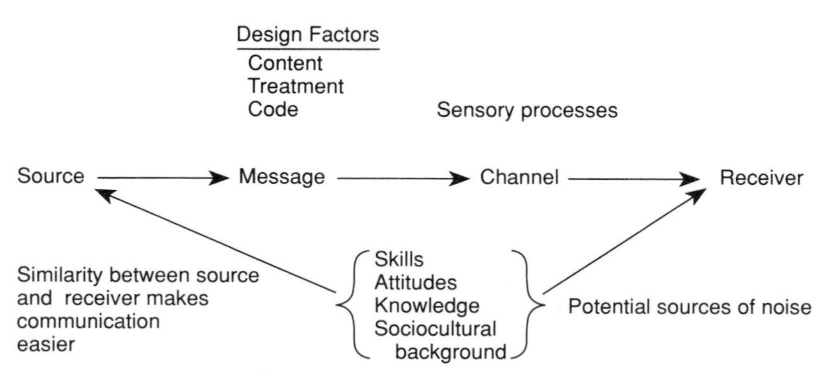

FIGURE 2.2 **Berlo's "ingredients" of communication. [Data from Berlo, D. (1960). *The process of communication: An introduction to theory and practice.* New York: Holt, Rinehart and Winston.]**

Bateson developed a system of concepts called *relational communication* (Ruesch & Bateson, 1951). He maintained that every exchange had two messages—a report message (content) and a command message (statement about the relationship). Most of the time, we concentrate on the content of a message; however, we are just as influenced by the more subtle nature of the message or the way in which it's communicated. For instance, if you come into my office, and after 5 minutes I say, "you may go now," not only have I told you that I have concluded my message, but I have made a statement about my control over our relationship. In saying this, I have indicated that I am in charge; I decide when and for how long we talk. If an organization decides to inform people that they're being laid off by putting pink slips in their paycheck envelopes rather than by holding individual or group meetings, the executives are sending a clear command message.

Within communication systems, relationships can be complementary (one dominant, one submissive) or symmetric (dominance met by dominance, submissiveness

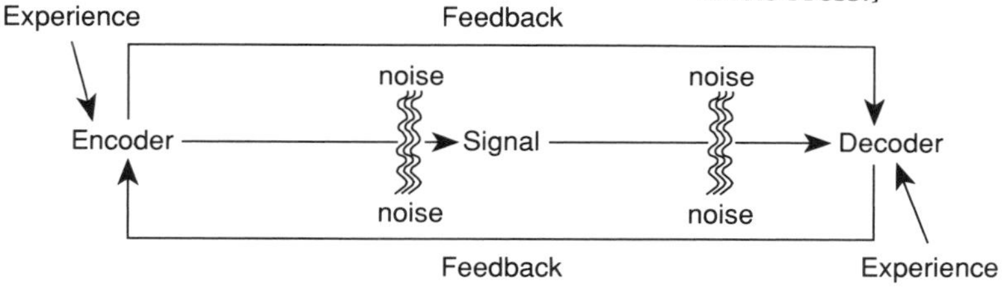

FIGURE 2.3 **Schramm model of communication. [Data from Schramm, W.L., ed. (1954). *The process and effects of mass communication.* Urbana: University of Illinois Press.]**

met by submissiveness). When designing messages or communication systems, it's important to consider the relationship or implied relationship of the participants in any exchange. In contemporary organizations, the movement is turning away from complementary relationships between dominant managers and submissive employees and toward fostering symmetric relationships among equally assertive, active colleagues.

When comparing communication models, it's tempting to evaluate them based on which is better or more current. However, each approach is valuable in explaining a particular type of communication situation. Linear models are good ways to describe one-way, singular announcements or orders. A newsletter story featuring employees' recent awards falls into this category. Interactional models describe mediated, asynchronous communication in which messages are alternately exchanged among parties, such as in electronic mail (e-mail) systems. An example of this situation is the use of e-mail to share information between two researchers. Transactional models are better at describing face-to-face communication in which both parties are actively and simultaneously exchanging information.

Clampett (1991), a contemporary organizational communication scholar, believes communication is like a dance. It involves skill, practice, a coordination of meanings, and a mutual and simultaneous interplay. You don't dance at people; you dance with them.

Although we use a lot of metaphors to describe the communication process, it's important to emphasize the fact that we're dealing with an imprecise process of attempting to share ideas and feelings. We use a lot of words like "deliver" (deliver training) and "get across" (get a point across), but communication is not transportation. We're trying to share ideas, not move them. As Berlo (1960) points out, "meanings are in people," not in the external world. So, it's very difficult to communicate.

While many communication theorists have looked at the general process of communication and the nature of information, some scholars focused on the effects of the medium used to present a message. McLuhan (1964) offered the viewpoint that the "medium is the message;" that is, certain inherent characteristics of media provide meaning in and of themselves, regardless of the content. For example, he argued that radio forces people to imagine images, while television requires less imagination and effort. Salomon (1984) conducted research on the effects of various kinds of media in providing children's educational material. He discovered that some media, such as TV, are considered easy, while others, such as print, are considered hard. Salomon explained this through the concept of *invested mental effort,* or how much concentration is required to decode messages transmitted by means of different media.

Throughout this book, we'll deal with communication from an organizational perspective, keeping in mind that the exchange of information may occur among individuals or groups, may take place instantly or for a period of time and at many locations, and may involve the use of a number of media technologies. When we talk about messages within a corporate communication context, we generally mean more than a phrase—we mean an idea, fact, or theme that we want to share in

FIGURE 2.4
Multimedia model of communication.

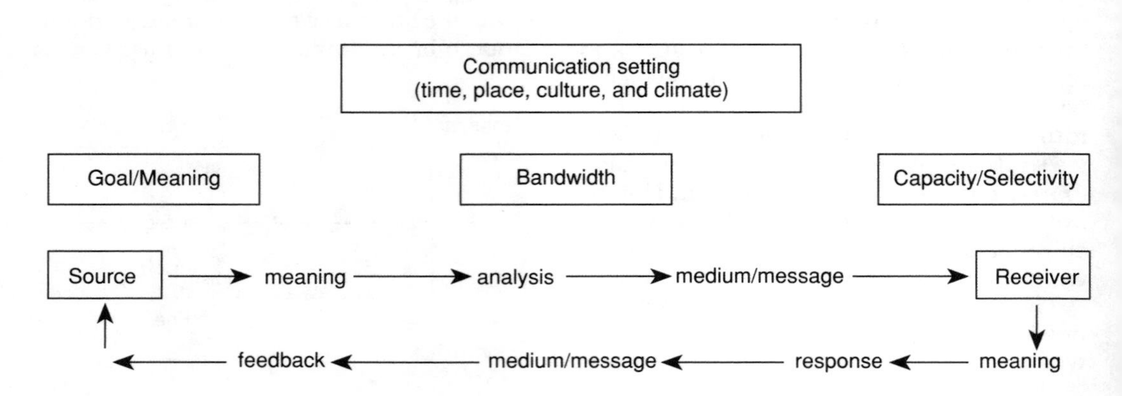

order to meet a particular goal. The multimedia model of communication describes this context. (See Figure 2.4.)

The source, or sender, of a message has a desired goal or outcome. This formulates into meanings that the source wants to share with the receiver. Through an analysis of the receiver(s), communication setting or environment, and practical constraints, a message is developed and a medium through which to communicate that message is chosen. The receiver has a given capacity for receiving that message at any particular time, and selectively attends to and intepretes it. After perceiving the data (message) through the chosen medium, the receiver creates his own meaning from it, responds in some way, and provides feedback through another series of messages and media choices. Obviously, no two people will attach the same meaning to a message. Therefore, since media are limited, to some extent, in *bandwidth*—capacity to transmit or display certain modalities like audio, text, or motion video—and all receivers are limited in their capacity to attend and to remember, no communication system is perfect. It is not as simple as moving a physical object from one place to another intact. Furthermore, the entire process takes place within a context—the communication setting. This setting consists not only of time and place, but also of the emotional climate and the national, regional, and organizational cultures.

It is becoming increasingly more difficult to communicate within contemporary information-age organizations. Although people have never been able to transplant their ideas directly into the heads of others, in the past, most people who talked to each other had common experiences. They primarily spoke in person to others from the same geographic area, social class, and gender. They usually talked about concrete things in the environment. Today, people communicate with others from all corners of the globe, often without ever meeting face-to-face. Coworkers may have had radically different backgrounds and may pursue incredibly divergent interests and lifestyles. In addition, most of our work is with intangibles—ideas and concepts. We can no longer point at an object in order to explain what we're talking about.

Wiio (1978), a well-known Finnish communication scholar, commented that if communication can fail, it will! Although managers and clients might blame the skills of their professional communication staff or a newly used medium for communication failure it is important to realize that it's not technology or technique at fault; rather, it's usually the built-in limitations of human beings.

"It is quite clear that man is a miserable component in a communication system. He has a narrow bandwidth, a high noise level, is expensive to maintain, and sleeps eight hours out of every twenty-four. Even though we can't eliminate him completely, it is certainly a wise practice to replace him whenever we can" (Miller, 1967, p. 50). With Miller's tongue-in-cheek commentary on humans aside, we are left with many questions about the nature of communication. As you think about the models and definitions just presented and reflect on your own use of the word communication, consider these questions:

- What is the scope of behaviors or interventions you consider to be communication?
- Does the scope include just individuals or groups?
- Must there be intentionality or can communication be involuntary?
- Must there be a criterion of effectiveness?
- Is the theory prescriptive or descriptive?
- Does communication have to be two-way?
- How complex is the situation that a model can describe in terms of individual messages and the communication setting?

Persuasion

Although a good deal of communication within organizations consists of just providing information or reporting, most of the messages that professional communicators design are intended to influence. Through a training program, we hope to encourage production personnel to follow safety precautions; through a graphic depicting company earnings over the past 5 years, we intend to support stockholder inclinations to invest more money in the company; through a rousing product launch meeting, we convince sales representatives that this new product will provide opportunities for big commissions; through a series of newspaper articles, a city government tries to dissuade residents from swimming in the reservoir.

Theories of persuasion began with Aristotle's *Rhetoric*. Aristotle said that persuasion was based on

- *ethos*—ethical or personal appeals, such as credibility and the perceived character of the source
- *pathos*—emotional appeals to our desires for status, security, and love
- *logos*—logical appeals based on scientific facts or rational conclusions

Most persuasive communications are still built on these principles. When we look at what influences people, several factors always emerge as important:

- *source credibility*—the creator or bearer of the message, and her perceived competence, honesty, motives, and charisma
- *receiver motivation*—the likelihood that the receiver of the message wants to change her ideas or behavior to meet some personal need or goal
- *psychological or physical barriers*—the internal and external factors that may inhibit a person from changing her attitudes or behavior

A number of interesting experiments have been undertaken to try to pinpoint the effects of source credibility. Researchers have conclusively found that the person who delivers a message strongly influences how much is remembered and believed. Of course, the credibility of the source depends on the context and the message; for example, your grandmother might be a highly credible source when it comes to gardening, but she might not be very influential in choosing a presidential candidate or a computer. People also try to assess the source's agendas, hidden or otherwise, when interpreting a message. They'll often ask themselves why the source is trying to convince them of something. Finally, the attractiveness of the source is important. Attractiveness, however, doesn't equal physical assets alone. A person is attractive to us if she represents something we'd like to be or someone with whom we can identify. A physically unattractive but highly successful entrepreneur may be an attractive source to potential investors.

In today's increasingly mediated communication environment, the bearer of the message is not always the source of the idea or data. For example, a professional actor on videotape may provide information on a new manufacturing technology or the company spokesperson may report on the CEO's condition after a heart attack. Although many communicators tend to choose message-bearers based upon how professional they look and sound, they may be the least credible sources. I have seen bank tellers laugh at interactive video programs in which an actor who is obviously not an experienced teller tries to convince them to use active selling techniques. I've also seen those same tellers listen intently to less polished but authentic bank employees. A colleague of mine involved in training for a major utility once gave me a tour of their training facility, pointing out rows of videotapes they'd purchased on various aspects of power plant operations. He said that they were quite effective until one of the employees pointed out that he'd seen the videotape instructor on a coffee commercial the night before.

No matter how convincing the source may be, people won't change their attitudes or behaviors unless they have some incentive. If they are perfectly happy the way they are, there is no motivation for them to modify their situation. Many organizational development specialists contend that unless an organization or individual is experiencing severe problems, they won't change. This is the reason that many effective commercials highlight shortcomings that the audience can usually find within themselves, such as bad breath, a boring life, or a dingy kitchen floor.

Luckily for communicators, most people have no shortage of needs. The most famous categorization of these was developed by Maslow (1970) in his *hierarchy of needs* (see Figure 2.5). Maslow stated that people have a series of needs, ranging from basic physiological needs (food and shelter, which are required to remain alive and safe) to higher order psychological needs (to maximize one's human potential). Unless the basic needs are satisfied, it's pointless to appeal to the other needs. However, if the final need of self-actualization is not realized (doing what one is ultimately capable of and fulfilled in pursuing), a person will never be truly satisfied.

An important step in planning communication programs and designing messages is to identify the audience's level of need so that you can develop an argument that appears to provide a satisfying solution. Common motivational appeals used in corporate communication are fear, the aspiration for power and affiliation, a drive for achievement, and a desire for financial gain.

Another factor in persuasion involves the reduction of obstacles. One major obstacle in influencing people to change their attitudes or actions is their existing opinions and patterns of behavior. The stronger the opinion or the more often they behave in a particular way, the less likely it is that they can be persuaded to think or act otherwise. If a message is consistent with existing beliefs and activities, it is likely to be adopted and acted upon. A series of consistency theories were developed over the past decades that tried to describe and predict these situations.

Osgood and Tannenbaum (1955) proposed the *congruity model.* When two elements are associated and they are perceived to be consistent, then there is congruity. If not, there is incongruity. People naturally want to make sense out of the seemingly random events that occur in the world around them and, therefore, they seek consistency or congruity. Messages designed to help people achieve this cognitive order will most likely be accepted, while messages that appear confusing or incongruent with people's past experiences and beliefs will likely be ignored or rejected. For example, when the popular children's TV host Pee Wee Herman was arrested for indecent exposure in a Florida theatre, many of his fans found it hard to believe that his arrest wasn't a mistake. They found it incongruous that a person could appear so wholesome and funny but could apparently be leading quite a different personal life.

Festinger (1957) developed a theory of *cognitive dissonance,* which is a concept similar to Osgood and Tannenbaum's congruity model. Festinger maintained that people try to reduce tension or uncertainty that is caused when there is a mismatch of information with their existing beliefs or experiences. He believes that the greater the dissonance, the greater the need for change.

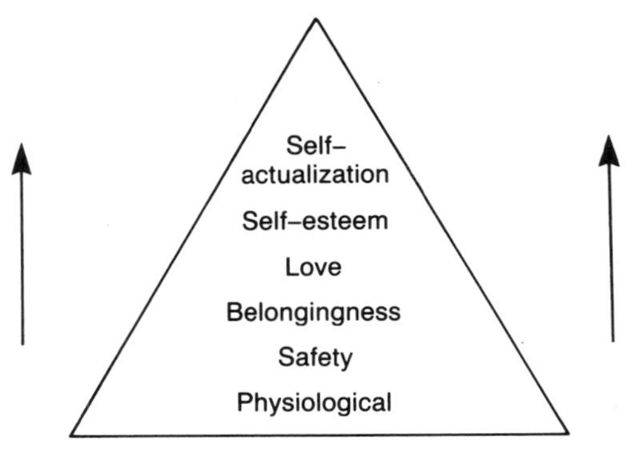

FIGURE 2.5 **Maslow's hierarchy of needs. [Data from Maslow, A. (1970).** ***Motivation and personality.*** **New York: Harper & Row.]**

Fishbein (1967) contrasted the concepts of attitudes and beliefs. He sees belief as a probability statement. When a person has a belief in something, he infers that something exists (e.g., a school superintendent may believe that the parents in his district support his policies), or that a certain outcome will result from his belief (e.g., a policy of silence in the halls will lead to better discipline in the classroom). In contrast, attitudes are evaluative. A person's attitude toward something is the sum of the strength of his beliefs about the thing and his evaluation of those beliefs. For instance, the school superintendent may believe that he has the backing of parents, but he may have an attitude of "I don't care" because he's more interested in the opinions of teachers and of his peers.

Understanding consistency theories helps us to recognize that it is very difficult for us to change attitudes, especially in the following instances:

- when they are held for a long time
- when they are rewarded and reinforced
- when they are associated with strong personal concerns
- when they are expressed publicly
- when they are consistent with logic or experience
- when they are supported by one's peer group

Although it may be tempting to use fear appeals, it's been shown that strong fear appeals make attitudes even more difficult to change. For example, if you'd like to encourage older women to have a mammogram done each year, it's important not to emphasize any particularly anxiety-arousing aspect of the situation. If women are too fearful of cancer, they'll be afraid to find out if they have the disease in the first place. If, however, the possible outcomes don't seem too frightening, the message may be more acceptable. People naturally guard themselves against fear, and they will most likely discount strong fear appeals as being exaggerated or just plain false.

One useful heuristic in developing persuasive messages is the Rule of Two (Bassett, 1968). This theory stresses that single, isolated communication programs often fail, and that they may even have a negative effect on attitudes or productivity; however, when isolated efforts are put together, results begin to show. Bassett claims that people wait for confirmation of the legitimacy and seriousness of the original communication before acting on it. If this confirmation is not soon forthcoming from some independent source the orginal communication is rejected. This is especially true when the audience is either indifferent to a message or is reluctant to respond to it. For example, if a coworker says a boss is unfair, a new employee will probably store that information for future reference. However, if a second source provides the same information, the person may accept that opinion as fact and start acting on it. Bassett warns that this Rule of Two is unconscious and passive. "People do not systematically seek out double inputs. Indeed, the fact that both come by accident appears to be an important element of their credibility" (Bassett, 1968, pp. 58–59).

According to the Keep America Beautiful organization, the single most memorable message ever sent to Americans about littering is the classic public service announcement with a Native American chief paddling his canoe in a river full of industrial waste and trash. As he watches a bag of garbage thrown from a passing car, the slogan "People Start Pollution, People Can Stop It" appears on the screen and a tear runs down his cheek. However famous this spot, critics such as Robert Cialdini, a professor of psychology, comment that the message may have contained counterproductive elements. "In addition to the laudable (and conceivably effective) recommendation in the ad urging viewers to stop littering, there is an underlying theme, as well, that a lot of people do litter." Cialdini and some of his students found that people littered less in clean areas than in dirty areas, and more if they witnessed other people littering. "Were we to advise the Keep America Beautiful organization on how to revise the PSA, then, it would be to make the procedurally small but theoretically meaningful modification of changing the depicted environment from trashed to clean" (Cialdini, 1989, pp. 222–223).

Many messages within organizations are simple and intended to persuade; vehicles such as public relations releases, employee newsletters, and motivational programs fall into this category. The theories of persuasion can also be applied to the design of appropriate strategies. However, increasingly, there are more complex messages that need to be provided, which involve the teaching of complex physical and conceptual skills. In this case, another set of theories is the appropriate base for development.

THEORIES OF LEARNING

In order for an organization's goals to be met, it's often necessary to provide instruction to its members. With rapidly changing technologies and practices in almost every industry, many companies spend hundreds of thousands or even millions of dollars on training and education. In fact, many psychologists have studied how people learn and have developed prescriptions to enhance the efficiency and effectiveness of learning. The three major schools of learning theory are *behaviorist, cognitive,* and *constructivist.*

Behaviorist approaches

The behaviorist approaches to learning were begun by Watson (1913), who published his view that psychologists should abandon their study of consciousness and instead focus on humans' observable actions. He and other psychologists who subsequently followed this line of thought developed *connectionist* theories, which stated that learning takes place when a stimulus is often associated or connected to some response.

Watson maintained that most human actions are learned through either conditioning or experience gained in the environment, rather than resulting from instinct or heredity. He explained learning in terms of frequency and recency: the more often or the more recently we have made a particular response, the more likely it is that we'll make it again. Thorndike (1913) added to this notion the concept of *reinforcement*, which is a satisfying consequence that occurs after a response to a stimulus. This model of learning is often depicted as S R r (see Figure 2.6).

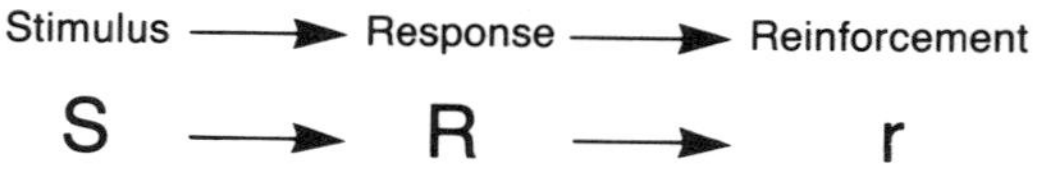

Perhaps the best known of the behaviorist theorists is B.F. Skinner. Although he agreed with Thorndike's theory of stimulus, response, and reinforcement, he believed that not all behaviors were a response to a particular stimulus. His theory says that such actions, called *operant behaviors*, are built into organisms; these include walking, playing, and reaching for food. In contrast, *respondent behavior* is elicited by a particular kind of stimulus. Skinner also talked about *positive reinforcers* (pleasant or satisfying consequences, such as money, attention, or food) and *negative reinforcers* (removing an aversive stimulus, such as stopping an electric shock). This is different than *punishment*, which is the use of some unpleasant consequence (Skinner, 1953).

Basically, the behaviorist approach to learning is that an organism is presented with a given stimulus and responds. If the response is rewarded, the strength of the stimulus-response unit is increased. If the response is followed by negative reinforcement or punishment, the stimulus-response chain is weakened or even extinguished. If a rat presses a bar and receives pellets, it learns to press the bar whenever it wants pellets; thus, bar-pressing becomes a very common behavior. Similarly, if a subordinate finds that each time she expresses an opinion as to how her job might be redesigned, her boss reprimands her for being a complainer, she learns to keep quiet and not to contribute new ideas.

Bandura (1977) developed a theory of social learning. He argued that people are not randomly programmed by the external environment, but that they choose goals and behaviors that are influenced by feedback from the environment. He pointed out that people learn not only by their own direct experience, but also through their imaginations (i.e., by role playing) and by modeling behaviors that they see in others. Bandura and his colleagues conducted a series of well-known experiments in which children who saw film clips of other children playing in an aggressive manner were much more likely to model or repeat those behaviors themselves. Bandura said that reinforcement is informational (we learn unstated rules by getting feedback on our behavior) and motivational (we are given rewards for certain behavior).

Stimulus ⟶ Response ⟶ Reinforcement

S ⟶ R ⟶ r

FIGURE 2.6 Behaviorist model of learning.

Although behaviorists have been criticized for having a too simplistic and mechanistic view of human learning and for emphasizing external rewards for performance, there are other important aspects to their way of thought that should not be overlooked. Remember that Skinner didn't teach his rats to press bars; he designed an environment and chose a task that would maximize the probability that the rat would perform the desired behavior even before it received a reward. A rat in a Skinner box doesn't have much to do other than to paw at the large bar protruding into its cage. When the rat happened to press the bar, which occurred quite frequently, this behavior was simply encouraged by the reward. The rats were also kept a bit hungry, so that the food that dropped after each press of the bar was, in fact, a genuine incentive rather than a boring event.

For professionals in the business of maximizing human performance, it's crucial to examine not just how to teach someone, but also how to design the environment and the reward structure so that it's likely that he will behave, and continue behaving, in the desired manner. Quite often, communication and training programs fail not because a message was misunderstood or forgotten, but because the workplace environment doesn't support the newly learned behaviors or because the wrong people were chosen for the job in the first place.

Cognitive Approaches

Instead of focusing on stimuli and people's outward responses, cognitive psychologists are interested in the mental processes of people, such as decision making, perception, and information processing, which are all elements that make up learning. One of the earliest approaches in this school is known as *gestalt psychology* (Wertheimer, 1945). Wertheimer objected to the kind of consciousness analysis being done by his German contemporaries and thought that the focus should instead be on the whole, or gestalt, pattern of consciousness and perception. Some of the school's early experiments demonstrated that humans combined incomplete figures in order to make wholes; for example, a person would actually perceive a square rather than a series of broken lines when these lines existed in a square pattern, or a series of lights sequentially flashing from left to right would appear as one light moving across a space, rather than as many lights.

Another important concept in cognitive psychology is *insight*. Have you ever stared at a problem, only to have a new and creative insight about how to solve it flash into your mind? This phenomenon, cognitive psychologists would argue, cannot be explained by a simple connection of a stimulus and a response. Wertheimer believed that there was far too much emphasis placed on rote memorization in education, and instead, he advocated finding ways to stimulate original ideas. This concept led the way to the discovery method of teaching.

Bruner (1966) argued that learning comprises the development of the cognitive structures of categories and coding systems. Without these, we would be unable to simplify our perceptions enough to make sense of the world and, ultimately, to learn. These processes of categorization and coding (and subsequently, decision making) have been the focus of cognitive psychologists' descriptions of learning. They have used the common technologies of their day as metaphors to describe the human mind. Some theorists have described the mind as a filing cabinet full of drawers, folders, pages, and words. More contemporary interpretations of information processing have used the computer as a metaphor of input, processing, memory, and output.

Memory is another significant concept in cognitive psychology. It is made up of *sensory memory* (a momentary lingering of a sense, such as a taste or a touch), *short-term memory* (the working memory of unprocessed data that quickly fades), and *long-term memory* (where virtually unlimited information—in terms of processed, coded, information—is stored for immediate retrieval). Our short-term memory is quite limited; if I recite my license plate number, you might remember it for a few seconds but unless you really attended to it, rehearsed it, coded it, and made some decision about where you'd store it in your mental hard disk, it would fade rapidly. (See Figure 2.7.)

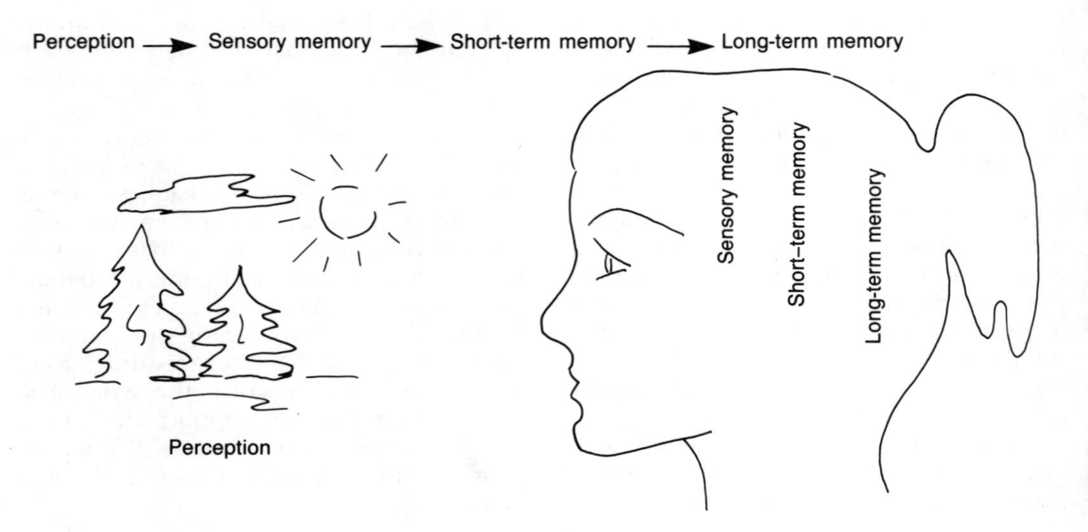

FIGURE 2.7 A map of human memory.

Miller (1956) found that the capacity of short-term memory is approximately "seven, plus or minus two," or five to nine items. One of the ways in which we process (simplify and code) data is by chunking, or clustering information into larger wholes. Instead of seeing a chessboard as a series of many pieces, chess masters chunk the data into just one whole image.

Integrating Behaviorism with Cognitivism

As you read through the behaviorist and cognitive approaches to learning, you may wonder which one is best. As in communication theory, no one model is necessarily superior, and each approach is better in explaining and controlling certain situations. So it is, too, with learning theory.

Gagné (1970), an important integrator of these two schools, has developed one of the most influential prescriptive theories of instruction. In it, he presents five categories of intellectual skills, and with them, various conditions for optimal learning. The first category is *simple learning,* which includes various forms of stimulus-response learning, such as learning to say "sorry" when you bump into someone or associating the Italian word "mangia" with "eat." Gagné says that the optimal conditions for simple learning are contiguity between stimulus and response, as well as reinforcement and repetition. The second type of learning is *discrimination,* or learning to distinguish and to respond appropriately to two or more similar items. For example, a child learns to discriminate between a dog and a cat and to play with each animal in different ways. In order to optimize discrimination, Gagné recommends that you present both stimuli, repeat them, and emphasize what makes them different. The third level of learning is *concept learning,* which involves learning what makes two or more things the same. Like discrimination learning, the stimulus objects should be presented, the ways in which they are different and alike should be emphasized, and reinforcement and practice should be provided. Higher up on the chain is *rule learning,* which involves chaining two or more concepts that help an individual to respond to new and different situations. Examples of rules include not only grammatical structures but also such responses as how to behave if you smell smoke in the middle of the night. In order to enhance rule learning, Gagné says you should tell the learner what is expected of her when the learning is complete, question the learner in order to stimulate recall of the concepts that make the rule, cue the learner to combine the concepts of the rule, and have the learner demonstrate concrete examples of the rule. Finally, the most complex form of learning is *problem solving,* which involves thinking out solutions to complex problems by combining rules. Similar to rule learning, the conditions for problem solving are to present the rules required for the problem's solution, to give verbal instructions or questions to elicit the

TABLE 2.1
Gagné's Conditions of Learning

Type of Learning	Example & Optimal Conditions
Problem solving	Calculate time to travel to Mars—present and elicit rules; guide thought process toward novel combination of rules
Rule learning	Round things roll—provide verbal cues for sequencing concepts properly; make application to specific examples; practice
Concept learning	Recognize "peopleness"—present a variety of stimuli differing in appearance but belonging to the same class
Discrimination	Which key fits your office door—contrast similar and dissimilar stimuli; gradually reduce the differences between stimuli until learner can make the correct discriminations
Simple learning	"Canine" means "dog"—present stimuli in conjunction with responses; provide feedback and practice

Data from Gagné, R.M. (1970). *The conditions of learning,* 2nd ed. New York: Holt, Rinehart and Winston.

learner's recall of the rules, and to guide the direction of thought processes. (See Table 2.1 on page 35.)

Constructivism

Both behavioral and cognitive approaches to learning theory emerge from the objectivist view that the physical world exists in one correct and complete form, and that the job of learning is to come to an understanding of these external objects and processes. Although objectivists acknowledge that different people see the world differently, this is considered to be a shortcoming in their experiences or learning. In contrast, constructivism argues that people impose their own structures on the physical world; therefore, there are different meanings and perspectives for every object or event. Constructivists feel that there is not any one correct meaning that we should be striving to discover or to teach. (See Table 2.2.)

The constructivist approach has many implications for the way that we design instruction and view the learning process. Rather than uncovering the facts and teaching them, the role of instruction within the constructivist framework is to present concepts within authentic experiences. Too often, concepts learned in school or training sessions don't coincide with the way they're experienced in the real world; therefore, much instruction is said to fail. Constructivist approaches seek to help learners to develop their own plans and concepts, rather than to try to force a shared reality that will simply never exist.

One example of a constructivist approach to training was produced by the Advanced Learning Technologies Project at Carnegie Mellon University. Researchers developed an interactive video simulation of corporate software engineering.

The trainee interacts with this virtual world in the role of a just-hired software engineer who is learning the profession. Through direct instruction and simulated experience, the student practices the process of formal code inspection. The learner can access various rooms in the virtual software company, including an auditorium, library, office, training center, and conference facility. Machine-based agents (knowbots) that simulate people, such as a trainer and a librarian, facilitate the use of resources to learn about the code inspection process. . . . Without using multimedia worlds and knowbots, this type of authentic experience is very difficult to simulate in classroom settings. Not only is the instructional environment dissimilar from the corporate context in which software development skills will be used, but also students do not know how to roleplay exemplary, typical, and problematic participants in code inspections. (Eddy, 1992, p. 58)

TABLE 2.2
Objectivism versus Constructivism

Objectivism	Constructivism
The real world exists outside of individuals' interpretations.	Reality is dependent on individuals' interpretations.
The mind processes symbols. Our thoughts reflect and are governed by reality.	The mind builds symbols. Our thoughts are grounded in our own perceptions and interpretations.
Meaning is the external world.	Meaning does not necessarily correspond to the external world; it is dependent upon understanding and is individually determined.
Symbols (communication) represent reality.	Symbols (communication) are tools for constructing reality and represent individuals' meanings.

Adapted from Jonassen, D.H. (1991). Objectivism versus constructivism: Do we need a new philosophical paradigm? *Educational Technology Research & Development, 39*(3), 9.

Theories of Communication and Learning: Parallel Paths

We can see many parallel paths of development among the theories of communication and the theories of learning. Starting with rather simplistic models of processes in which humans are acted upon by their environment or by others' messages, these theories have gradually become more complex and include a fuller description of cognitive processes involved in coding and decoding messages, storing them in memory, and finally using concepts to solve problems. Of course, communication and learning are intertwined; we cannot learn unless we have decoded some message and made sense of it. We have nothing to communicate unless we have learned something. As we move toward an examination of how these ideas fit into organizations, similarities in the development of theoretical perspectives and various assumptions about communication and learning, which have influenced the practice of organizing and managing human interaction, can be seen.

THEORIES OF MANAGEMENT AND ORGANIZATION

Although the theories of communication and learning we have reviewed so far have focused on the individual, we're now going to shift to the group context, specifically work groups and whole organizations.

The Scientific/Classical School

In the early 1900s, managers and scholars began to look at the large organizations that were emerging and sought to prescribe better means for managing them. Some of the earliest writers in this area were Taylor (1919) and Fayol (1949); they attempted to develop a scientific or rational approach to management that was not dependent on the whims of managers, nepotism, or bonds of friendship.

This early school of management developed such practices as time and motion studies, in which workers were carefully observed and timed as they performed each small task that made up their jobs. Management engineers then would redesign the work to reduce the number of discrete motions and time needed to produce an item, and workers were held to strict standards of output. This approach to managing also created ways of selecting personnel scientifically, devised incentive plans, and created organizational structures in which there was a clear division of work, reporting structure, and orderliness. The assembly line approach to creating products and managing people that emerged from this school fit well into the manufacturing technology and social structure of the era.

When you conjure up images of what it would be like to work in an organization run by scientific principles of management, it's likely that the word *bureaucracy* comes to mind. Weber (1949) developed this design for organizations in the 1940s, and it includes the notions of

- hierarchy
- division of labor
- rules and regulations
- impersonality

Most organizations today show a strong influence of this school; there are clear rules and regulations for hiring and firing, well-delineated job descriptions, and an organizational chart that makes it clear who the boss is. Employees are seen as prone to laziness, lacking the ability to think for themselves, and apt to make decisions based on personal preferences rather than on what's best for the company. Of course, in the typical manufacturing plants of the day, strict adherence to procedures, the ability to quickly replace a worker on the assembly line, and clear channels of authority were necessary to keep the system functioning. In addition, the only existing models for management of large groups were found in the military and in the church, both very hierarchical organizations.

Smither (1991) claims that authoritarian management is not outdated and may, in fact, be superior to more humanistic or participatory forms of management, especially in the following cases:

- when employees are poorly educated or uninterested in psychological growth or personal fulfillment from their jobs

- when productivity is more important than employee satisfaction
- when the focus of performance is short-term rather than long-term
- when the manager is comfortable with a directive style
- when there is a good relationship between workers and managers so that they will not resent a directive approach

In these systems, communication is a tool for centralized authority. It is used to give orders in a top-down flow of information through specified channels. Interpersonal communication and behavior are downplayed—the emphasis is on the task and the structure.

The Humanistic School

The humanistic school of management reflects the view that structure and rules alone cannot make productive organizations; it emphasizes the softer aspects of leadership behaviors and organizational climate. Humanists say that productivity is not based on strict rules, but rather on social norms. While the scientific school emphasized money as an incentive, the humanistic school emphasizes the importance of noneconomic rewards. This approach focuses on how workers react as a group, and stresses communication and shared decision making.

Perhaps the most famous studies that served to provide a foundation for the humanistic school were conducted by Elton Mayo in the 1930s. Collectively called the *Hawthorne Studies*, these consisted of a series of studies conducted at Western Electric Company's Hawthorne Plant. Researchers were trying to determine what physical environment or job structure would lead to the greatest productivity. The surprising results were that no matter how the researchers manipulated the environment (i.e., light, compensation, schedules), the productivity increased. The only explanation was that the workers were being paid more attention, knew that they were being studied and, therefore, were more productive. Mayo (1947) also found that there were strong group norms concerning productivity and how work should be done; no matter what the environment or regulations, the group had unspoken standards about how much work was appropriate.

Chester Barnard, an influential writer on management, was the President of New Jersey Bell Telephone Company in the 1930s. Instead of focusing on bureaucratic order, he emphasized cooperation among managers and their subordinates and believed that organizations must have a moral purpose. Barnard (1938) wrote that the three essential activities of a manager are these:

- to provide a system of communication
- to secure essential efforts from individuals
- to formulate and define the organization's purpose

Later management theory turned to managerial styles rather than just their duties. One of the most famous typologies was developed by McGregor (1960), who described two mind-sets of managers that he called *Theory X* and *Theory Y*. Theory X is the concept that people have an aversion to work so they need to be coerced to do their jobs. Theory Y holds that work is as natural as play, and that workers want to realize their potential and contribute to the organization. McGregor encouraged managers to adopt a Theory Y perspective in order to treat workers more humanely as individuals, and to reap the benefits of their energies and creativity by valuing their active participation. Theory Z is an extension of this concept by Ouchi (1981), who believed that managers must realize that people not only have a drive to be productive, but also to generate trusting interpersonal relationships and foster worker participation. Other typologies continue to be constructed.

In General Electric's 1992 Annual Report, Chairman/CEO Jack Welch and Vice Chairman Ed Hood made clear their expectations for managerial styles. They delineated four styles of managers. The first is one who delivers on commitments and shares the company's values. His future is onward and upward. The second type neither meets commitments nor shares their values; in other words, she doesn't belong. The third type misses commitments but shares the values; she usually gets a second chance. The fourth type of managers, those who deliver on the numbers but don't share the values, were warned that their future was not too secure.

This is the individual who typically forces performance out of people rather than inspires it; the autocrat, the big shot, the tyrant. Too often all of us have looked the other way—tolerated these 'Type 4' managers because 'they always deliver'—at least in the short term. And perhaps this type was more acceptable in easier times, but in an environment where we must have every good idea from every man and woman in the organization, we cannot afford management styles that suppress and intimidate. Whether we can convince and help these managers to change—recognizing how difficult that can be—or part company with them if they cannot will be the ultimate test of our commitment to the transformation of this Company and will determine the future of the mutual trust and respect we are building. (Welch & Hood, 1992)

Likert (1961), similar to McGregor, developed a typology of management styles. His four types are

- exploitative-authoritative
- benevolent-authoritative
- consultative
- participative

These forms of supervision view the manager in a variety of ways, from a rather selfish tyrant to a benevolent, paternalistic dictator to one who asks his subordinate's views to one who works with a team to guide them in self-directed work. Likert also developed the *linking pin* concept, which insists that a manager is a link to other groups rather than the ruler of his own group.

The Managerial Grid (Blake & Mouton, 1964) is still a popular tool used to describe five managerial styles. It is often used as an instrument to analyze an individual manager's tendencies and preferences within the context of organizational analysis projects and management training courses. Using a questionnaire, a manager's predisposition for concern for people versus concern for production is mapped. Blake and Mouton have advocated that managers see these as complimentary rather than as mutually exclusive goals. A good manager cares both for his subordinates and for the production goals of the unit.

Since the humanistic school emphasizes that employees be stimulated to participate and to contribute to the organization, theorists have been asked to develop concept systems for motivation. The *motivation-hygiene theory* (Herzberg, Mauener, & Snyderman, 1959) is one of the most commonly cited works in this area. Herzberg claims that certain work factors are in the category of *hygiene;* in other words, they eliminate dissatisfaction but themselves do not lead to satisfaction. Factors such as salary or work conditions and policies fall into the area of hygiene factors; a poor salary will cause dissatisfaction, but a great salary will not necessarily lead to a happy employee. In contrast, *motivators* actually increase satisfaction; some of these include achievement, recognition, responsibility, and the opportunity for advancement.

Also interested in the factors that led to satisfaction, Lewin (1948) is widely regarded as the founder of the field of group dynamics, or *field theory.* He stated that people operate within a complex energy field and that the group is one important factor in this field. A group is more than the set of its members; for example, groups evolve their own goals and life space, and provide security and a context. However, each person must also have freedom outside the group. Dissatisfaction will occur from two extremes: (1) if individuals don't have enough freedom from the group, or (2) if the group is too weak to be functional.

A more contemporary view of motivation centers around factors that can enhance the motivational potential of jobs. Hackman and Oldham (1980) have suggested five core dimensions:

- skill variety
- task identity
- task significance
- autonomy
- feedback

These factors should be considered when designing jobs, rather than finding artificial and external means of increasing motivation among workers.

A number of practices have emerged from the human relations movement. These practices center on fostering honest and open relationships among peers and between supervisors and subordinates, on group decision making, and on the plasticity of the roles and relationships among individuals. In order to train people to behave in this manner, *T-groups* (training groups), a common technique in the 1970s, were used to expand interpersonal sensitivity, self-awareness, and collaboration. Facilitators led work groups in exercises that caused people to disclose themselves, to speak more frankly than they were accustomed to, and to tell others how they were feeling. Although there were many misguided approaches to such exercises that resulted in people being embarrassed and annoyed, these approaches also served as attempts to break people out of their very rigid, bureaucratic molds.

More recently, the *quality circle* approach has been embraced by many organizations. This approach involves holding regular meetings of work groups or informal teams of people from different departments who work on ways to continually improve the organization's function. Self-directed, or high performance, work teams are another similar approach to management in which groups of workers broadly share the functions of production as well as typical management decision making, such as performance appraisals, training, planning, and budgeting.

The humanistic school has moved from a model of *human relations* (it's nice to treat people decently) to a model of *human resources* (good organizational climates lead to enhanced productivity). Although the influences of this school are powerful, it would be a mistake to assume that the bureaucratic school has been replaced by the humanistic school.

Critics of humanistic approaches say that they promote dysfunctional practices (i.e., insensitivity to rules). Furthermore, it has not been demonstrated that a participatory management style is always preferred by subordinates, or even if it were, that it leads to a more effective organization. It is simplistic to assume that satisfaction equals productivity. Conflict avoidance is not necessarily good, as conflict can often make for a vigorous organization. Participatory decision making is only valid if everybody wants to participate and has something valuable to contribute. Contemporary magazines are heralding the return of the authoritarian or charismatic manager who can break all the rules about modifying their styles to adapt to their subordinates, and yet can inspire them to "forsake their own goals and adopt those of the leader" (Smither, 1991, p. 41).

Organizational theorists have generally moved from advocating one particular management approach or organizational structure to advocating a *contingency approach* in which certain sets of environmental variables would be examined in order to recommend the optimal strategy. Lawrence and Lorsch (1976) developed such a contingency-based model. They found the major environmental variables to be the nature of the human inputs and tasks to be performed and the predictability of the organizational environment. Lawrence and Lorsch (1976) believed that the greater the uncertainty in an organizational environment, the greater the need for differentiation among members, as well as the greater the need for integration or collaboration.

Unpredictable environments need a cadre of specialists who can make rapid decisions, and systems that promote effective and rapid communication among those specialists. An example of this type of organization is a research lab. These kinds of organizations generally need flatter structures with broader spans of control and decentralized decision making. As we'll examine in Chapter 3, an increasing number of organizations now fit this picture.

What does this mean for organizational communication? As Perrow (1972) stated, we have moved from "survival of the fittest" to "cooperation" in 60 years. Handy (1989) calls it a "culture of consent" because people like to work and want to contribute. Since an organization's main activity is communication, it should be enhanced and should flow in all directions. A manager's job is to create a good environment and to elicit participation. However, organizations still need structure, or chaos and frustration reign. Someone still has to make the decisions, and roles do have to be defined. Although we want to take into account individual needs, organizations cannot operate without some rational means for selecting and managing personnel. The interplay of leadership styles, communication patterns, and organizational structure is crucial in building the concept we'll discuss next—corporate culture.

Corporate Culture

Almost every executive I've talked to in the past few years has been concerned with maintaining or changing the culture of his or her organization. This metaphor of culture has been the key, these executives believe, to creating the kind of environment they need to move the organization in the direction they desire.

Applying the term "culture" to organizations is rather new; most people think of it as a characteristic of ethnic or national populations. However, a number of communication theorists have defined and applied it more specifically to organizations as

> patterned communication. (Hall, 1959)

> culture acts like music for dancers. (Clampett, 1991, p. 52)

> a common frame of reference for interpreting and acting toward one another and the world in which they live. (Bormann et al., 1982, p. 147)

> a pattern of basic assumptions—invented, discovered or developed by a given group as it learns to cope with its problems of external adaptation and internal integration—that has worked well enough to be considered valid, and therefore, to be taught to new members as the correct way to perceive, think and feel in relation to those problems. (Schein, 1985, p. 9)

> the way we do things around here. (Burke & Litwin, 1989, p. 277)

The authors who brought the term culture to every managers' lips are Deal and Kennedy (1982) through their immensely popular book, *Corporate Cultures: The Rites and Rituals of Corporate Life*. Deal and Kennedy assert that an organization must build and maintain a strong corporate culture in order to prosper. "A strong culture enables people to feel better about what they do, so they are more likely to work harder" (Deal & Kennedy, 1982, p. 16).

Among other case studies, Deal and Kennedy (1982) describe the culture of Tandem computers. Tandem supports a widely shared philosophy of working together; this is seen by slogans, such as "It takes two to Tandem," on T-shirts, posters, and other objects throughout the corporation. But beyond this sort of motto, Tandem differentiates itself from many similar companies in more fundamental ways. The company has no formal organizational chart, and few meetings and memos. The president is widely known and believed to be a hero. The culture is open and informal; communication is enhanced, for example, by holding their famous Friday afternoon beer busts.

How can you tell what an organization's culture is like? Generally, researchers figure this out by extensively interviewing employees. The culture consists of elements such as

- inside jokes
- slogans
- rituals
- myths
- beliefs
- language
- values

In many ways, culture is an organization's look and feel, to borrow a term from the software industry. Today, when employees consider prospective employers, they look not only at salary, benefits, and location, but at the organization's total work environment and philosophy. A large part of the culture is communication; it is not just what the organization communicates, but how it does so. Therefore, the communication professional's role in establishing and fostering culture is to share not just facts, but what the organization is like and what it stands for.

As we examine culture in light of the communication models we've presented, you'll remember that frame of reference is important in those models, even in the early linear ones. A common culture makes it easier to communicate. One of the ways in which many new quality initiatives within organizations strive to improve functionality is by providing various departments with a common language or set of terms to use in sharing their ideas and requesting information.

Many aspects of an organization need to be examined and brought into consistency if a strong, identifiable culture is to be created. These include

- *socialization*—recruiting, orienting, and mentoring employees
- *the physical setting*—what the environment says about the company's values and the role of people in it
- *graphic identity*—the look of the logo, print materials, and so forth
- *communication channels*—how people can communicate with each other, including aspects of formality and accessibility

Today, many managers believe that they need to change their organization's culture. Internal departments and external consultants, such as the accounting firm, Arthur Andersen, specialize in *change management*. Perhaps the best-known example of changing a culture is what Lee Iacocca accomplished at Chrysler in the late 1980s. Iacocca himself became a visible disciple for changing the way that Chrylser operated and for the way that the public and its employees thought about the company. To facilitate this changed image, assembly line workers coined slogans like the "New Chrysler." The company not only altered its image, but within a few years, significantly altered its profitability as well.

Most writers and practitioners in the area of change management believe that an organization needs one powerful leader to spearhead the effort. Deal and Kennedy (1982) talk about the concept of the *symbolic manager:* Symbolic managers see themselves as players—scriptwriters, directors, actors—in the daily drama of company affairs" (p. 142). As work life becomes increasingly symbolic and as reality is indeed being constructed in people's minds, the ability of executives to shape an organization through communicating the right messages in a powerful manner has never been more needed.

APPLYING THEORIES

We've reviewed many theories and concepts. Now, how do they apply to the way we design communication systems? Certainly, any simplistic answers to this question will fall far short of providing foolproof recipes, there are nevertheless some heuristics, or rules of thumb, that can help us in our decision making:

1. *People are not passive recipients of messages, as some of the linear approaches to communication models might have implied.* In contrast, they are involved in choosing, processing, and responding to information in unique ways. Therefore, communicating, learning, and managing are transactional and participative processes in which greater involvement generally leads to a more effective process and a more satisfactory outcome.

2. *More communication is not necessarily better, either in instruction, motivation, or coordination.* Although, generally, people within organizations report that they desire more information, this does not necessarily imply sending more messages. It is important to make a distinction here between data and information; additional memos, meetings, or media may only provide facts that are either already known or are irrelevant to the individual's goals. Rather, it is the job of communicators to provide information, or that which is not known but is relevant to one's particular situation. This obviously means tailoring messages for individual audiences. Dissatisfaction can occur when people are given too much data, or when their time is taken up with seemingly endless meetings.

3. *No matter what you do, you will not always be able to change people's opinions of their organizations or managers.* Some managers believe or hope that a new communication campaign or program will turn their organizations around. Although well-designed interventions can indeed help foster a good organizational climate, it is no substitute for good decision making, a marketable product, competent production and financial management, and honesty.

4. *Organizations need to seek a balance between authoritarian, decisive, top-down,*

and rule-based communication and participatory, open, creative, and discovery-based communication. Neither the older scientific approaches nor the newer humanistic approaches to management are necessarily correct; thus, a situationally based set of theories will prove more useful.

Identifying and Testing Assumptions

Most professional communicators will not admit to subscribing to any particular theory of communication or of learning. However, everybody makes decisions based upon some notion as to what accounts for certain consequences. Let's examine some actual cases (with details changed to protect the guilty!), and try to discern the theories under which the individuals involved were operating.

Case 1: The Mind Dump. An organization hired a group of marketing representatives to work in a new business unit. Some had spent more than 10 years with the company in another capacity and, basically, understood the product and the company's operating procedures, while others were fresh out of college. The new training manager decided to offer a 9-day seminar on both the technical and sales skills aspects of the job. Well-versed outside consultants were brought in to present the seminars, which were supported (or perhaps weighed down!) by approximately 30 pounds of manuals and books. The seminars provided almost no opportunity for practice of or feedback on any of the skills; therefore, most of the 9 days were spent watching, listening, and occasionally taking notes.

My firm was asked to evaluate this method of training and to recommend ways in which this content might be packaged in the future so that the information could be delivered to one or two reps when hiring for this position slowed down. In our observation and interviews with participants, we found that most of them understood little of what was being presented. They were literally overwhelmed with thousands of facts, formulas, and procedures. However, the training manager felt that the seminars were quite comprehensive, effective, and relatively inexpensive for all that was being taught. Much was taught, but little was learned.

What were the assumptions that led the training manager to design such a seminar, and to think that it was successful? He believed that the message was delivered. Since he was operating from a linear communication perspective, no attention was paid to feedback or receiver responses. The more-is-better theory of communication was the underlying principle. The manager never thought about the limits of human memory. His approach to improving performance was to tell it all, and hope that everything would be learned and memorized. Since there was no formal evaluation of the course, it seemed like a good way to prepare the new sales reps. However, a 30-minute videotape or 50-page manual, covering a few important facts or procedures that a rep would need in her first few days on the job, would have resulted in as much or more learning at a fraction of the cost.

Case 2: Let's Talk. A marketing department needed to develop a new product brochure. The executive in charge requested input from every regional manager, a process that lengthened the process by about 6 months and wound up dissatisfying many regional managers, as not every manager's input could be incorporated into the final design. In the end, the brochure looked a lot like what the communication department had designed in the first place, but the client was sure that the brochure would be better accepted because everybody had bought into the process. Actually, the regional managers were angry at being asked to spend time on something that they felt wasn't in their job description or within their areas of expertise.

What were the client's assumptions here? She had no trouble recognizing that communication is two-way; unfortunately, she didn't know where to stop. She believed that more input is better and that her subordinates wanted to share their opinions. While a participatory decision-making style is quite effective, not every decision needs to be made by the group. The organization was facing a serious recession. The regional managers were more worried about making their sales quotas than in what form the new brochure would take. In this case, she was too focused on human needs and not attentive enough to the production needs of the organization.

Case 3: The Videodisc that Spun Off Course. A large hotel chain commissioned the production of a slick interactive videodisc for front desk managers on customer relations. The organization realized that it was important for people in this job to uphold an image in the eyes of the customers, yet these people were often overworked and underpaid. So, they spared no expense in selecting the newest technology and the most professional actors and production crews available. However, when the program was released, they found that the intended audience was amused at the rather phoney situations and movie-star actors, or even worse, they were downright angry at the hotel's obvious outlay of this kind of money on a program. They felt that the situations and models were unrealistic and that the company would have done better by giving them raises or better working conditions with the money they had spent on the videodisc hardware and software.

What were the faulty assumptions here? The hotel's training manager correctly acknowledged that tailored messages are generally more effective and that opportunities for testing and feedback promote learning. Both of these elements are supported by interactive videodisc. However, he incorrectly assumed that the beautiful actresses would be good role models; in fact, they were so far from people whom desk managers could or would aspire to be, that they became objects of ridicule. The program wasted all of its potential since the senders of the messages were obviously not credible. The training manager also assumed that the audience had some incentive to change their behaviors, or that their shortcomings were due to a lack of training rather than a lack of motivation or other problem inherent in the job and in the environment. Although the videodisc could certainly be shown to teach some facts and behaviors effectively in a research situation, it did nothing to improve the performance of that hotel's front desk clerks.

By the way, the hotel decided to persist with the technology, but cheaply produced subsequent programs using in-house employees as talent. Those programs proved to be extremely well-received and effective.

With an overview of communication, learning, and management theories as a base, we'll now examine how communication can help to drive organizations as they emerge and renew themselves to meet the challenges that face them on the road ahead.

REFERENCES/ SUGGESTED READINGS

Bandura, A. (1977). *Social learning theory.* Englewood Cliffs, NJ: Prentice Hall.

Barnard, C. (1938). *The functions of the executive.* Cambridge, MA: Harvard University Press.

Bassett, G.A. (1968). *The new face of communication.* New York: American Management Association.

Berlo, D. (1960). *The process of communication: An introduction to theory and practice.* New York: Holt, Rinehart and Winston.

Blake, R., & Mouton, J.S. (1964). *The managerial grid.* Houston: Gulf Publishing Co.

Bormann, E.G., Howell, W.S., Nichols, R.G., & Shapiro, G.L. (1982). *Interpersonal communication in the modern organization,* 2nd ed. Englewood Cliffs, NJ: Prentice Hall.

Bruner, J. (1966). *Toward a theory of instruction.* Cambridge: Harvard University Press.

Burke, W.W., & Litwin, G. (1989). A causal model of organizational performance. In J.W. Pfeiffer (Ed.), *The 1989 annual: Developing human resources.* San Diego, CA: University Associates, 146–147.

Cialdini, R. (1989). Littering: When every litter bit hurts. In R. Rice & C. Atkin (Eds.), *Public communication campaigns* (pp. 221–223). Newbury Park, CA: Sage Publications.

Clampett, P. (1991). *Communicating for managerial effectiveness.* Newbury Park, CA: Sage Publications.

Cormon, S.R. (1990). "That works fine in theory, but . . . " In S.R. Corman et al. (Eds.), *Foundations of organizational commu-*

nication: A reader (pp. 3–10). White Plains, NY: Longman.

Daniels, T.D., & Spiker, B.K. (1991). Perspectives on organizational communication. Dubuque, IA: Wm. C. Brown.

Deal, T.E., & Kennedy, A. (1982). *Corporate cultures: The rites and rituals of corporate life.* Reading, MA: Addison-Wesley.

DeVito, J.A. (1982). *Communicology: An introduction to the study of communication,* 2nd ed. New York: Harper & Row.

Eddy, C. (1992, May). The future of multimedia: Bridging to virtual worlds. *Educational Technology,* 54–50.

Farace, R.V., Monge, P.R., & Russell, H. (1977). *Communicating and organizing.* Reading, MA: Addison-Wesley.

Fayol, H. (1949). *General and industrial management.* Translated by Constance Storrs. London: Sir Isaac Putnam.

Festinger, L. (1957). *A theory of cognitive dissonance.* Stanford, CA: Stanford University Press.

Fishbein, M., ed. (1967). *Readings in attitude theory and measurement.* New York: John Wiley & Sons.

Foltz, R.G. (1961). Communication in contemporary organizations. In C. Reuss & Silvis D. (Eds.), *Inside organizational communication.* New York: Longman.

Gagné, R.M. (1970). *The conditions of learning,* 2nd ed. New York: Holt, Rinehart and Winston.

Hackman, R.J., & Oldham, G. (1980). *Work redesign.* Reading, MA: Addison-Wesley.

Hall, E.T. (1959). *The silent language.* Garden City, NY: Doubleday.

Handy, C. (1989). *The age of unreason.* Boston: Harvard Business School Press.

Herzberg, F., Mauener, B., & Snyderman, B. (1959). *The motivation to work.* New York: John Wiley.

Jonassen, D.H. (1991). Objectivism versus constructivism: Do we need a new philosophical paradigm? *Educational Technology Research & Development, 39*(3), 5–14.

Kreps, G.L. (1986). *Organizational communication.* New York: Longman.

Lawrence, P., & Lorsch, J. (1976). *Organization and environment: Managing differentiation and integration.* Boston: Harvard University School of Business Administration.

Lewin, K. (1948). *Resolving social conflicts: Selected papers on group dynamics.* New York: Harper & Row.

Likert, R. (1961). *New patterns of management.* New York: McGraw-Hill.

Littlejohn, S.W. (1983). *Theories of human communication,* 2nd ed. Belmont, CA: Wadsworth Publishing Co.

Maslow, A. (1970). *Motivation and personality.* New York: Harper & Row.

Mayo, E. (1947). *The human problems of an industrial civilization.* Boston: Harvard Business School.

McGregor, D. (1960). *The human side of enterprise.* New York: McGraw-Hill.

McLuhan, M. (1964). *Understanding media: The extensions of man.* New York: McGraw-Hill.

Miller, G.A. (1956). The magical number seven, plus or minus two: Some limits on our capacity for processing information. *Psychology Review, 63,* 81–97.

Miller, G.A. (1967). *The psychology of communication.* New York: Basic Books.

Osgood, C.E., & Tannenbaum, P. (1955). The principle of congruity in the prediction of attitude change. *Psychological Review, 62,* 42–55.

Ouchi, W. (1981). *Theory Z.* Reading, MA: Addison-Wesley.

Perrow, C. (1972). *Complex organizations.* Glenview, IL: Scott, Foresman & Co.

Rokeach, M. (1969). *Beliefs, attitudes, and values: A theory of organization and change.* San Francisco: Jossey-Bass.

Ruesch, J., & Bateson, G. (1951). *Communication: The social matrix of society.* New York: Norton.

Salomon, G. (1984). Television is "easy" and print is "tough": The differential investment of mental effort in learning as a function of perceptions and attributions. *Journal of Educational Psychology, 76,* 647–658.

Schein, W.H. (1985). *Organizational culture and leadership: A dynamic view.* San Francisco: Jossey-Bass.

Schrage, M. (1990). *Shared minds: The new technologies of collaboration.* New York: Random House.

Schramm, W.L., ed. (1954). *The process and effects of mass communication.* Urbana: University of Illinois Press.

Shannon, C., & Weaver, W. (1949). *A mathematical theory of communication.* Urbana: University of Illinois Press.

Shockley-Zalabak, P. (1988). *Fundamentals*

of organizational communication. New York: Longman.

Skinner, B.F. (1953). *Science and human behavior.* New York: Macmillan.

Smither, R. (1991, November). The return of the authoritarian manager. *Training,* 40–44.

Taylor, F.W. (1919). *Principles of scientific management.* New York: Harper & Row.

Thorndike, E.L. (1913). *The psychology of learning.* New York: Teachers College.

Watson, J.B. (1913). Psychology as the behaviorist views it. *Psychological Review, 20,* 158–177.

Weaver, W. (1949). The mathematics of communication. *Scientific American, 181,* 11–15.

Weber, M. (1949). *The theory of social and economic organizations.* Translated by A.M. Henderson & T. Parsons. New York: Oxford University Press.

Weick, C. (1969). *The social psychology of organizing.* Reading, MA: Addison-Wesley.

Welch, J.E., & Hood, E.E. (1992). To our share owners. *General Electric 1992 Annual Report.* Fairfield, CT: General Electric Corporation.

Wertheimer, M. (1945). *Productive thinking.* New York: Harper.

Wiio, O. (1978). *Wiio's laws—And some others.* Espoo, Finland: Welingoos.

3 Communication and the Information-Age Organization

We've established the importance of communication in organizations and taken a quick look at the various shapes it takes, including employee information, public affairs, managerial communication, training, and management information systems. In Chapter 2, we examined the various theories and assumptions that underlie the practice of corporate communication. Now, we're going to shift gears and move toward actually applying these ideas within organizations.

This chapter focuses on the bigger picture of where we're going in corporate communication—not in terms of our specific tools or terminology—but more important, in terms of where we need to take our organizations. Corporate communication does not exist for its own sake; rather, it is an essential tool for shaping organizations. Too often, though, professional communicators talk only to each other and are overly concerned about the specific techniques or equipment they use to produce their interventions. Instead, we need to be more concerned about where we go, not just how we get there. We need to move into the role of tour director, rather than auto mechanic, in terms of driving organizational change.

"Change" is the word on the lips of most executive officers in this era. As the first world industrialized nations of North America and Europe face increasingly stiffer competition from other parts of the globe, and as large, established corporations see their market shares being eaten away by small, aggressive, and flexible young start-ups, there is, as one of my colleagues put it, terror in the boardroom.

Let's look at some startling facts about life in North American business. Twenty years ago, most people worked with tangible products, and more than half of the workers were native-born Caucasian males between the ages of 16 and 34 who were supervised by a manager-to-subordinate ratio of 1 to 7. Today, most work involves interfacing with continuous, symbolic data streams in an environment that changes constantly. The average worker is 40 years old and female. Much of the available work force for entry-level production positions is foreign, poorly educated, and lacking in the basic skills needed to do the jobs required. The ranks of middle management have shrunk in the face of technology, which can replace them within companies that are often on the edge of bankruptcy or have been bought out by their former competition. The manager-to-subordinate ratio now averages 1 to 37 (Braden, DeWeaver, & Gillespie, 1991). Managers can no longer motivate employees by simple incentives, such as raises or titles. The work force today is more interested than ever in the quality of their life, in balancing family and professional needs, and in shaping the directions and values of the organizations in which they spend their time.

COMMUNICATION DEPARTMENTS: FROM GATEKEEPER TO GATEWAY

What is the role of communication in these new (or perhaps, renewed) organizations?

Unfortunately, corporate communications departments are generally considered—for good reason—the tool of the status quo in large bureaucratic organizations. They are often seen as gatekeepers of the news; that is, they make sure that no leaks get out to the press, that employees hear the company's side to every story, and that all meetings are carefully orchestrated so as not to embarrass upper management. Its formalized channels of communication, such as the house organ (newsletter), are designed to circumvent the grapevine or other ad hoc means of lateral communication. Training programs are often little more than corporate indoctrination. Indeed, after interviewing many corporate communications professionals, I think the name of their departments should be changed to "communication prevention"!

Today, most organizations are attempting to move from their highly bureaucratic forms to more flexible, participatory styles of management. As we've observed in Chapter 2, bureaucracy is not necessarily bad or outdated and all of its aspects should not be overthrown. It grew from attempts to make management scientific—to eliminate bias and nepotism and to make policies clear and fair. Taylorism, the set of philosophies and techniques developed by early management theorist Fredrick Taylor in 1919, advocates monitoring employee behavior using quantitative techniques (i.e., time and motion studies). Newer adaptations of these ideas are still seen in today's quality programs, which emphasize statistical measures in describing and improving performance. Many high performance teams are failing to live up to expectations and even some Japanese firms like Honda are moving backward toward centralized management (Smither, 1991).

Structures are changing whether we like it or not. In the 1980s, more than 4000 U.S. companies were merged, acquired, or changed their names and their cultures. Many American companies were bought out by foreign counterparts or domestic competitors. More than 100,000 organizations have requested copies of the Baldrige Award guidelines to help them revamp their internal procedures so that they can compete for this coveted prize for total quality management. Communication expertise has been a major component in helping organizations transform themselves, but professionals have had to move from their old roles of managing or containing communication to facilitating communication and helping organizational members adopt new policies and practices.

In 1986, Owens-Corning Fiberglass Corporation began a restructuring program to avoid a hostile takeover. In less than a year, the work force was cut from 28,000 to 17,000. Exactly 1 year later, employees gathered at company headquarters for their own celebration, called Pink Pride. They wore pink T-shirts to match the color of the company's product, and "laughed, and hugged and toured special exhibits of company products." A few employees without executive support (in fact, despite some opposition) planned it themselves, encouraged by a new-age management consultant who had worked on empowering employees and unleashing their creative energies. This celebration was a concrete example of the toppling of bureaucracy, the need for the establishment of a new corporate culture, and the role of communication in change. After this experience, employees learned to take charge themselves without waiting for management direction or even permission, and to develop ideas and systems without waiting to be asked. One such project was Pink Link, a computer system envisioned and designed by employees that helps customers to place orders. It was completed in just 8 days, with no committees, no reports, and no permissions asked (Cook, 1988, p. 52).

More and more, communicators are being asked to play the role of *change agent* within their organizations. Our creative and persuasive talents and powerful tools may make us the best qualified people to lead organizations into the information age.

Before you, as a communications professional, can comprehend the potential problems and challenges of a given enterprise and eventually design effective communication systems, you need to be able to identify the organization's current status with regard to its internal and external contingencies. Goldhaber et al. (1979) developed a model that describes four organizational states:

1. *Proactive relaxation* (the "honeymoon"). In this state, the environment is stable and internal communications are effective. There is relatively little need for new information, but the organization's communication system is ready (proactive) to cope with an unexpected information overload.
2. *Proactive coping* (the "marriage"). In this state, the environment is unpredictable, but the communication system is effective. The organization is placed in a more challenging situation, but is able to handle rapid changes.
3. *Reactive hibernation* (the "time bomb"). In this state, the environment is stable, but the communication system is ineffective. The organization may be lulled into a state of inactivity by its current predictable environment and, thus, does not prepare an effective way to deal with surprises. Such organizations may survive temporarily, but are in a potentially dangerous situation. Preventive measures need to be taken to bolster the organization's ability to survive a future threat to its stability.
4. *Reactive stress* (the "explosion"). In this state, the environment is unpredictable and the communication system is also ineffective. Organizations in this state are extremely vulnerable. An example is when changes occur quickly and important knowledge rests only in the minds of a few key individuals. If they leave, the enterprise may be left in a state of chaos, unable to produce its product or manage its affairs. Facing an uncertain future, key members may lose faith and leave (or worse yet, mentally leave). Emergency communication interventions are needed in such a case.

As organizations grow and adapt to new internal and external contingencies, communication policies and tools must change as well. Although each enterprise is, of course, unique, there are some global trends in business that need to be kept in mind.

EMERGING AND NEW-AGE ORGANIZATIONS

What does the information-age organization look like? There is no one answer; rather, it's like looking through a kaleidoscope—the images we see are intriguing, have some common characteristics, are in a state of continual change, and are beautifully varied. Most organizations today find themselves operating in increasingly unpredictable environments. This section discusses some of the emerging characteristics.

Flat Structures

In general, organizations will have fewer levels of management; in other words, they will have a flat organizational chart. Instead of tall, rigid layers of managers responsible for separate departments, organizations are moving toward *matrix management* in which employees may have to answer to several project managers for their activities in support of different work responsibilities, rather than reporting to just one boss. Workers will increasingly be responsible for managing themselves. In this simpler structure, the hierarchy is softer and any penalties for ignoring the formal chain of command are generally not invoked (D'Aprix, 1977). In fact, Handy (1989) observes that the word "manager" is beginning to disappear, and in its place, people are using the terms "team leader," "project head," "coordinator," or "executive." "Managers, after all, imply someone to be managed, they suggest a stratified society. An organization could not logically be staffed only by managers, but it could by executives" (Handy, 1989, p. 153). (See Figure 3.1 on page 50.)

Management Based on Information, Not Power

Drucker (1988) believes that most businesses will become knowledge-based, in that they will consist largely of specialists who direct and who discipline their own performance through organized feedback from colleagues, customers, and top management. People who are close to the information, rather than generalist managers or

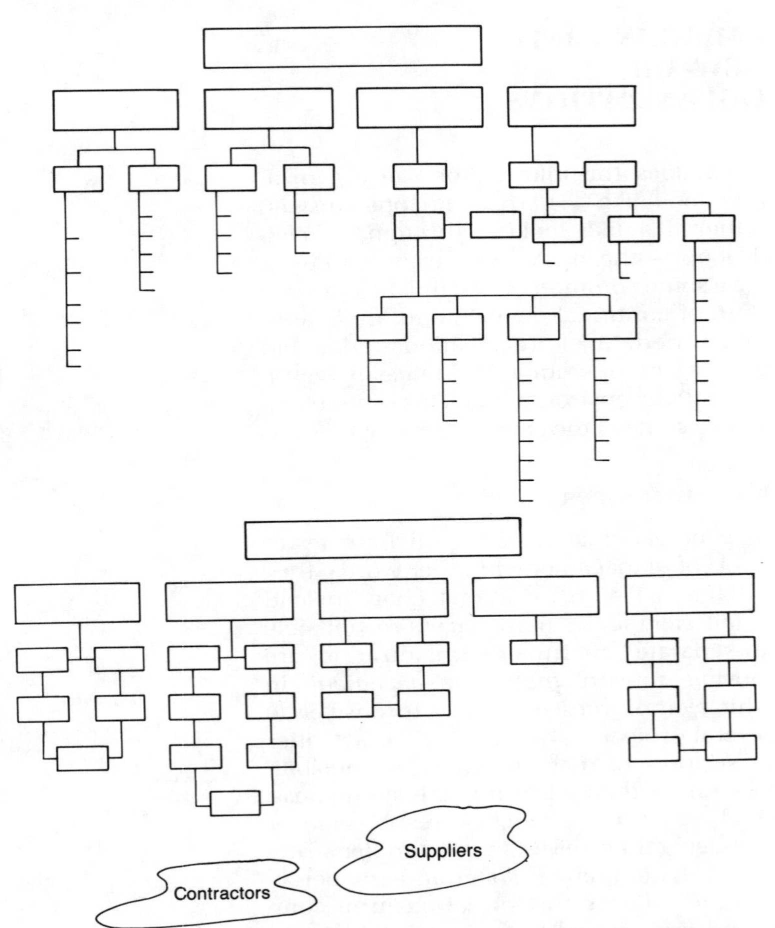

FIGURE 3.1
Traditional versus "flat" organizations.

people with seniority, will make the decisions. He says that information technology makes this transformation inevitable. Sharing decision making and power is one of the most fundamental and frightening changes facing corporate executives today.

Handy (1989) has coined a new formula for success: I^3 = AV where *I* stands for *Intelligence, Information,* and *Ideas,* and *AV* means *added value.* "In a competitive information society, brains on their own are not enough, they need good information to work with ideas to build on if they are going to make value out of knowledge" (Handy, 1989, p. 141). Increasingly, he notes, organizations are talking about their intellectual property.

Faster and More Flexible Structures

Traditional organizations had rather stable products and departments. Today, customers are no longer willing to order merely what's in stock or buy into a vendor's idea of what they need; rather, they will expect their suppliers to design and build products that fit their particular requirements. *Just-in-time manufacturing* is a new production paradigm that responds to customer specifications and builds products that will be shipped out immediately, eliminating expensive stockpiles of raw materials and inventory. The fast and flexible organizations will be the only ones that thrive—or even survive. Close communication with the customer will be necessary to design products that sell, and efficient internal communication will be necessary to efficiently bring those designs into reality. Therefore, "coordination and communication will replace command and control" (Reich, 1977, p. 77).

Coleman Company used to need 2 months of inventory to supply large chain stores with its camping stoves and lanterns. Today, using direct data links to the cash registers of major customers (i.e., Kmart and Wal-Mart), Coleman can produce and ship an order within a week. Two years ago, it manufactured just 20 models of camping gear; today, it sells 140 models in 12 color combinations, while cutting inventory costs by $10 million, reducing scrap by 60% and raising productivity by 35%. Coleman did this through a vision, assistance from a consulting company specializing in speed, employee training, and enhancing communication within the company and with its customers (Dumaine, 1991).

Emphasis on Human Capital

While industrial-age organizations' most important resources were financial capital, buildings, and production equipment, today's organizations' most vital assets are skilled and loyal people. The whittling away of middle management (through self- or computer management), labor shortages, and the movement toward information organizations means that people are not merely replaceable cogs in the wheel that grinds out products (Naisbitt & Aburdene, 1985).

This can be a frightening notion to company owners who suddenly realize that their most valuable resources put on their coats and walk out the door every day—some never to return. Some companies have found themselves in major crises when employees resigned and retired, leaving them without the knowledge to perform certain vital processes.

More Small Companies

Communicators do not always find themselves in the position of changing taditional bureaucratic organizations. In fact, there have never been more business start-ups, and many professional communicators will have the opportunity to build communication systems and practices from the ground up. In 1991, 90% of U.S. firms had fewer than 20 employees, and these small organizations were creating 90% of the new jobs. Helping an organization grow from a spare-bedroom operation to a larger and more complex entity, and shaping the company's image, policies, practices, and communication networks will provide many challenges.

Small companies are attracting much of the talent that used to be siphoned by corporate giants. In an era when even the trusted empire of IBM lays off tens of thousands of people with little notice, more people in the work force are developing a high level of distrust toward large organizations. Since there is less bureaucracy and often an owner or manager who actually runs the typical small organization, they can move faster, adapt to individuals' unique needs and contributions, and more easily accept the advice of communication professionals. Kanter (1989) calls these young, dynamic organizations *newstream,* as compared to the traditional mainstream.

Comparing the Old with the New

Let's examine more specifically the ways in which tomorrow's organizations will be increasingly different from their predecessors

TABLE 3.1
Contrasting the Corporation in the Present and the Future

Traditional	Future
Organization structured and orderly	Organization flexible at all levels
Control based on supervision	Control based on feedback
Position- or status-oriented	Person-oriented
Rule-oriented	Results-oriented
Pay for status	Pay for performance
Tall hierarchies	Flat hierarchies
Duties specified by organization chart	Responsibilities on organizational chart crossed frequently
Centralized	Decentralized
Short-term planning	Long-term planning
Numerous raw materials	Basic raw material is information
Production based on repetition	Production based on creativity
Relies on exhaustible information	Relies on inexhaustible information
Value increased by labor	Value increased by knowledge
Emphasis on achievement	Emphasis on self-actualization
Foundation for work is profit	Foundation for work is systematic acquisition of information
High profit margin is wealth	Access to information is wealth
Information acquired as needed	Information central to operations
Communication based on the need to know	Communication is basic function
Communication tools scattered	Communication core combines telephone, radio, television, computer, and fax
Meetings are a major consumer of time and money	Media systems reduce communication costs

Adapted from Ruch, W.V. (1984). *Corporate communications: A comparison of Japanese and American practices.* Westport, CT: Quorum Books, and Kanter, R.M. (1989). *When giants learn to dance.* New York: Simon and Schuster.

(see Table 3.1 on page 51). As you think about these factors, keep two perspectives in mind: how can I as a professional communicator help the organization deal with these changes, and how will these changes affect my role within the organization?

THE ROLE OF NEW COMMUNICATION SYSTEMS

A communication system in an organization is like the nervous system in an organism. It's what detects and responds to the environment and coordinates the functions of the various specialized organs and limbs. Many organizations I've worked with want to grow and change, but don't realize the dramatic transformations that this requires. Imagine an amoeba trying to evolve into a Rhodes scholar—just making the existing nervous system larger or giving it a fashionable new name won't do the job. It's got to develop a completely new, more sensitive and complex infrastructure. This takes time and expertise.

Schrage (1991) "an organization without an adequate nervous system is a cripple. Media technologies—the memos, telephone calls, fax transmissions, and personal communications—are the neurons, synapses, nerve pathways and spinal cords of the organization." This nervous system shapes what the organization sees, hears, and communicates. Organizations with few receptors for external stimuli now manage complexity by acquiring expertise, that is, by hiring more specialists. However, in doing so, the underlying communication structures are unchanged and the perceptual biases of the organization don't shift. Other organizations respond by buying more technology. "We need to shift away from the notion of technology managing information and toward the idea of technology as a medium of relationships" (Schrage, 1991, pp. 111–112).

There are no simple prescriptions or even radical surgical techniques that communication doctors can use to cure all the ills of their organizational clients. Simple products, or quick fixes, often act merely as a Band-Aid™, covering up more serious underlying problems and wasting valuable time and resources. However, there are some general techniques and technologies that are useful in dealing with some of the major challenges that face us.

Erosion of Hierarchies and Bureaucracy

The Situation. Three factors are forcing organizations into adopting a more participatory, open, and flat structure. First, today's managers are generally more liberal and have been influenced by the humanistic theories popular during the 1960s to the 1980s, the years when most of these managers were in college. Therefore, their inclinations are toward less formality and control. Second, as information becomes the most precious commodity in organizations, it deprives managers of some of the control they were once able to impose on systems in which objects were the most important assets.

Information is unlike other resources because you can't own it—old means of control just don't work anymore (Cleveland, 1990). Information leaks out, and more informed and educated people demand openness. Technology creates more opportunities for communication through channels that cannot be easily formalized and regulated. More knowledge means more power. Therefore, today, it is almost impossible to be a dictatorial and secretive manager.

In the late spring of 1991, things were not going well for IBM, and CEO John Akers let loose with a string of biting criticisms at a meeting of his top managers. Akers told his executives that he wanted his message to get out to IBM employees, and one of his branch managers took that literally. He wrote a detailed report on Akers' comments and sent it out to his coworkers via IBM's e-mail network, called PROFS. From there it sped through offices worldwide—about 25,000 miles in 3 days—with IBM employees adding comments as they passed it along. A few days later, unidentified IBM employees mailed and faxed copies of that memo to major U.S. newspapers, where the story made headlines on May

30. No longer are the official channels of communication, or even the grapevine, the major means of information exchange (Rebello & Schneidawind, 1991).

Employees are increasingly questioning managerial decisions and can't be counted on to automatically endorse new policies. Managers are being forced to no longer make unilateral proclamations, but to facilitate the creation of shared visions. Organizations have seen that it is slow and inefficient to have decision making occur solely at the top of the organizational chart. The people who are closest to the information and situation need to be empowered to resolve problems and to institute new practices themselves. You have undoubtedly been in the situation of asking a clerk or customer service representative to solve a problem only to be told by that person that she didn't have the information or authority to do anything about it. This sort of unresponsiveness to the customer has cost many companies valuable business, and they are quickly coming up with new mechanisms that free their employees to do whatever is necessary to resolve problems and to satisfy the customer.

Another factor in the erosion of bureaucracies is a matter of survival. Organizations must pare down their ranks and procedures in order to work more quickly and cost-effectively. In tight economic times, middle managers are the target of massive lay-offs. It seems that the value of their role is questionable, especially in the face of changes that empower line-level employees to make their own decisions. In addition, new computer and communication systems allow managers to receive reports and share information with more people, so a typical manager's span of control can be widened. Physical proximity is being replaced by proximity in time.

General Electric (GE) has been engaged in transforming its century-old corporate culture, and one of the major results has been a dismantling of the bureaucracy.

We've been trumpeting the removal of bureaucracy and layers at GE for several years now—and we did take out 'Sectors,' 'Groups' and other superstructure—but more remains, and unfortunately, it is still possible to find documents around GE businesses that look like something out of the National Archives, with five, 10 or even more signatures necessary before action can be taken. . . . Layers insulate. They slow things down. They garble. Leaders in highly layered organizations are like people who wear several sweaters outside on a freezing winter day. They remain warm and comfortable but are blissfully ignorant of the realities of their environment. They couldn't be further from what's going on. (Welch and Hood, 1992, p. 4)

Perhaps Toffler (1990) summarizes these dramatic organizational imperatives best: "Bureaucracies, with all their cubbyholes and channels prespecified, suppress spontaneous discovery and innovation. In contrast, the new systems, by permitting intuitive as well as systematic searching, open the door to precisely the serendipity needed for innovation." Toffler predicts that we are moving toward communication systems that are "profoundly antibureaucratic" as information "spills out of the formal channels." Although corporations "spend billions to construct electronic alternatives to the old communication structures . . . all these require enormous changes in the actual organization, the way people are ranked and grouped" (Toffler, 1990, p. 177).

The Role of Communications. Goldhaber et al. (1979) have contended that power in organizations will be increasingly dependent on obtaining, processing, and disseminating (or withholding) information. They contend that successful managers will need to blend what is known about communication with what is known about organizations to build an intelligence network.

Such a network in a flat, decentralized, and primarily worker-managed environment will need to look much different from the typical formalized, multilayered, and centralized communication systems now in place in most organizations. Today's communication tools are primarily used by upper management to give orders and to selectively provide information. Tomorrow's network will need to have a number of nodes or sources of input and support collaboration, rather than the simple dissemination of mandates and data. In many cases, the existence of new multiuser communication networks are accelerating the downfall of bureaucracy since it's quite easy and common through media, such as e-mail, for a person at the lower ranks of an organization to communicate directly with the CEO.

More emphasis will be placed on group decision making. Organization members will often need training on how to do this, and they will require access to new support systems. Some companies use professional facilitators to run crucial meetings. New technologies that can promote idea-sharing include software systems that a group can use to enter and to refine their thoughts during brainstorming sessions, and rooms wired with individual pads for anonymous responses and immediate display of the voting results on a computer screen. Some specially designed meeting rooms, called *decision support centers,* have been shown to generate more than $50 per person/hour savings in meeting time and a 92% reduction in time required to complete a project. Other collaborative meeting environments include Colab at Xerox PARC and the Capture Lab at Electronic Data Systems (Schrage, 1991).

New management information systems will need to capture and to process data from remote sites quickly and easily distribute digested information to the ranks of workers who will be increasingly responsible for making decisions. In place of the ranks of middle managers who served as linking pins to communicate information among various work groups and divisions at regularly scheduled meetings, we'll now see new media systems, such as electronic bulletin boards, e-mail, and even newsletters and video reports, which can quickly disseminate information generated by anyone on the system.

Finally, intelligent systems, such as artificial intelligence-based expert systems and performance support tools, will need to be used to multiply and to broadcast an organization's collective knowledge throughout the work force. Expertise will no longer reside in just a few key individuals; rather, the role of experts will be to build communication and information systems through which they can vicariously provide their skills and knowledge to others as the need arises. If a machine breaks down on the production line, a worker will be able to touch a nearby computer screen, respond to some simple questions about the situation, and receive a diagnosis of the problem including video, audio, and text explanation of how to fix it. A spreadsheet containing crucial financial information will be disseminated throughout the organizational local area network. Users can click on a cell to get an audio explanation of a particular figure or formula, or access an on-line tutorial on budgeting to help them interpret it and create their own financial models.

It is important to note that technology is certainly not the only solution to supporting communication in the new lean and mean organization. Meetings, planning sessions, and team-building exercises will be needed to bolster human relationships, provide a forum of consensus-building, and reteach employees new skills and patterns of interaction. When companies restructure themselves, they are increasingly acknowledging the need to build in an understanding of and commitment to the new organization. Sometimes they'll take everybody away for a retreat for a few days. Many respond by offering team-building sessions or even by sending work groups to outdoor education sessions, such as "Outward Bound," where teams climb mountains, build bridges, and canoe through rapids together to build trust and encourage collaboration.

Even the physical architecture of organizations is changing to communicate on a very basic level the changes in the managerial architecture. One client organization in which OmniCom conducted a workshop actually had a three-tiered dining room. Employees on the lowest level walked through cafeteria lines to choose their food; the second tier offered a slightly better fare for middle management; we were separated from our workshop participants and taken to the top level where a chef carved a huge roast beef served on linens and fine china. Contrast that to another of our clients, Consolidated Diesel, which built an entirely new facility to house its new-age production environment. At this facility, there was one door through which everybody entered, no reserved parking spaces, and only one attractive and open cafeteria in which everyone from the assembly technicians to the plant manager ate.

Think of communication systems as the wiring of an organization. Yesterday's skyscrapers needed a rather simple means of providing a few outlets fed from a central source to each isolated office. Today's flat, spread-out style, as well as an increasing need for power and data, require more sources as well as differentiated channels for different kinds of energy and information. Just adding a few more outlets from the same old source or painting them to match the new color scheme won't do the job. A completely different design and system of conduits is needed to support the new organizational architecture.

Increasing Diversity

The situation. If you look at a 1950s movie or television program depicting typical office scenes of that era, you'll immediately become aware of the dramatic changes in just the way today's organizations look on a superficial level. The work force composition of most organizations is becoming much more diverse. The traditional office scene of middle-class, middle-age, white male managers is an image of the past as women and ethnic minorities increasingly enter the workplace. Facing a shortage of traditionally qualified applicants in the labor pool, organizations are hiring more women, immigrants, ex-convicts, older workers, individuals with disabilities, and temporary or part-time workers (Odiorne, 1990). Within organizations, more native languages, life-styles, and backgrounds are represented. This diversity makes communication much more difficult because there is much less shared experience. While it presents challenges, it also can be an asset to organizations whose multiplicity of viewpoints can help them compete with and often beat out larger, stodgy, monolithic organizations. (See Figure 3.2 on page 56.)

Many organizations are moving from being regional to national to *transnational,* which means they have a network of global semiautonomous locations. In North America, we used to feel that diversity meant the appearance of foreign workers in domestically owned and managed firms. The challenge was to learn how we could best manage them. But today, many European and Asian firms are taking over North American companies, and employees are now in the position of learning to work for foreign-born managers and owners. The changes don't just mean mastering a few words in a new language; rather, the very culture of organizations change under foreign control.

Although North America had characterized itself as the melting pot into which various national languages and cultures blended together to create a homogenous American culture, the more apt metaphor currently is a salad bowl in which a mixture of distinct entities retain their original characteristics, while grouping together to create a larger and more interesting whole. In organizations today, the emphasis is on fostering and adapting to diversity rather than attempting to overcome it.

Michael Cooley, a prominent European technology adviser, has led a group of researchers in defining the new business posture of Europe in the face of the common market introduced in 1992.

> The Japanese and American business cultures have emphasised focus and size respectively whereas the European business culture has emphasised collaboration and variety. . . . It is, of course extremely hard to assess the respective economic merits of different cultures in different situations and at different times. It is clear, however, that collaboration, cultural sophistication and creativity in its widest sense are advantageous in the post industrial society. It is for these reasons that Europe must not become a 'melting pot' as the United States has been. Its regional and cultural variety will provide the basis for addressing diversified markets and responding to the demand of product variety. (Cooley, 1990, p. 2)

Handy (1989) has developed a new symbol—the shamrock—that represents the organization of today, with three different groups of contributors:

- A core of full-time workers without whom the organization would cease to exist. These people are lured to stay with the corporation by large salaries and fringe benefits, and in exchange, work long hours and are willing to bend the rest of

FIGURE 3.2
This internal bulletin describes the increasing diversity in the workplace. (Courtesy of Westinghouse Electric Corporation.)

Westinghouse Corporate Internal Communication — WENS Bulletin 20, April 1991

The changing face of the U.S. workforce

As baby boomers begin retiring about the year 2010, who will fuel America's future productivity machine? The answer — to a large degree — is women, minorities and immigrants, who together will comprise over 80 percent of net new additions to the workforce in the waning years of this century.

Traditionally, industry leaders have had little need to consider the social implications of a radically different domestic labor force. But the U.S. Department of Labor's 1987 publication of a study conducted by the Hudson Institute changed all that. Called "Workforce 2000," the study opened many eyes by establishing a direct link between changing demographics and its potential impact on America's competitiveness.

The study attributed the makeup of tomorrow's workforce to several profound and measurable trends: 1) a slowdown in population growth and, therefore, a shrinking number of workers; 2) an aging labor pool with fewer younger workers standing in the wings to replace them; and 3) the growing share of women, minorities and immigrants working in the nation's stores, factories and offices.

What is the magnitude of these trends? Let's take a closer look.

The winds of change

During the 1990s the country's labor force will be expanding by less than one percent per year. That's significantly lower than the 2.9 percent annual growth rate of the 1970s, and it represents the slowest rate of increase since the 1930s. One reason is that, beginning in the late 60s, America's baby boom turned into a baby bust as birth rates plunged. The aging of the baby boom generation (one in three Americans are boomers born between 1946 and 1961) will push the workforce's median age from today's 36 years to about 39 at the turn of the century.

Besides this gradual aging, the 21st century's labor force will have an appearance quite unlike today's. Although native-born white male Americans comprise nearly half of our current workforce, they'll contribute only 15 percent of its net new additions between the years 1985 and 2000. The workplace will take on a completely different look as more women, minorities and immigrants work alongside white male Americans than ever before in our nation's history.

For example, by the year 2000 women will total 47 percent of the workforce; 61 percent of all women aged 16 or older will be at work; and 75 percent of working women will be in their childbearing years. Obviously, business must be prepared to better meet the unique needs of working couples — needs such as pregnancy leave and child care.

In addition, blacks, Hispanics and other minorities will account for 29 percent of all job seekers between now and the year 2000 — a level that's twice their current share of the workforce. And 600,000 immigrants, largely from Asia and Latin America, are projected to enter the country in the final 13 years of this century — the largest numbers since the first world war.

Jobs will be different, too

As the age, physical appearance, lifestyle and culture of our workforce change, so, too, will the types of jobs being created. The "Workforce 2000" study says that the service sector — industries such as health, education, trade, finance, insurance, real estate, transportation, retailing and government — has grown from less than a quarter of all jobs in 1850 to about 70 percent today, and continues to expand. Manufacturing, which produced approximately 30 percent of all the nation's goods and services in 1955, will account for only 17 percent by 2000. And, although manufactured goods will still be produced in great quantities, goods production is likely to generate an increasingly smaller share of the country's income.

Unfortunately, America seems ill-prepared for this emerging job market. Most of the jobs being created in the 90s will require a higher degree of skill and education. In the meantime, a widening gap is emerging between the advancing skill requirements of the new economy and the relatively low education and skills of many new workers. Today, only about one-half of our high school graduates go on to college, and one in four students doesn't even complete high school.

Bridging this gap will require increased emphasis and improvements in education, training, automation and other strategies both in the classroom and on the job. Such actions are costly but necessary if U.S. companies are to maintain their competitive edge.

The cumulative impact of the changing ethnic, racial and sexual composition of the labor force will be dramatic. Steps must be taken now to help build a workforce that's sufficiently staffed and educated to provide continued growth and expansion of the nation's economy. The years ahead will require more companies to examine and adjust the way they manage this diverse workforce, to provide a new mix of necessary benefits, to adopt new hiring practices that look beyond traditional sources of personnel, and to better educate and train both new and existing workers.

Our next article will take a closer look at how the new workforce will impact American companies.

their lives around their organizational commitments.

- A ring of contractors who do specialized work for the company.
- A flexible part-time and temporary labor force who can provide more help during busy times or extend the operational hours of a business.

Increasingly, as much as 80% of the elements of some organizations' products and services are produced by contractors outside the organization. Furthermore, more people will be interested in working part-time, which aligns well with organizations' flexible needs. However, new contractual arrangements, incentives, and communication systems will be needed to run such a shamrock organization (Handy, 1989).

Diversity brings with it different work styles, career paths, job descriptions, reward systems, and, of course, communication strategies. What was appropriate for male heads-of-households working full-time in paternalistic organizations will not necessarily be right for single mothers, part-time semiretired workers, or self-employed contractors who work for a group of clients out of a home office.

So far, we've just looked at the increasing

diversity within organizations. Today, we're also seeing more diversity among organizations. Companies today are taking on more distinct cultures and creating unique environments, and these differences are often important criteria for potential employees and customers. Not every enterprise looks and behaves the same way. While some of our clients still work Monday through Friday from 8:30 A.M. to 5:00 P.M. in typical office settings with standard equipment and decor and choose their attire based on the prevailing conservative dress code, there are many exceptions.

In 1979, when we started OmniCom Associates' operations from our home (a small townhouse) and later from an electronic cottage situated on a country road with offices overlooking a stream, we were considered a novelty. One of our favorite clients, who for all appearances was the epitome of the conservative older manager, learned to love our style, and when he recommended us to others he always added, "they're different." Although we've moved to slightly larger quarters and have vehicular access via a real paved road now, we have consciously made the decision to maintain our offices in a separate two-story wing of our home. People no longer ask us why we choose to work in this setting, even though we are successful enough to have a traditional office. Most of our meetings with the CEO of our major client organization are held at the kitchen table on evenings or weekends since his company's headquarters are located 5 hours from his home and he spends a great deal of time on the road. Many clients work in different time zones and primarily from various field offices, so our work with them doesn't necessarily end, or even begin, at 5:00 P.M. We find the fax spitting out memos late at night and on weekends. In fact, one of our associates, who also works from her home office, once answered a phone call from Australia in the middle of a Sunday night. Our flexibility and atmosphere are important ingredients of our service to clients.

Our chief programmer, Jim, worked for us for 3 months while living in Japan when this travel opportunity presented itself to him. We stayed in touch by means of an electronic information utility. He programmed while I was sleeping and then shipped updates to me for my comments in the morning. I reviewed his progress during my day and his night. It was one of our most successful and efficient projects. We now comment that we've learned how to work 24 hours a day! Besides the favorable outcomes of the particular situation, Jim also became more valuable to us because of his experiences, and we didn't each have to cope with his simply quitting the job because of a temporary relocation.

The Role of Communications. Many managers simply aren't prepared for the diverse work force and work styles that today's environment demands. It's relatively easy to communicate with other people who are similar to you and who work in the same place at the same time. Trying to convey your ideas to people whose backgrounds and languages are different and who may not even be physically accessible to you on a regular basis is much more challenging.

Of course, all members of organizations must become much more skillful and sensitive communicators in order to succeed, but moreover, the law imposes requirements for the treatment of women and minorities. For example, a recruiter can be liable for a lawsuit if she asks a female applicant how her husband might feel about relocating, and a manager might bring on real trouble by commenting negatively on a male technician's unusual African-American hairstyle.

Organizational publications and media materials must be carefully designed to accurately represent and appeal to the diverse work force. Training materials and sessions need to be designed so that they can be translated into different languages and so that exercises don't conflict with cultural expectations. For example, in Japan, people are accustomed to working in teams and find the notion of individualized study very strange; in doing workshops in Finland, we found that it's considered uncomplimentary to publicly ask questions of a presenter; in Holland, it's rude for a computer-based training program to address you by your first name. In assisting Lederle International to provide computer-based training to its pharmaceutical sales representatives throughout the world, we found that it was not sufficient to merely translate lessons into different languages. Styles of interaction with physicians, medical protocols, and even the legality of various drugs and procedures meant the programs often had to be substantially redesigned to fit individual markets.

Many organizations such as Corning, Digital Equipment, the Bank of Boston, and GTE are offering training on cultural differences. Some companies have found that key women and minority members of the work force were leaving the company because they believed that they couldn't form the necessary relationships. Being sensitive to how formal and informal communication takes place within organizations is very important for corporate communications professionals. It's easy to say that all members of an organization get information in the same way because they all get the newsletter and e-mail accounts. However, what you may find is that important informal training and information-sharing takes place in other settings, such as a golf course or in the men's locker room at a health club, where only a select subset of people are included. While communication managers certainly can't mandate who can socialize with whom, they can try to make their clients aware of some of these relatively invisible inequities.

Modern communication systems are needed for those organizations whose constituents are not necessarily in the office everyday from 9:00 A.M. to 5:00 P.M. Eastern Standard Time. Voice mail, fax, and accounts on electronic information utilities can allow people to send messages and data quickly even if the intended recipient is not present. Today's catalogs feature briefcase computers and printers, modems, fax machines, and cellular phone systems, and there's even a magazine called *Portable Office*. Managers will also need more sophisticated means of appraising subordinates' work than merely walking around to see if they're at their desks.

While mediated forms of messaging may depersonalize communication, this is not always detrimental. Research projects have found that people are much more open and honest when communicating by way of e-mail, conveying thoughts that they'd never say in person. In large systems where users may not be acquainted on a face-to-face basis, ideas are not confounded with the characteristics of the person who generated them in terms of their age, gender, or appearance. These factors may present a real advantage to organizational members who traditionally are at a social disadvantage in the workplace.

F International is a freelance programming firm started in Great Britain by Stephanie (Steve) Shirley in 1962. By 1988, it had 1100 workers and brought in nearly 20 million British pounds. Seventy percent of F International's staff work from home and over 90% of them are women. They believe that their company is 30% more productive than their competition, which is based in traditional offices. Individuals can meet with each other or use specialized equipment in the main office or in one of several branch offices or work centers. E-mail, newsletters, seminars, and a series of meetings with executives keep each worker linked to the group (Handy, 1989).

Having a more flexible communication system means that organizations will be able to tap into a much broader work force. Instead of losing a talented engineer when she decides to spend more time with her newborn baby, a company can provide her with telecommunications equipment to allow her to work at home. People homebound because of illness or disability can become productive workers. In addition, representatives in other regions or nations no longer need to feel isolated from their home offices or removed from communication and training activities.

Increasing Participation

The Situation. Today's workers want to know and to contribute more. We have moved from the situation in the late 1800s and early 1900s, when the North American work force was primarily made up of immigrants who often didn't have formal educations, didn't even really speak English, and were just happy for a job, to the current climate in which employees are generally much more vocal. Now that more companies are publicly owned, financial information and decision factors can no longer be kept as top management secrets. In fact, in many organizations, employees are also stockholders and can (and do) demand explanations about the operations.

As you'll recall from Chapter 2, the human relations and human resources schools of management emphasize that organizations need to increase participation in decision making among members. Everybody needs to contribute, not just follow orders. In line with the erosion of bureaucracy, a higher and more sophisticated level of participation is required of most employees.

More often, we're seeing less division of labor, even in production environments. Factories are moving from the assembly line model in which one person did one routine job to more sophisticated and demanding structures, such as cell manufacturing, where a small team is responsible for many tasks in building a product.

At Consolidated Diesel in Whitakers, North Carolina, groups of salaried technicians are responsible for the self-managed production of major diesel engine components. Each worker is able to operate and maintain a number of highly automated production systems (many of them robotic) and is responsible for making suggestions for improving the system, budgeting, scheduling, ordering supplies, and training other team members on how to do their jobs. When we worked with their human resources staff there a few years ago, we found that one line was working 10-hour days and Saturdays during November so that they could meet their production goals and then take off a whole month around the holidays. Technicians are promoted when they can verify that they have learned a new skill and taught someone else their skills. The idea is to create a pool of highly qualified and flexible people, so that if one person is out, others can step in and do the job. A similar model of compensation based on increasing skills is in effect at Delta Business Systems and General Foods.

One might quite understandably question whether there's an adequate return on the obvious investment needed to support these increasing levels of participation and training. Kanter (1989) cites companies that can document a 30% to 50% reduction in manufacturing costs, elimination of an expensive layer of management, and a very low number of labor arbitration cases that are realized by their investment of 10% to 15% of their employees' total work hours in communication and in training.

Many organizations lose valuable contributors when they decide to take their ideas outside and to form their own companies. To help to retain these key people and to capitalize on their creativity, some enterprises are encouraging *intrepreneurship*, a system by which people can create their own mini-businesses within the larger organization. Other companies set up separate departments, such as the audiovisual department, as profit centers, independently accountable for their own operations and earning the money necessary to support themselves from internal and external clients. Ultimately, some of these are totally spun off as independent companies. Reich (1987) calls these various new ways to capture and to build upon the knowledge of workers collective entrepreneurship.

The Role of Communications. Kanter (1989), a business scholar and popular writer, comments as follows:

> Many opportunities for synergy come in the form of information sharing; thus, the channels need to be established to enable managers and professionals from different business units and different parts of the world to communicate. Post-entrepreneurial companies tend to be characterized by more frequent events that draw people together across areas—executive conferences, meetings of professionals assigned to different businesses, boards and councils that oversee efforts in diverse places and transmit learning from them. Training centers and educational events are a potent means for increasing communication. (p. 113)

Expanding the roles of people who yesterday were unionized hourly workers and today are salaried technicians with managerial responsibilities is not a task that can be done overnight. Companies, such as Consolidated Diesel, have had to set up extensive recruiting and screening procedures, and invest time and money in training. One of the things our organization did with them was to develop simple media production tools and teach workers to design and produce their own video and interactive media programs to document unusual conditions (i.e., equipment malfunctions) and to provide a record for others on how to perform various tasks. The company also provided courses on managing and budgeting and hired in-house consultants and facilitators to assist work groups in their meetings.

As Toffler (1990) documents, the new GenCorp Automotive plant is using similar strategies, providing each worker with $8000 to $10,000 worth of training, encompassing not only technical material, but also problem solving, leadership skills, role playing, and organizational processes. This will enable them to switch jobs and minimize boredom.

Executives are increasingly feeling the need for people to buy into change. No longer can a division manager simply announce a switch in operating procedures or philosophy and expect everyone to agree, follow, or even understand. Handy (1989) says that in this new culture of consent, organizations will behave more like universities in which collegiality and persuasion are the means for change, not authority and orders. Many managers are frustrated when they feel that they've persuasively communicated a vision and a direction, only to find that their subordinates don't seem to be doing anything about it. Since many changes in management are extremely complex and symbolic, it's not as easy as just saying, "use this new machine" or "fill in the form this way." The changes that management are often trying to incorporate may seem vague and elusive to the workers, and even if they agree philosophically with the changes, they may not know how to implement them.

In order to facilitate management communication and employee participation, many organizations are mounting extensive communication campaigns to introduce new notions and provide opportunities for feedback. For example, Corning Incorporated's education and training department helps units facilitate change by organizing large-group interactive sessions involving everyone in that unit. The manager explains his rationale and the decision-making process, and small groups of employees discuss the information in break-out sessions and return to the large group where spokespeople present their opinions and ask questions of management. They leave those sessions, says Ed O'Brien, vice president for human resources at Corning Asahi Video Products, feeling as if they've understood and shared in the process of change.

As communication technologies become cheaper, easier to use, and more accessible to each person in an organization, we're seeing decentralization of the production of the vehicles (i.e., newsletters and videos) that used to be the province of specialized communication departments. Since a central role of management is communication, as management becomes diffused within the corporation, so does the production of various communication interventions. More people will become involved in the design and the development of training, in marketing, and in news materials that directly relate to their own units.

Increasing Speed and Flexibility

The Situation. As Toffler (1970) predicted more than two decades ago, the future is rushing at us at an increasing speed. Organizations are becoming transformed internally, their industries and the economic environment are often unstable, and society as a whole is undergoing radical changes. One of the most powerful underlying themes in all of this is speed. Products and services are being designed, developed, and marketed on a continual basis, and often there is no such thing as a stable product line.

Companies simply move faster and across larger distances. Dow Chemical uses computer systems to continually analyze foreign currency rates, transportation systems, and local production costs to identify the most efficient producer of each of its products, selecting among factories on three continents. Two years ago, people would ask if we had a fax machine. Today, a fax is assumed to be as ordinary and necessary to an office as a rest room. Even overnight express delivery of letters is now considered too slow. However, when we still need to use a courier, Federal Express, with their scanner system, can tell us where our package is at any minute.

An important aspect in an organization's success is its efficiency. When profit margins are tight, it's imperative for workers to produce their product or service as quickly as possible. Decision making is moving faster, too, as on-line management information systems provide access to up-to-the-minute data on production and sales. While older batch-style computer systems produced reports only once a month, modern systems now offer continuous monitoring of important indicators and can digest large amounts of data into accessible statistics and analyses. Toffler (1990) goes so far as to assert that speed will be *the* crucial factor in separating the successful countries and companies from those that will take a back seat. He maintains that the United States may very well regain some of the manufacturing jobs lost to the Far East because even though they can do it cheaper, we can do it faster.

The Role of Communications. Earlier, I used the metaphor of communication as the nervous system in an organization. Toffler (1990) believes that primitive organisms and societies have slow nervous systems while more advanced ones process signals faster—speed will separate the less developed organizations and nations from the leaders.

In order to cope with rapid changes, people need more information and training. Unfortunately, in today's corporate climates, the level of uncertainty and distrust is often high. Increases in efficiency can only happen if people feel secure enough to pay attention to their jobs rather than spending their time and energy on speculation and feeding the rumor mill. They also need to be trained in the most effective processes and technologies.

When New York State Electric and Gas (NYSEG) reorganized into two internally competing divisions, there was obviously much concern and confusion among employees. Their human resources and organizational development specialists produced newsletters and videos, conducted meetings, and established an e-mail hot line to receive questions about the situation.

As design, production, and sales occur more quickly, so does communication. Many organizations are struggling to provide training programs, but find that spending several weeks or months to develop a professionally executed course may be too long—by then the content is outdated. Communication programs are quickly becoming perishable products. No longer is it possible for producers to assume that an organization can amortize a program or even the hardware on which to play it back over several years. I distinctly remember editing a video orientation program for a major bank, having had the script approved only days before. In the middle of the edit session, a call from a vice president came in for our client, who learned that about a third of the video was now incorrect because the bank had just decided to reorganize its operating units. The people, managerial structure, and names of departments changed so often in that bank, which was in the middle of being bought out by a foreign holding company, that I sometimes joked that we could have a full-time job just rewriting and reediting that one video program!

Obviously, we need faster and more easily updateable communication tools. The problem is that many high-tech vehicles, such as computer-based training and interactive multimedia, are often produced by outside vendors and are neither easily nor inexpensively edited. Many organizations have paid over $100,000 for slick videodisc programs only to find that they don't have access to the source code of the computer program that controls the videodisc and displays text. When they try to go back to the original producers to make updates, it's not uncommon to find that they're out of business.

Not all leading-edge technologies have this problem, though. Systems like e-mail and teleconferencing allow people to communicate almost instantly. When Eastman Kodak found that the company was losing money because their technicians were commonly replacing a large part in their copiers rather than making a smaller, less costly repair, their business TV group quickly got together a national satellite video conference where experts explained the more efficient procedure. The business group at Kodak paid for the teleconference in a matter of weeks based on the money saved by thousands of less expensive repairs.

Another firm involved in the analysis of scientific data for large government projects has their on-line management information system hooked up to a computer graphics program in their executive conference room. Whenever the CEO decides to make a presentation to the press, he has up-to-the minute charts and graphs that are continually and automatically updated and plotted by the graphics software and displayed by a data projector. Information can't wait for slides to be processed anymore; by then, it's outdated. Today's computer-literate managers are often creating their own handouts and slides for presentations rather than waiting for audiovisual specialists to do it. Software now makes it easy to create professional-looking results, and up-to-the-minute information is more valuable to them than are artistic renderings. This shift in emphasis from appearance to speed is having a major impact on corporate media departments who have in the past relied on their aesthetic talents and high-end equipment, and on management's inability to find any other alternatives.

Organizations can't operate quickly unless their communication systems can detect, process, and transmit information quickly. We're not talking about just sending messages rapidly, we're referring to continually sensing the environment and processing external data. A wealth of important information can come directly from the consumer, and today it can be done faster than through the old means of focus groups or surveys. Supermarkets today are using scanners at the checkout to do much more than merely speed up the lines. The data on customer purchases is linked directly by satellite to the stores' home offices and to many of their major suppliers. Some companies provide their customers with direct links to their computers. No longer is there guessing about what customers are buying, the information is available daily.

For example, some businesses in the fashion industry, such as the Italian clothing manufacturer Benetton, collect data on fashion trends, such as what color is hot, and can change manufacturing plans and ship new designs within a matter of weeks. Communication professionals are finding themselves working closely with computer and data transmission experts to bring new systems on-line.

However, finding ways to quickly update and display information is only the tip of the iceberg when it comes to dealing with the continual acceleration of our lives. Change, especially rapid and continual change, often leads to stress. Even when changes are positive for companies and individuals, it is taxing on their systems. As companies engage in downsizing, the nature of work for many people is radically different. Some managers may be forced to go back to carrying tools. On the other end of the continuum, former line workers are now expected to take on managerial responsibilities.

To cope with this stress and prevent it from negatively affecting the organization's productivity, there is a need for such communication interventions as employee assistance or counseling programs. Stress can lead to absenteeism and even to workers' compensation suits. After incidents like bank robberies or suicides, professional counselors often engage in debriefing. Allowing people to talk through the situation can help them get back on track.

When Western Airlines merged with Delta, a consultant designed a mock funeral for Western, which included having managers write down their worst fears about the merger, throwing them into a casket, and having a paver flatten the casket and its contents. After the funeral, managers marched into an auditorium where they were given caps and gowns and participated in a graduation ceremony. The name of Western's employee newsletter was changed from *Update* to *The Best Get Better*, a toll-free hot line was set up to provide updates on the merger and record callers' questions, and its Health Services Program developed seminars and videotapes on dealing with anxiety and stress. "Overall, the success of the merger process at building commitment was a result of Delta's policy and style of implementation on the one hand, and Western's proactive communication and counseling on the other" (Kanter, 1989, p. 87).

Balancing speed with quality, thoughtful decision making, and increasing participation is becoming an important challenge for us all.

Emphasis on Quality

The Situation. One of the most prevalent new themes in organizations is quality. Although most manufacturing organizations have always had quality control departments and the notion of quality would appear to be a common goal for any organizational process or product, American management has embraced the Q-word with an almost religious fervor. It became apparent in the 1980s that the United States had lost its edge to foreign competition. Not only could foreign corporations do it cheaper, they could do it better. Increasingly, U.S. organizations were characterized by sloppy work and rude, incompetent service. In the early 1980s, several individuals began advocating fundamental remedies for these organizational ills, and focused their strategies around the term quality. Today, a large percentage of organizations have started quality programs, using terms such as total quality management, continuous improvement, and zero defects.

There are three prominent names in quality consulting. W. Edwards Deming is an American who started the quality notion in Japan in 1950 and was instrumental in the turnaround and eventual triumph of the Japanese economy. The foundation of his model is SPC, *statistical process control,* a method of examining work processes. In SPC, analysts use statistics and charts to plot quantitative deviations from the ideal in a production process. Deming has also developed a 14-point program for managing productivity. Joseph Juran's focus on quality is the customer, whether that customer is a consumer or another internal department for whom services are provided. He urges organizations to open their lines of communication, listen to each others' needs, and examine the entire company for problems and find ways to eliminate them. Philip Crosby is known for several popular buzzwords in quality, such as "zero defects," "conformance to requirements," and "quality is free." His emphasis is on measuring the cost of nonconformance to standards (Oberle, 1990).

Many organizations have adopted one or more of these strategies for achieving quality. Most have launched into the major management reorganization that is necessary to carry off these principles by implementing quality programs. Generally, a department or staff has been assigned to oversee the quality effort, in many cases, headed up by a vice president.

Adding to the excitement that generally accompanies organizations' launches of quality programs are two prestigious awards for which hundreds of companies apply each year, the Deming Prize and the Baldridge Award. In 1989, Florida Power & Light Company (FPL) became the first non-Japanese company ever to win the Deming Prize, which is Japan's prestigious award for total quality. FPL says their new procedures improved their reliability and customer service while saving them $300 million. The Malcolm Baldridge National Quality Award is sponsored by the U.S. Department of Commerce and named after a former secretary of commerce intent on improving the competitiveness of American business. Every year since 1988, two companies in manufacturing, service, and small business, are awarded this prize. Xerox won this award and in the process improved unscheduled maintenance by 40% and reduced service response time by 27%. Milliken, another Baldridge winner, has reaped a 42% increase in productivity. However, these changes didn't occur quickly or cheaply. For example, FPL spent over $5 million and Xerox spent $800,000, plus the salaries of 17 full-time employees who did nothing but prepare the application for the Baldridge Award (Quality Awards Aren't Free, 1990).

Although a goal of most quality programs is to try to eliminate red tape, many of them seem to have actually spawned more bureaucracy. Some people think that rigid sets of procedures and endless meetings to brainstorm solutions to sometimes trivial problems do nothing to help an organization become more productive. In fact, after FPL won the Deming Award, they disbanded their Quality Improvement Department, Quality Improvement Promotion Group, and their Quality Support Services, and eliminated three levels of reviews. Their management says that they have not abandoned their focus on quality, but rather have learned about it, incorporated it into their routines and, therefore, no longer need the formal support of individuals whose only responsibility is to quality (The post-Deming diet, 1991). Not everyone buys into the tenets of rigid conformance to standards and reduction of deviation; for example, the President of Southwest Airlines says he hopes that his employees regularly break the rules because that's the only way they can adequately respond to unpredictable demands and the environment.

The Role of Communications. An integral part of all quality programs is enhancing communication. In fact, as I talk to many managers who are in charge of quality programs, I feel that their title should be director of communications. All quality programs involve large investments in training; that is, teaching managers and line workers the concepts of quality, how to measure it, and how to creatively solve problems. After people have learned the basic ideas and skills, they need a forum in which they can share ideas. Often, meetings or quality circles are formed in which interdepartmental teams can tackle problems and move toward continuous improvement. Facilitators help departments develop a common language that they can use to communicate their processes and needs to each other. Finally, incentive systems are needed to help change people's behavior. Most quality programs have instituted special columns in newsletters that recognize outstanding contributions to quality and other events that spotlight employees who have made unusual efforts or creative suggestions. (See Figure 3.3.)

Many prescriptions for achieving quality directly relate to organizational communication. Deming's famous Fourteen Points include driving out fear, breaking down barriers, ridding the organization of slogans, and encouraging people to educate themselves and work in teams.

FIGURE 3.3
The Guthrie Healthcare System fosters their quality initiative through communication vehicles, including the newsletter, a brochure, and the "Caught in the Act" program, which recognizes exceptional service. (Courtesy of Guthrie Healthcare System.)

Examples of exceptional service are:

1. Escorting a visitor directly to an area rather than simply providing directions.
2. Providing personal attention beyond job requirements to assist co-workers, patients, families, and visitors (i.e., "not passing the buck")
3. Helping a co-worker change a flat tire.

What do I do with the completed form?

Please place the completed nomination in the slot located in the Cafeteria Quality Awareness Center.

What happens after I nominate someone?

Twice per month, nominations will be reviewed and those individuals selected will be notified and recognized for their exceptional service. Names of successful nominees will be posted in the Cafeteria Quality Awareness Center twice per month.

Guthrie Quality Initiative
"Caught in the Act"
NOMINATION

The following employee was "caught in the act" of providing exceptional service at Guthrie.

Name ____________ Department ____________

What was done: ____________

Why was it exceptional: ____________

Submitted by:
Name: ____________
Department: ____________

Wallace Company in Houston, Texas along with IBM, Cadillac, and Federal Express, was awarded the Malcolm Baldridge National Quality Award in 1991. But only 10 years before, the company was on the brink of failure. They were literally pushed into the quality movement by one of their major customers, the Celanese Chemical Company, who itself was essentially required to adopt total quality management by its customer, the Ford Motor Company. Wallace retained a training and development consulting firm, which conducted a needs assessment. The survey found that there was fear and mistrust in the company, that employees weren't sure that management was sincere about quality, and that many associates said they wouldn't recognize the CEO if he walked past them.

Embracing the quality movement, Wallace virtually reinvented itself. They began to call employees associates, provided more than 19,000 hours of training at a cost of more than $700,000 over a 3-year-period, revised their computer and communication systems, and conducted weekend retreats. While they always had an open door policy, people didn't really believe that their suggestions would be taken seriously. One way of driving out fear was to change the suggestion system. They now respond to all suggestions within 48 hours and implement about 80% of them. Senior managers participate in weekend training retreats and the CEO himself conducts a lot of the quality training. Besides winning the Baldridge Award, Wallace's sales rose 69% in 3 years. They went from near failure to becoming an international model for quality.

However, the quality movement can become an obsession rather than a philosophy. After winning the Baldridge Award, Wallace lost $1 million in the first half of 1991. A consultant hired to diagnose the problem found that company executives were spending too much time giving speeches on quality rather than doing their work (Anderson et al., 1991; Galagan, 1991).

The approach that most organizations have taken to quality is to launch it as a program. In some cases, this has involved large meetings, pep rallies, slogans, logos, training sessions, and mission statements. For example, when AT&T began their quality campaign, they contracted with a multimedia communication company to produce a program that would motivate employees in an imaginative, innovative, and inoffensive way. The producers designed puppet characters based on their study of the company's corporate culture, language, and corporate archetypes. One of the main focuses of the launch was a 17-minute video, "Quest for Quality," which is a journey inside the AT&T logo. "The story's heroes are the crew of the 'Starship Resolution' who have been given the mission to seek out the evil forces undermining the quality of the Alliance" (Quest for Quality, 1991, p. 16).

Many critics of the quality with a capital Q notion maintain that quality shouldn't be a separate program managed by a separate department. Rather, they believe it should be infused with every sector and every individual in an organization. Doing this may involve special interventions that raise awareness and provide incentives and skills; however, the goal for most quality departments should be to make themselves obsolete. In order to establish new behaviors and patterns, communication systems must be set up to facilitate them.

One powerful concept is that *quality is truth* (Handy, 1989). In order for people within organizations to accurately assess their situations, make appropriate deci-

GUTHRIE *Clinic* **"Today"**

JULY 1991

New Employees Welcomed

There are three new employees arriving on July 1, 1991, whose names will become very familiar to you.

Debbie Bush is a Cytotechnologist who is joining our Cytology Department. Debbie worked with United Health Services in Johnson City before coming to Guthrie Clinic.

Marcia Kelly comes to the Guthrie Clinic as Director of Patient Data Collection and Management. Marcia comes to Sayre from Berkshire Medical Center/University of Massachusetts, Pittsfield, Massachusetts.

Lonson Nash is a Pathologist Assistant. Lonnie comes to us from Salisbury, Maryland.

A warm welcome is extended to all.

Several employees have been nominated through Quality's Caught in the Act system to be recognized for their exceptional service. We congratulate these individuals:

Don Stevens, Maintenance, was nominated by Pam DePrimo, Human Resources. Don gave directions to an elderly patient, and when she seemed confused, he took the time to personally escort her to her destination.

Emily Kelley, Administration, was nominated by Floyd Metzger. Emily observed patients in a local pharmacy trying to have prescriptions filled. The pharmacists were struggling because they were unfamiliar with all the physicians and residents' names. She suggested to administration that pharmacies in the Guthrie market area receive a Physician Directory. This suggestion was implemented and will be a standard practice when new directories are printed.

Susan Hall, Clinic Administration, was nominated by Russell Knight, RPH President. Sue encountered a patient seeking information about Guthrie Courtesy Transport. Rather than just refer her to someone else or some other department, she tracked down the requested information to ensure the patient's requirements were fully met.

Check the Quality Awareness Center in the cafeteria. Names of Caught in the Act nominees from throughout the Sayre campus are posted there.

New Physicians

Our medical staff will be increasing in the near future. The following is a list of the new physicians; some of them are old friends and some will be new friends.

Dr. W. Donald Cooke
Allergy .... Sayre
Dr. David Cowger
Emergency Medicine .... Sayre
Dr. John Dawsey
Radiology .... Sayre
Dr. Louis Dubois
Pulmonary .... Corning
Dr. James Fine
Anesthesiology .... Sayre
Dr. Thomas Gustafson
Family Practice .... Pine City
Dr. Elizabeth Ho
Internal Medicine .... Horseheads
Dr. Mal Homan
Endocrinology .... Sayre
Dr. Jacqueline Huntly
Women's Health Center .... Sayre
Dr. Alexander Lalos
Gastroenterology .... Sayre
Dr. Dean Markham
Infectious Disease .... Sayre
Dr. Kirk Musselman
Cardiology .... Sayre
Dr. Michael Najarian
Surgery .... Sayre
Dr. Christine Phillips
Rheumatology .... Corning & Wellsboro
Dr. Todd Phillips
Family Practice .... Mansfield
Dr. Daniel Sporn
Cardiology .... Sayre
Dr. Timothy Sullivan
Vascular Surgery .... Sayre
Dr. Barbara Wiseman
OB/GYN .... Sayre
Dr. Jeffery Wiseman
Surgery .... Sayre
Dr. Joseph Yarze
Gastroenterology .... Sayre

Look To The Sky

The new Guthrie One helicopter has arrived.

The Bell 412 SP (special performance) has increased cabin space for patient care. This helicopter will hold three patients, has an extended fuel range of 400 miles, and is equipped with color radar for flying in adverse weather. The Guthrie Flight Team will be training in the new helicopter, doing mock runs and emergency procedure training for the first few weeks of its arrival.

With the arrival of our new helicopter, Guthrie reaffirms its commitment to providing the region with the highest level of prehospital care.

FIGURE 3.3
(*Continued*)

sions, and communicate effectively with one another, their information must be accurate and clear. Some fundamental problems in people-oriented systems are progressive cover-ups of mistakes and bad news, withholding of essential data, white lies, faulty assumptions, and avoidance of asking important questions of employees and customers. Truth is more than just a matter of not lying—rather it's knowing and telling what's accurate. When looked at this way, truth is an essential ethical standard and a crucial concept that professional communicators must deal with.

While quality efforts may be introduced as a special program or communication campaign, for this concept to take root in organizations, it must become a way of life. Communication departments need to put into place regular channels for promoting and rewarding the behaviors and standards that their companies have defined as their goals. Many of the special practices, such as meetings, suggestion boxes, and publicity for outstanding efforts, that characterize quality campaigns should become standard procedures and systems managed by communication professionals. They should gradually move from being identified with the quality department or effort to being ordinary parts of the management and communication system (Landes, 1992).

Social and Ecological Consciousness

The Situation. While social responsibility used to be thought of as an individual's prerogative, it has now become a major criterion for measuring organizational effectiveness and attractiveness. Ecology has moved from being identified with 1960s hippies and doomsayers to playing a major part in the consciousness of individuals and the missions of organizations. With increasing public pressure to reduce pollution and the bad publicity of some major environmental disasters like oil spills and toxic waste dumps, corporations have become much more sensitive to their impact on and responsibility for the environment. The slogan, "think globally, act locally," has become a major theme in corporate communications as organizations realize that they not only have to alter their practices, but they have to change their images as well. And the role of corporations in supporting social goals, such as child care, public education, and health, has burgeoned.

"The final new reality in the world economy is the emergence of the transnational ecology. Concern for the ecology, the endangered habitat of the human race, will increasingly have to be built into economic policy" (Drucker, 1989, p. 133). Information-age organizations need to minimize their intrusion on and consumption of natural resources. They need to heighten management's awareness of their impact on their communities, not just the bottom line. In fact, many contemporary economists urge organizations to measure their contributions to the economy and the environment of their locations. And they need to become good citizens, contributing not only money, but expertise and time to their community organizations.

Ben and Jerry's, a successful ice cream company, gives a percentage of its income back to a foundation that is used to support not-for-profit groups in the community (see Figure 3.4). In Ithaca, New York, they support the Learning Web, which is a means of providing mentoring of local professionals to high-school students. Their store in Ithaca is run by Youth Scoops, an organization made up of teenagers who learn to run businesses ethically and soundly. The company also sponsors a traveling circus, promoting the use of solar energy and attention to the needs of children. Large organizations like AT&T and IBM give their executives sabbaticals during which they contribute their time and expertise to not-for-profit groups, and Georgia Power Company devotes a large percentage of its resources to promoting industrial development and tourism in its region.

The Role of Communications. Most people recognize the role of corporate communications departments in promoting a positive, socially responsible image for their corporations. In fact, they are often accused of covering up or actually lying about undesirable corporate practices, touting the company's role in cleaning up the environment when it is really polluting the local water supply, or bragging about its modern child care policy while systematically pay-

FIGURE 3.4
Ben & Jerry's mission includes improving the quality of life in the broad community, and their annual report includes a social audit, as well as a financial audit. (Courtesy of Ben & Jerry's Homemade, Inc.)

ing women less than their male counterparts.

However, communicators must move out of the job of talking about what other people in the organization are doing and increasingly assume leadership roles to promote socially and environmentally responsible actions. Communication policies and technologies can play a vital role in controlling pollution and waste. New communication systems, such as e-mail and teleconferencing, can allow employees to telecommute from their home offices instead of commuting to the office every day. Almost one million people conduct business, at least partially, from home offices in the United States today. These same systems can eliminate much of the wasteful use of paper by sending messages across phone lines and computer networks instead of using paper and transportation systems. Moreover, they can accommodate the human and personal needs of workers with disabilities, illnesses, young children, or other situations that would benefit from working at home. Finally, corporate communications departments are taking leadership roles in advocating the use of recycled paper and nontoxic inks in company publications and correspondence.

Many organizations tap their communication professionals as important resources, which they offer to local civic and charitable organizations. They coordinate corporate donations, fund-raising drives, and community campaigns. For instance, the director of communications at Pet, a large food company, was responsible for focusing its sponsorship of programs that attempt to alleviate hunger. Media production departments frequently produce promotional and motivational programs for local groups (i.e., the United Way), and video professionals often donate the production of public service announcements that air on broadcast stations. Georgia Power Company donates the resources of its media production group to produce videotapes and interactive media programs, which aim to attract businesses to locate in their area. They were even instrumental in producing an interactive multimedia system that helped Atlanta win the bid to become the site of the 1996 Olympic Games. Even small companies support the professional and civic groups in their communities by providing help in developing newsletters, posters, and concert programs.

Often, communicators have unusual access to the needs and goals of the communities in which they work. They can become leaders in directing management attention and in encouraging company resources to address those issues, and then can ensure that those actions are given appropriate recognition by employees and the public.

Communication as a Value-Added Commodity

The Situation. Most products these days are complex; skill and information are needed to use them correctly. Even simple food products need nutritional labeling and recipe ideas. When people, or especially corporations, buy, they are not looking for a one-night stand, but a long-term relationship. What often makes a product stand out over its competition is the service or information that goes with it. This may include packaging, instructions, assistance from a vendor in setting up or using the product, and formal customer training programs and seminars.

One of the most lucid descriptions of the role of information in our economy has been developed by economist Paul Hawken (1980), who calls this the *informative economy.* He says:

> In every product there is a relationship between the amount of mass and the information that product contains. An ingot of lead has a high ratio of mass to information; a video cassette recorder has a high ratio of information to mass. . . . In order to put more information into products, companies have to put more information into their employees through education, communication, and honesty. (Hawken, 1980, p. 74)

Hawken explains that the U.S. economy is not destined to grow at the rate it did when we were expanding rapidly as a country, and that we will have to learn to do more with less. This means changing from an affluent society with a wealth of goods and services to an influent society in which the information contained within goods flows into you.

> An affluent society may possess an opulent and abundant amount of goods, but that does not mean it will be able to utilize, appreciate, and maintain them. An influent society will have less, but its relationship to what it has will be more involved and concerned; people will take care of what they have, and what they have will mean more to them. In other words, an affluent society amasses goods, while an influent society processes the information within goods. (Hawken, 1980, pp. 74–82)

The economy in developed nations today is increasingly based on service rather than products. We move more information than goods. The ways in which companies can provide those services can be enhanced greatly by what's called the secondary service package. Secondary service adds value to the basic product—it's not just an extra. It provides a leverage in that it boosts the primary service (Albrecht & Zemke, 1990).

Genigraphics was one of the first providers of computer graphics systems to create business slides. Although they always provided training for their customers since the early 1980s (they had to because virtually no artist knew how to operate this type of mini-computer system). As their competition increased so did their need to provide even better information to customers. The business graphics business is dominated by entrepreneurs operating on a small margin. For them, however, $100,000 for a powerful computer and software was a major expense, and so was the down time of their artists until they got up to speed.

As the market got more competitive and their basic product was imitated by other firms, Genigraphics saw training as a major value-added segment of their offerings. They redesigned their 2-week classroom course to consist of an initial introduction to the system, which was provided on diskette and was run on the system itself. That system-based training was later followed up by a 1-week coaching session conducted by a training specialist. This allowed artists to learn the system immediately without waiting to be scheduled for training, and also gave them more hands-on practice. Genigraphics used the training system at trade shows and demonstrated how easy and efficient it was to learn their system by actually allowing prospective customers to try it out. Not only did the new training system reduce Genigraphics' costs, but it provided an important new feature that set them apart from their competition.

The products you buy often have a significant information component or add-on. When you buy software, you often get a free newsletter with tips on how to use it more productively. Many software companies support on-line bulletin boards where customers can call in with questions and share tips or actual programs with each other. In addition, routine information provision can be done through interactive phone services, where customers call in and use their touch-tone phones to select the content they're interested in. Some companies offer free or inexpensive seminars or users group meetings to customers (see Figure 3.5).

The Role of Communications. Information must be recognized as an essential element of basic products by communication professionals. Other business conditions have imposed the role of change agent or facilitator on communicators; however, the fact that information and communication

FIGURE 3.5
The communication component of products and services is an important aspect of their value. Many organizations hold client meetings or users' groups to provide additional instruction and networking opportunities for customers. (Courtesy of OmniCom Associates.)

OmniCom Associates
Sixth Annual Client Workshop
Friday October 12, 1990

OmniCom Associates cordially invites you to our Sixth Annual Client Workshop. As in the past, this complimentary event will provide you with an overview of current and emerging corporate information technologies as well as an opportunity to network with colleagues in the field. David and Diane will present synopses of their conference presentations at NSPI, SALT, and ASTD and discuss what's new in the field.

Schedule:

9:00 Coffee and Welcome

9:30 - 12:00 New Issues in Corporate Information Systems:

- **Update on New Information Delivery Systems:** digital video (DVI, CD-I), hypermedia, NEC's new "PC-VCR"
- **Instructional Design and Development Tools:** discussions and demonstrations of ID expert systems, flowcharting / storyboarding programs, and OmniCom's new development tool, *DesignStation*
- **Rapid Prototyping:** discussion and demonstration of a new method and philosophy of instructional design and development using existing software tools
- **Mobile Communications:** we'll focus on portable and cellular communication devices and give you a tour of *our* new office on wheels

12:00 Lunch

1:00 - 3:00 Client Presentations

- Recent and Current Projects: NYSEG's new teller orientation video; Lederle International's *LederLearn* authoring system and CBT programs, digital video for land planning by Cornerstone, Alltell's new interactive media effort, Espar's new training system and PR/advertising strategies, Radford University's Oral Communication Program individual instruction programs, Amway's trade show videodisc, and more.

Please respond by September 15. We are holding a block of rooms for Thursday night at La Tourelle and will be happy to add your name to the list. Housing in Ithaca is very tight on Friday and Saturday night (fall weekends tend to be "sold out") so if you plan to stay on, please make your plans as soon as possible!

We look forward to your participation.

systems are now key features of so many products and services means that finally communicators are becoming mainstream line managers instead of staff support. Communication is no longer a frill, it is inseparable from the merchandise.

Not many marketing and production managers are yet aware of the powerful competitive edge that excellent communications can provide; however, there is a growing number of good examples. For instance, several of my associates went to a free presentation offered by Federal Express on international shipping. Baxter International [formerly American Hospital Supply (AHS)] provides customers with on-site computer order entry systems. This means that salespeople don't have to call, and hospitals can just key in their orders at their convenience direct to Baxter. Customers can also use the system to enter their hospital's in-house inventory; this add-on feature has helped Baxter to better predict demands and serve their customers (McGowan, 1985). The well-known jean manufacturer, Levi Strauss & Company, uses a system called *LeviLink* to allow retailers to send orders electronically, and goods can be shipped within 4 days (as opposed to about 40 days for paper-based methods of ordering). Sales increased 20% to 30% in accounts using *LeviLink*, reportedly because fewer customers left without finding the size and style they needed. Moreover, stores were able to keep lower inventory levels because ordering is so tightly linked to sales (Alter, 1992, p. 559).

Ron Martin is the vice president of employee communication at American Express Company in New York, and is also past chairman of the International Association of Business Communicators (IABC). During an interview with him in December 1991, he described a new multimedia system he and his staff were pilot testing. It enables employees at their desktop computers to access typical data and text files, as well as enabling them to see cable or broadcast TV, in-house video programs, and excerpts from new ads that American Express will be airing in the future. He explained that it was currently a new and proprietary system that the computer manufacturer was beta-testing in their facility and wasn't quite sure how much he could divulge about it at that point, or if he could invite people in to see it. In fact, some people had found out about the system and were asking to see it. Martin explained that his first reaction was "why not?"; he was accustomed to sharing ideas and products with colleagues, especially because of his high level of involvement in IABC, which promotes such interchanges. But when the computer vendor asked to bring guests from a brokerage that competes directly with Shearson Lehman, a subsidiary of American Express, Martin suddenly reconsidered his posture. This was not just any communication vehicle, this multimedia system provided a new competitive edge to American Express. Martin says he's reevaluating their policies for letting people learn about such new communication systems and admits he hadn't really realized the impact of this kind of technology, which probably should remain proprietary because it's such a strategic tool. As communication and information become imbedded in products and processes, it suddenly takes on a new character; it's no longer something we can afford to give away and may soon be put in the category of trade secret.

Communicators can enhance their companies' products by developing excellent packaging and labeling, by providing clear and engaging manuals, by developing efficient systems to keep customers informed and in the loop, and by developing sophisticated systems to track and to process customer accounts. When I place an order to my favorite clothing catalog, I only have to tell them my ID number. They usually say something like, "Well, Ms. Gayeski, would you like this order shipped to your Coddington Road address? And how did that pink blouse you ordered last month work out?" When I call Federal Express to see if a package I sent was delivered on time, they can tell me who signed for the package and at what exact time. And one of the factors that interested me in buying an automatic bread-making machine was the booklet of several dozen recipes that came with it.

Thought of in this way, communication

can become a valuable commodity, not just an expense. In today's market, sometimes there are few ways in which to distinguish a company's products—especially when the company is also trying to hold costs down. Communicators may be able to make those products attractive and keep close ties with customers so that they continue to buy and to develop a strong relationship.

EMPOWERMENT THROUGH COMMUNICATION INTERVENTIONS

What is Empowerment?

Empowerment is a buzzword (perhaps an unfortunate one) in business today. It means many different things to different people. Kizilos (1990) says that it's been described as "participative management or resource and information sharing or downright delegation of authority or enabling" (p. 48). Alter (1992) defines it as "giving people the ability to do their work; the right information, the right tools, the right training, the right environment, and the authority they need" (p. 509). Generally, the word is used to describe the management practice of distributing the power and tools to make decisions throughout the organization. Whatever you might call it, enabling people to do their jobs the best way they know how and freeing them of artificial barriers to excellent performance is not just a noble humanitarian goal, it's a business imperative. In order to remain competitive, companies must allow and encourage every member of the firm to contribute and to excel.

Putman (1991) describes empowerment as "increasing the behavior potential" of individuals and groups within organizations (p. 4). He lists several inadequate paradigms often associated with this au courant term:

- *vitamin enrichment*—Empowerment is seen as an object that can be added to the current impoverished workplace by a simple intervention, such as sending workers to an image-building seminar. The question here is "why set up unempowering work cultures in the first place, and try to add some artificial empowerment later?"
- *Robin Hood*—The paradigm here is take power from the rich (upper and middle managers) and give it to the poor (the line workers). This approach suffers from the fallacy of a zero sum model in which there is assumed to be a fixed sum of power and the objective is to take it away from the bad guys.
- *healing*—In this model, the approach to improving the workplace is to remove personal barriers to effectiveness by self-awareness, image improvement, and self-affirmation techniques. While some of these interventions may help the individual, they do nothing to change the workplace to which she must return. Even if people want to be empowered, the organization may not support their actions.
- *learning*—Similar to the healing approach, the learning paradigm of empowerment stresses giving the individuals the things (in this case, concepts and skills) they need to increase their decision making abilities. Again, the organizational environment is neglected.

Putman (1991) advocates the behavior potential paradigm to empowerment by viewing it as a product of the individual and the organizational system. He notes that people who are normally quite capable in some situations become incapacitated in other environments. Managers often try to shelter employees from bad news or micromanage their subordinates by checking and commenting on trivial matters. However, these seemingly helpless employees go home after work and sit on school boards, volunteer in suicide prevention groups and fire departments, manage households, and may even run second businesses of their own. How can companies tap into that creativity and competence? What happens to make people so powerless at work?

Although empowerment has generally been associated with helping rather low-level employees to make decisions and contributions without a lot of red tape, this notion is limited. Top managers often say they feel powerless—unable to drive their organizations to make them move where they want them to go. Clearly, the focus should not be on power, but rather on the ability to provide, access, and act on information.

In 1990, the Royal Bank of Canada began to implement broad changes that centered around a new strategy for increasing the authority and prominence of field representatives and branch managers. According to an article in the *Harvard Business Review*, their plan read like a primer in translating empowerment into operational management principles. Among the interventions introduced to facilitate this shift were formal communication mechanisms (i.e., conferences, newsletters, and computer systems), and the establishment of an ambitious training program, called Area Manager University. Communication technologies were even used to help teams decide on the strategies they'd use in this new initiative. A computer system allowed 350 attendees of a management meeting to anonymously vote on the proposals (Charan, 1991).

The term "empowerment" is actually a poor choice for what most people are talking about. Do we really want everybody to have power or do we want to do away with the concept of power? Maybe more appropriate words are "free," "robust," "capable," or "vigorous." For organisms and organizations to be robust, they need to be appropriately designed and highly developed internal systems that are fueled by fresh and well-balanced input, conditioned by exercise, and unburdened by regular output. Communication systems, in particular, are a source of vigor for organizations. They keep corporations in touch with the outside world and coordinate activity within themselves. Through the sharing of information, individuals can increase their capabilities and can add to the creativity and general knowledge pool of the group. Many organizations are impotent simply because they have not exercised their communication systems, to the point where they have atrophied. And, of course, to supply those systems with honest communication, individuals must be free to do so without the restraint of some other source of power.

IMPEDIMENTS TO CHANGE

Many executives say they'd like to change their organizations from their former bureaucratic, slow, homogeneous forms, but I wonder how many of them really want to change or know what it takes to change. They realize that their organizations are becoming stagnant, but secretly hope that they'll retire before they actually have to change their behavior, share their power, or use any new forms of communication technologies. They feel confused, awkward, and afraid.

Argyris (1976), a management theorist, maintains that many managers hold an espoused theory that they articulate and defend, but behave according to another theory-in-use. He describes a *Model I theory* in which the goal of an organization is to achieve the purpose that an individual executive has defined, to win and never lose, to suppress conflict, and to stress rationality. A *Model II theory* is centered around the value of valid information based on shared and intelligent decision making, an internal commitment to consensus, and a respect for intuition and emotions. Argyris contends that most managers espouse a Model II theory, while actually operating under Model I assumptions and rules.

Futurist Toffler (1990) describes the way in which most organizational change occurs:

- a new problem or opportunity presents itself that doesn't fit into anyone's existing bureaucratic niche
- someone suggests setting up a new unit to deal with it
- other existing units see it as a threat and torpedo the idea
- a compromise of an interdepartmental committee is attempted
- it moves too slowly; the CEO appoints a "czar" to cut through red tape
- the czar doesn't have the information so the CEO assigns it to her personal troubleshooter
- existing departments are outraged and work to assure failure

> Any hope of replacing bureaucracy, therefore, involves more than shifting people around, laying off "fat," clustering units under "group vice-presidents," or even breaking the firm into multiple profit centers. Any serious restructure of business or government must directly attack the organization of knowledge—and the entire system of power based on it. (Toffler, 1990, p. 171)

Restructuring, of course, is more easily said than done. Organizations move slowly, and executives are usually not the owners—they must answer to many people, including the union, the board of directors, and the stockholders. In addition, existing communication systems don't readily support change. Although communications professionals try to facilitate reorganizations and new management philosophies by producing newsletters, holding meetings, and creating motivational video programs, our tools are generally inadequate. Traditional media are one-way, top-down. We need to develop new collaborative systems and means for sharing ideas and information laterally and from the bottom up. Organizations in the midst of change need people to help coach managers on new styles. The bottom line is that communication systems cannot stay the same when the whole company is changing around them.

There are communication systems, however, that more appropriately channel information broadly and efficiently. These systems include participatory and accessible media such as e-mail, desktop publishing, small-format video, and teleconferencing, to name a few. Unfortunately, the availability of new technologies themselves have not historically led to their adoption, and their adoption has not necessarily led to a rethinking of the basic communication culture of organizations. The concept of *cultural lag* helps us recognize that even when technologies are proven and available, institutions generally lag in their adoption of them and in benefiting from the changes that they can offer.

For example, Yates (1989) points out that the Illinois Central Railroad had virtually unlimited access to the telegraph before the Civil War, but used it very unsystematically. It was not until the Civil War brought on urgent situations and scheduled irregularities that the railroad's management converted to telegraphic dispatching. Other firms, like Scovill, adopted technologies such as circular letters, before it adopted the duplicating and filing technologies that were needed to manage this new medium. Currently, many organizations have introduced e-mail systems, but executives still have their secretaries print out hard copies of each message sent to them and read them in that form, obliterating any benefits of the original medium.

Darryl Connor, a popular speaker in performance technology and training circles, says that organizations will not change until they experience pain. The pain of not changing must be greater than the pain of changing in order to overcome the natural fear of shaking up the status quo. Sometimes the change agent's role is to make people focus on and admit that pain. Only then can new communication or training systems make sense.

The single factor that has led to the systematic application of most new communication techniques and technologies has been the intervention of one top manager who has championed new approaches. While in many cases, that top manager has been a communications professional or has been strongly influenced by a communications professional, it would be incorrect to assume that this is the norm. Unfortunately, many corporate communications practitioners are reactive, rather than proactive, in advocating change. Too often, they and the profession as a whole do not have management's ear. And generally, the professional communicator doesn't have the power or budget to make change happen quickly.

It's a common belief that people resist change. Actually, they don't resist change, they resist the personal and political impacts of change. If it makes them work harder, if it disturbs their friendly grouping of colleagues, or if it means less money in their personal or department budgets, then the walls go up. A lot of research has been done on the adoption of new communication and educational technologies. Overall, potential adopters were not afraid of new media systems per se, they often welcomed them as new tools to learn to use. However, users were often afraid of what the systems would mean to their jobs. Teachers, in particular, have been worried that educational TV or computers would take away their job security, or at least, the freedom to design their own classroom experiences.

The word "resistance" is often used when people talk about change. Putman (1985) cautions us that we need to make a distinction between *resistance* and *opposition*. He maintains that the word "opposition" should be used when there are clear and straightforward reasons for people opposing change. Resistance should be reserved for those times when there is no such apparent or logical ground for opposing change. For example, if a communication consultant recommends that an in-house video department be closed down, those video staff members will likely oppose that change. It's easy to understand why they would not support such a decision. However, if managers are asked to show a video program on financial projections for the company within their department meetings and don't do so, we may say they are resisting the use of this new technology. Putman has concluded that people resist change when they feel that they are being coerced. Understanding this, we can see that resistance is a reaction to the perception of being coerced, and is fundamentally a problem in human relationships.

In order for communications professionals to avoid the problem of resistance to change, we need to adopt new philosophies and practice different communication behaviors ourselves. We need to establish ourselves as helpers, not dictators. We can't make other people change and grow and learn; they must do it themselves.

1. Put yourself in their shoes and make sure they understand you can adopt and reflect their viewpoints. Don't let people think you're trying to make them look stupid or old-fashioned.
2. Don't be possessive of your ideas; in fact, as long as they're perceived as your ideas, you've lost. The ideas have got to become their ideas.
3. Don't push. When people are in the process of changing, they will predictably feel ambivalent about the process. The more you try to sell your ideas, the more they'll be forced to think about the opposite position. One successful consultant even goes through some theatrics to point out that, although he can advocate a position strongly, he is open to thinking about others. He tells clients, "let's pretend that I were trying to sell you this idea." Then, he physically moves to another place in the room, changes his demeanor, and emphatically and enthusiastically presents a point-of-view. After his soliloquy, he returns to his place and to a more placid bearing, and discusses the situation with his clients.
4. Be open to new information. Don't feel like you have to win the situation by imposing your point-of-view. Remember that your goal is to improve the general situation, not inflate your ego by proving yourself right. Collaboration rather than indoctrination is the key.

We need to recognize that new tools and policies are not often the ideal solution to communication problems, but they provide some experience that is better than what presently exists. Employees generally say they want more information and they want it provided face-to-face. Of course, this is inefficient, if not impossible.

During a conversation about this book, a vice president of communications told me that company employees were generally dissatisfied with the present process of executive visits during which the executive management team went to each site once a year to share their thoughts and plans with the people there. Employees believed that they learned nothing new in these meetings. They said that they would rather have top managers spend time with them and see what it is actually like to perform their jobs. They want to be listened to, not talked at. However, if the management team did this, it would occupy almost all of their work days during a year.

I suggested that they use simple camcorders to let employees put together a location overview, complete with vignettes from typical work locations and questions from employees that the executive team could review prior to their visit. In this way, management could address the needs and concerns of the workers at each site and probably learn some more about the company's basic operations in the process.

While this may not be as satisfying as having the chairman help answer the phones or dig trenches with each person, it's a far cry from what they're doing now.

In order for changes to occur, we must do more than announce reorganizations and install new communication technologies. Long (1989) suggests that several crucial human resources issues need to be dealt with:

- training and support of employees
- recruiting and selection of job candidates
- job design
- health, safety, and ergonomics
- compensation policies
- performance evaluation
- industrial relations

Long (1989) suggests that human resources professionals are seldom involved in the implementation of new technologies early enough to do anything other than react to problems after they've emerged. Professional communicators can be effective change agents, but we must step up and make our influence felt much more vigorously and promptly.

NEW TOOLS AND THE NATURE OF COMMUNICATION

We've already seen many examples of how new communication tools can support new models of management and communication within organizations. Although these new tools are powerful means for supporting change, they don't bring it about just by being plugged in. *Technological determinism* is a perspective that posits direct cause-and-effect relationships between technology and social or organizational change. Technology is seen as a cause, as something that is so powerful that its very presence can change organizations without necessarily a plan. Another related deterministic viewpoint is that technology is a consequence of organizational change.

Many well-meaning people assume that technologies will bring about the social and organizational revolutions they are capable of influencing. In 1982, Steve Wozniak, one of the inventors of the original Apple Computer, sponsored a rock concert attended by about half-a-million people that predicted a new age of democracy and community to be brought about by the widespread use of personal computers. Some people quite literally make communication technologies heroes; for example, *Time* magazine named the personal computer "Man of the Year" in 1983!

However, technology doesn't necessarily bring about any particular type of organizational change or, in fact, any change at all. Many people are afraid that new communication tools will depersonalize jobs, make it easier for managers to monitor their behavior, and make workers even more powerless. Others feel that these tools will empower workers, make organizations more democratic, and introduce more freedom into the workplace. Obviously, all of those scenarios are plausible. What is the purpose of introducing the technology? Is it to tighten control, deskill jobs, and eliminate people or is it to broaden their sources of and access to education, information, and collaboration? We can see numerous examples of both points of view and of organizations in which each of those extreme and opposite consequences of technology can be found.

One problem of current research on new information technologies is that we cannot predict the causes or effects of a technology based on our studies of previous technologies because, as Nass and Mason (1991) put it, "we do not know whether relevant features have been changed" (p. 46.) They claim that there is currently no specified set of attributes that vary across technologies and that can be demonstrated to be either the causes or consequences of certain organizational behaviors.

> For example, consider a study that demonstrates that personal computers have led to de-skilling in the workplace. If having a value of 'very rapid' on the characteristic 'processing speed' is the key to the de-skilling of workers, then ceteris paribus, technologies that process information rapidly, such as adding machines, and technologies that preprocess information, such as sorting machines, should also lead to de-skilling. . . . However, if the key feature is having the value 'keyboard' for the variable 'input device,' then typewriters, type-setting machinery, and push-button phones should lead to de-skilling of workers. If the pivotal criterion is 'fits on a desktop' for the variable 'size' . . .

then telephones, staplers, and reading lamps should lead to de-skilling (Nass & Mason, 1991, p. 49).

Obviously, a given technology has many characteristics, and the combination of them plus the way in which they are applied in an organization makes the difference between an oppressive system and an empowering one.

Zuboff (1988) has studied organizations that have introduced new information systems into their core operations. She distinguishes between *automating* systems, which seek to cut out or reduce people, and *informating* systems, which increase individuals' and the organizations' intellective capacity. She also cautions readers to consider the potential negative impact that technology may have on individuals and, ultimately, what they can and will do for their organizations.

An alternative to determinism in looking at the introduction of technology is to characterize it as a process and a complex intervention. The intervention is *strategic* in so far as some participants have a vision and strategy for linking work and the introduction of new tools. The intervention is also *sociotechnical* in that it effects social, as well as technical, systems within organizations (Kling & Iacono, 1989). People in organizations need to be taught to use new tools productively—not just how to push buttons but how to strategically employ new systems to attain higher quality output and more desirable working conditions.

Fernando Flores, a developer of organizational communication systems, applies broad theories to the invention and application of software systems. "In using the word 'technology' people are generally concerned with artifacts—with things they design, build, and use. But in our interpretation, technology is not the design of physical things. It is the design of practices and possibilities to be realized through artifacts" (Flores et al., 1988, p. 153). He and his colleagues have developed a theory of language as social action. They maintain that people in organizations don't merely send messages back and forth, but rather have a series of requests and promises. They have developed an on-line communication system called *The Coordinator,* which manages conversations, including word processing, a calendar and appointment manager, and messaging. It coordinates messages in terms of requests and responses to requests, making the intent of a message crystal clear. This helps in large organizations where there is less cultural commonality by creating a new shared culture or tradition of explicit language acts.

Technology can be viewed as existing along two continua. These are global versus local and digital or symbolic versus iconic or pictorial. While people may fear that computer systems will imprison people in their homes or office cubicles, allowing them to interact only through flickering screens, these same families of tools allow us to share ideas and make friends with people all over the world. Although communication systems can allow us to process large quantities of data in digital or symbolic form, they also may display rich pictorial or iconic images. (See Table 3.2.)

As with any tool, even a simple pencil and paper, it's not the medium but the message and underlying intent that lead to observable effects. We need to move from technocentric systems based on predictability, repeatability, and mathematical quantifiability to anthropocentric systems that include intuition, subjective judgment, tacit knowledge, imagination, and intentionality (Cooley, 1990). When employed without a concomitant change in management paradigms, new communication tools have had little impact on organizational life. Teleconferencing generally doesn't reduce travel; however, it enables more people to meet each other and more projects to be spawned, which eventually result in face-to-face meetings. Word processing hasn't yet significantly reduced the number of secretaries. Individuals in these roles now spend more time fancying up their bosses' simple word processed memos and creating even more pieces of paper. Even e-mail hasn't reduced the reliance on print. Large quantities of paper and toner cartridges for laser printers are still being used for printing e-mail memos for executives who don't like to use computers.

TABLE 3.2
Potential Advantages versus Disadvantages of New Media

Advantages	Disadvantages
• Demassified and decentralized communication and management	• Dehumanized workplaces
• Increased participation and influence throughout ranks	• Costs money and time to bring on-line
• Increased choices of messages, styles, careers	• Increased pressure and information overload
• Enables alternate work styles	• Increased control and monitoring
• Saves time and money	• Centralized control in information and media systems
• Improves decision making	• Focuses attention on media rather than message or product
• Builds bridges with customers and colleagues	• Short-circuits hierarchy/creates anarchy

Yet, the importance of tools should not be underestimated. Schrage (1990) talks about the monstrously cynical view that tells people they should improve and that they could be better if they were only someone else.

> The virtue of good tools is that they don't ask you to become someone else. They invite you to create an extension of yourself that, with a little time and skill, lets you be more than what you are. Tools are a medium for self-expression. . . . Tools are, literally and figuratively, the way people come to grips with their work. . . . To see technology primarily as a way either to reduce costs to accelerate output is simplistic: *technology is really a medium for creating productive environments.* (Schrage, 1990, pp. 66–67)

Another example of new communication tools is the emerging class of software systems that support instructional design automation. Since there is currently pressure to provide more instructional materials, especially within the military and corporations, the aim of these packages is to make the process of designing education or training materials more efficient and to provide some expert decision making to novice designers.

Some developers are approaching this task by creating expert systems that turn designers into clerks who type in simple responses to the software's questions. The system then creates the instructional plan for them. However, not all systems are designed with the goal of deskilling the job of instructional design or eliminating the creative process. Another approach, instructional design support software, can empower subject-matter experts (SMEs) to create their own well-designed training programs. (See Figure 3.6.) This class of software

> can assist work groups in collaborating on design projects, and can reduce the level of monotonous and clerical tasks inherent in instructional development. However, when applied without the mindset of empowerment, this same family of software can make instructional development into a rigid and boring process, confining the creativity of performance technologists and limiting their growth. (Gayeski, 1991, p. 34)

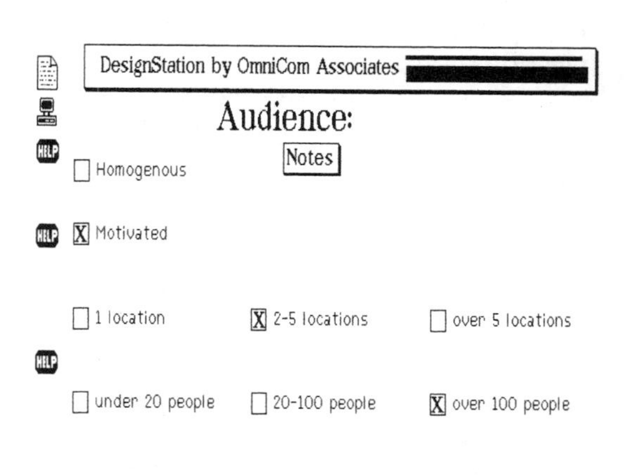

FIGURE 3.6
Electronic support systems, such as *DesignStation,* enable communication materials to be created more efficiently and consistently. This software creates a data base of content and strategies for projects, and includes a help system for novice developers. (Courtesy of OmniCom Associates.)

When assessing communication and management technologies, review the questions:

- Will the job be deskilled?
- Will changes in the organizational structure be implied?
- Upon whose model or theory will the tool be based?
- Will people resist using it?

Communication managers and developers face an embarrassment of riches in terms of new technologies. Many audiovisual and computer specialists tell me that they feel overwhelmed and confused about the options that face them, and don't know where to begin. Technologies and programs are being developed faster than we can keep up with them, and unfortunately, in some cases, they are also outdated almost as soon as we get them out of their boxes. Although it seems, at times, almost impossible to keep up with the many new communication technologies continually being introduced, it is imperative to do so. For example, I subscribe to approximately 25 professional magazines and journals and attend at least five conferences each year to keep up with new communication practices and tools. We need them to facilitate our own work, and our organizations need to be rewired to remain vital.

Many professional communicators are convinced of the merits of new tools, but don't know how to get started. Video, computer, and telephone systems are being adopted and managed through a variety of departments, including training, information systems, corporate communications, telecommunications, and marketing. This can lead to turf wars over whose responsibility it is to specify and control these new technologies.

Here are some approaches which organizations have found useful in exploring new communication tools:

- Budget for annual research and development projects that are aimed at using new tools to solve a few significant, but well-defined problems. These applications can provide a valid platform for assessing the new technology, can be used to demonstrate to future clients and sponsors, and can be used as models for future interventions.
- Contract with experts in new technologies to assist in the creation of model programs.
- Hire professionals to teach your organization's staff to develop and to manage new technologies.
- Visit other organizations who have applied (or attempted to apply) particular technologies in order to investigate successes as well as failures.
- Fund several staff members to attend professional seminars or enroll in college courses in electronic media technologies.
- Collaborate with colleagues in other industries, at universities, or who are consultants in order to develop innovative programs whose costs can be shared.

Our firm, for example, serves as a corporate communication research and development center where clients can learn to use and actually try out innovative technologies and techniques. They can borrow equipment and work with experienced staff while developing their first applications, and thus, avoid having to purchase equipment or hire staff until they are sure that their organization can really benefit from a particular system.

> The key is to remember that even as we use technology to shape our environments, technology is shaping us. Technology, like gravity has its own imperative. New technology rearranges our perceptual world and subtly redefines our relationship with our environment. (Schrage, 1990, p. 8–9).

THE IDEAL CORPORATE COMMUNICATION SYSTEM

What is the ideal corporate communication system? This is a question I've posed to a number of executives and communication professionals as I conducted research for this book. Very few had any answer, even for their own organizations. One vice president remarked that he felt that he'd never been given the opportunity to remake his company's communication system, so he'd never really given it any thought. Although the company was undergoing radical transformation and he had a long history of managerial successes within it, he somehow never recognized that he had the power to think and act so globally.

According to a survey on best practices in employee communication prepared for the Human Resources Policy Institute (Young & Post, 1992), the primary factors related to effective corporate communication include the following:

- the senior executive as a communication exemplar
- congruence between rhetoric and action
- two-way, face-to-face communication
- decentralized responsibility for employee communication
- dealing with bad news
- recognizing who is the customer of employee communication
- the right strategy
- the communication budget
- to which department or executive employee communication reports

Of course, there's no one ideal. Communication systems must be positioned within an organizational framework and theory, and must undergird the corporate goals and culture. However, there are a number of attributes that researchers and practitioners alike seem to generally agree upon as being characteristic of excellent and progressive communication practice:

- The system must foster a Theory Y approach in which messages can easily travel up, down, and laterally through an organization.
- Feedback and collaboration is supported.
- Communication within and between all levels and departments is frequent, honest, and symmetrical.
- Communication load is optimized (neither overload nor underload).
- Accuracy and speed of information is optimized.
- The system is sensitive to human and environmental needs (flexibility in time and mode, respectful of individuals' rights and feelings, ecologically sound).
- Policies and interventions revolve around a consensually determined and unified corporate culture and image.
- Rules and tools are articulated clearly and made available to all members.
- Communication practices and interventions add value to the organization's products and members' lives.
- New technologies and techniques are continually examined and employed strategically to support individual and organizational goals.

Unification

A major problem with current organizational communication systems is that they're unsystematic! Although there may be a company logo and a set of rules for dealing with the press, new technologies and human communication practices are not being dealt with in a unified manner. The lines between media are blurring, and the roles of trainers, information systems specialists, communication relations specialists, telecommunications managers, computer scientists, graphic artists, and writers are also muddled. There is usually no person looking at communication systems as a whole across organizational departments. The upshot of this is continual reinvention of wheels, a lack of consistency in policies and practices, duplication of technologies and services, and no standardization of tools.

> The rapid growth of technology affecting data, voice, image, and text processing demands a new organizational structure for the effective and efficient coordination of these various elements. The alert manager realizes that these diverse groups, which developed independently, serve one ultimate purpose—to process and produce information that will meet the organization's aims and goals. (Meltzer, 1981, p. 112)

Excellent modern corporate communication systems will have to be led by a visionary who can see beyond the archaic compartmentalization of rules and tools.

Collaboration

As management moves from authoritarian to participative approaches, communication systems need to move from an emphasis on reporting to a facilitation of collaboration.

> We have telephones, videocassettes, satellites, fiber optics, facsimile machines, paper, audiocassettes, overhead projectors, whiteboards, blackboards, and so on—we live in environments rich with tools that make it easier for people to communicate with each other. But where are the tools for collaboration?. . . If one truly believes that organizations succeed

> by profitably amplifying the talents of their workers, then the challenge is to create collaborative tools and environments—techniques and places that enable people to be themselves while they are being more than themselves. These tools and environments add value to relationships even as they empower individuals. . . .The ability to communicate ideas crisply and clearly in order to let colleagues and subordinates know what to do simply isn't enough any more—if it ever was. (Schrage, 1990, p. xxiii)

Most current media practices and technologies encourage indoctrination, not collaboration. Corporate media have a bias toward telling and selling. People feel that they have to fit their messages within a particular known medium, and often forget about ensuring understanding of the meaning. The ideal communication system would provide a better balance between linear transmission of messages and facilitation of relationships and work.

Helping Organizations Evolve

As we've seen throughout this chapter, many organizations are in a state of chaos. They are either young and developing or old and struggling to cope with rapid change and difficult social and economic conditions. The ideal corporate communication system is sensitive to the organization's status within its life cycle and its environment, and supports healthy change. Some specific approaches offered by Deal and Kennedy (1982) are

1. Position a hero in charge of the process.
2. Identify a real threat from outside to motivate change; recognize, or even create, an enemy.
3. Create pivotal transition rituals like mourning rites for the old culture. Then, establish a transition or temporary culture.
4. Provide transition training in new values and behavior.
5. Bring in outside shamans.
6. Try to build security during the transition.

Change is not easy; often, people are confused, have low morale, and resist change. People become preoccupied with change, not their job. They need time to plan for the change, to mourn the old ways, and to celebrate the new culture. During this time, they deal with ambiguity and stress.

It's important to communicate during periods of change; however, this is often a time when communication is poor. People may be working longer and harder and communicate less often or less well. Leaders may tend to be secretive. Then the rumor mill starts up. To counteract this, formal communication must be direct and frequent, otherwise, people assume the worst. It's crucial to also remember to communicate with customers. They may be quite aware of reorganizations and transformations and may wonder about how this will affect your company's product, service, personnel, or prices.

Decades ago, organizations didn't change very frequently. When they did, management would just announce the changes, and then wait for things to normalize. Now, people announce change, react with communication or training, and wait for normalization. Ideally, change should be planned for—training and communication should precede it. This means that communicators need to be close to top management in developing strategies rather than finding out about them once they've already been implemented.

Finally, communicators should not be merely advocates and support groups for organizations in transformation. It is external channels of communication that foster innovation—they are the eyes and ears of the enterprise, while internal channels foster stability (Kreps, 1986). Keeping these two in balance is the juggling act of a master electronic jester.

REFERENCES/ SUGGESTED READINGS

Albrecht, K., & Zemke, R. (1990). *Service America!* New York: Warner Books.

Alter, S. (1992). *Information systems: A management perspective.* Reading, MA: Addison-Wesley.

Anderson, K., Ballman, G., Eberlein, C., Geiger, N., McDonald, C., Pinto, P., Rossett, A., & Summers, M. (1991, December). Win a Baldridge, talk yourself out of a job. *Training,* 37.

Argyris, C. (1976, Winter). Leadership, learning, and changing the status quo. *Organizational Dynamics,* 29–43.

Braden, P. V., DeWeaver, M. J., & Gillespie, L. (1991). Keeping U.S. companies competitive: How performance technologists can and should contribute. *Performance Improvement Quarterly, 4*(4), 47–61.

Charan, R. (1991, September/October). How networks reshape organizations—for results. *Harvard Business Review,* 104–115.

Cleveland, H. (1990). Epilogue: The twilight of hierarchy: Speculations on the global information society. In S. Corman et al. (Eds.), *Foundations of organizational communication: A reader* (pp. 327–341). White Plains, NY: Longman.

Cook, K. (1988, September 25). Scenario for a new age. *The New York Times magazine,* 26–27, 52–53.

Cooley, M. (1990). *European competitiveness in the 21st century: Integration of work, culture, and technology.* Brussels, Belgium: FAST Programme.

D'Aprix, R. (1977). *The believable corporation.* New York: AMACOM.

Deal, T.E., & Kennedy, A. (1982). *Corporate cultures: The rites and rituals of corporate life.* Reading, MA: Addison-Wesley.

Drucker, P. (1988, January/February). The coming of the new organization. *Harvard Business Review,* 45.

Drucker, P. (1989). *The new realities.* New York: Harper & Row.

Dumaine, B. (1991, October 7). Earning more by moving faster. *Fortune Magazine,* 89–94.

Flores, F., Graves, M., Hartfield, B., & Winograd, T. (1988). Computer systems and the design of organizational interaction. *ACM Transactions on Office Information Systems, 6*(2), 153–172.

Galagan, P. (1991, June). How Wallace changed its mind. *Training & Development,* 23–28.

Gayeski, D. (1991). Software tools for empowering instructional developers. *Performance Improvement Quarterly, 4*(4), 21–36.

Goldhaber, G., Dennis, H., Richetto, G., & Wiio, O. (1979). *Information strategies: New pathways to corporate power.* Englewood Cliffs, NJ: Prentice Hall.

Handy, C. (1989). *The age of unreason.* Boston: Harvard Business School Press.

Hawken, P. (1980). *The next economy.* New York: Holt, Rinehart and Winston.

Kanter, R. M. (1989). *When giants learn to dance.* New York: Simon and Schuster.

Kizilos, P. (1990, December). Crazy about empowerment? *Training,* 47–56.

Kling, R., & Iacono, S. (1989). Desktop computerization and the organization of work. In T. Forester (Ed.), *Computers in the human context* (pp. 327–334). Cambrige, MA: MIT Press.

Kreps, G.L. (1986). *Organizational communication.* New York: Longman.

Landes, L. (1992, February). Down with QUALITY program-itis. *Communication World,* 29–32.

Long, R. (1989). Human issues in new office technology. In T. Forester (Ed.), *Computers in the human context* (pp. 327–334). Cambrige, MA: MIT Press.

McGowan, W. (1985). Information-age technology—The competitive advantage. In J. Rosow (Ed.), *Views from the top: Establishing the foundation for the future of business.* New York: Facts on File Publications.

Meltzer, M. (1981). *Information: The ultimate management resource.* New York: AMACOM.

Naisbitt, J., & Aburdene, P. (1985). *Reinventing the corporation.* New York: Warner Books.

Nass, C. & Mason, L. (1991). On the study of technology and task: A variable-based approach. In J. Fulk & C. Steinfield (Eds.), *Organizations and communication technology* (pp. 46–67). Newbury Park, CA: Sage Publications.

Oberle, J. (1990, January). Quality gurus: The men and their message. *Training,* 47–52.

Odiorne, G. (1990, July). Beating the 1990s labor shortage. *Training,* 32–35.

The post-Deming diet: Dismantling a quality bureaucracy. (1991, February). *Training,* 41–43.

Putman, A. (1985). Managing resistance. In C. Bell & L. Nadler (Eds.), *Clients and consultants: Meeting and exceeding expectations* (pp. 191–199). Houston, TX: Gulf Publishing Co.

Putman, A. (1991). Empowerment: In search of a viable paradigm. *Performance Improvement Quarterly, 4*(4), 4–11.

Quality awards aren't free. (1990, January). *Training,* 12–13.

Quest for quality: Is it perception or reality? (1991, September). *AVC Development & Delivery*, 16–17.

Rebello, K., & Schneidawind, J. (1991, May 30). Seeing red over IBM's blues. *USA Today*, 1B–2B.

Reich, R. (1987, May/June). Entrepreneurship reconsidered: The team as hero. *Harvard Business Review*, 77.

Ruch, W.V. (1984). *Corporate communications: A comparison of Japanese and American practices.* Westport, CT: Quorum Books.

Schrage, M. (1991). *Shared minds: The new technologies of collaboration.* New York: Random House.

Smither, R. (1991, November). The return of the authoritarian manager. *Training*, 40–44.

Toffler, A. (1970). *Future shock.* New York: Bantam Books

Toffler, A. (1990). *PowerShift: Knowledge, wealth, and violence at the edge of the 21st century.* New York: Bantam Books.

Welch, J.E., & Hood, E. E. (1992). To our share owners. *General Electric 1992 Annual Report.* Fairfield, CT: General Electric Company.

Yates, J. (1989). *Control through communication.* Baltimore: The Johns Hopkins University Press.

Young, M., & Post, J. (1992, March). Part II: Managing to communicate, communicating to manage: Best practices apply equally in good times and bad. *CCM communicator*, 1–5.

Zuboff, S. (1988). *In the age of the smart machine.* New York: Basic Books.

4 Communication Analysis and Design

We've looked at the big picture in corporate communications: what the field is about, its major theoretical foundations, and the ways it is changing and is changed by new organizational paradigms and management theories. This chapter marks the shift from what you need to do to how you should do it.

Communication analysis and design are processes used to get you from where you are to where you want to be. They are the nonproduction or planning phases of professional communication activity. They revolve around the following activities:

- documenting the current status
- developing goals
- specifying the ways in which your company falls short of its goals
- developing general plans for how to narrow those discrepancies
- writing specific, measurable objectives
- designing communication strategies and messages
- adopting assessment and on-going feedback methodologies

A crucial question to ask when analyzing and designing new communication systems is: "are we primarily concerned with maintaining the status quo or with changing it?" Does the organization need to rewire its communication system philosophically or physically, or is there the option of using the current system to provide new messages? Is there a reason to change basic management and communication practices and philosophies, or will it suffice to develop an individual program?

FRAMEWORKS FOR CORPORATE COMMUNICATION ANALYSIS

There are many types of analyses that professional communicators perform and many perspectives from which to examine organizations and their communication processes. Before we look at a long list of analytical techniques and program design models, we need to understand some of the basic principles around which they are created and used.

Communication by Objectives

Communication systems and individual communication programs need to be driven by objectives. Unless clients can specify where they want to go, communication professionals can't decide how to get there or when they've arrived. Very often, sponsors merely request a program or a new technology without a well-thought-out rationale. Although it may seem like it's more efficient just to respond to such a request, this leads to long-term deterioration of communication systems or departments. Unless communicators can demonstrate how they've helped their organizations achieve specific goals, they will always be considered frills, or the soft side, of management. Communication by objectives is an extension of the popular concept of management by objectives (Odiorne, 1965). (See Figure 4.1.)

Analysis starts out by specifying objectives at the most global, organizational

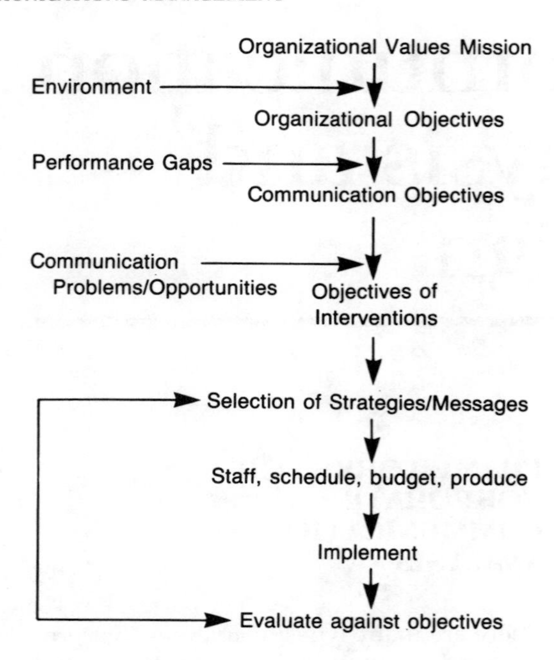

FIGURE 4.1 Communication by objectives. [Adapted from Haynes, G. (1981). Organizing and budgeting techniques. In C. Reuss & D. Silvis (Eds.), *Inside organizational communication* (pp. 62–76). New York: Longman.]

level—this is the *strategic level* of analysis and design. This perspective involves establishing objectives, such as profitability, sales, awareness, decreased staff turnover, public opinion, increased productivity, and so forth. From here, we can derive communication objectives. These may include the frequency and accuracy of internal communication, the provision of timely training, the number of times the company is written up favorably by the press, and the ability of employees and stockholders to articulate the organization's business philosophy and plans. After developing communication strategies and objectives, you then have the basis to design even more specific communication policies or interventions, through defining the audience, the messages and images, and the settings. This is the *tactical level* of analysis and design.

The most visible part of the Renaissance communicator's work is producing things (such as annual reports, meetings, video programs, and graphics) just as most organizations are known for the products they make or the services they offer to the public. However, we all realize that a lot goes on behind the scenes to bring those commodities to the public.

Before you can start producing products, you have to examine the general status of information in the organization. The following characteristics of information must be assessed first:

- accuracy
- amount
- speed
- access

Some of the typical challenges or deficiencies in organizational communication that you might look for include

- insufficient or unclear messages
- inaccurate or conflicting messages
- lack of timely information and training
- inappropriate spokespeople as primary message sources
- perceived risks in communicating openly
- information overload
- distortion of messages
- status differentials in communication access
- lack of incentives for good communication
- physical barriers to communication
- lack of communication productivity tools
- mismatch between communication rules and tools and management philosophy
- lack of consistency among messages and images
- inadequate knowledge on how to access information and communication channels
- lack of favorable public exposure
- widespread misconceptions about the organization
- inefficient organizational operation because of skill and conceptual deficiencies
- slow or inappropriate response to market and environmental needs

In this chapter, we'll examine the ways in which we can determine the problems, and then design appropriate systems and messages to remedy those problems.

What's the Real Problem?

As defined in the first chapter, corporate communications practice means developing and using rules and tools to enhance the dissemination, comprehension, acceptance, and application of information in ways that help to achieve an organization's goals. In doing this, communications professionals generally work on projects on behalf of clients or sponsors. For example, a human resources department may need a means for telling employees about the various benefit plans from which they may select. The marketing department may request a trade show booth, brochures, and a survey of potential customers. The office of the CEO usually needs an annual report and coordination of stockholders' meetings. Even traditional and ongoing activities like publishing an employee newsletter rely on information provided by other sources, as well as general guidance about the desired effects of the newsletter, which are provided by people outside the communication department. These other departments or individuals who request (and sometimes directly pay for) communication services are called clients or sponsors. The word *client* is generally used when outside consultants or agencies refer to their customers; *sponsor* is generally used for internal customers.

Communication projects are usually begun in one of two ways. A client may request a particular product or service after identifying (or assuming) the likely cause and solution to the problem. Or, communication professionals may use analytical tools to uncover potential communication problems, develop solutions, and propose projects. Depending on the skills, roles, and relationships between the client or sponsor and the communication service provider, one of these models may prevail. For instance, internal communication departments may be charged with ongoing assessment of internal communication satisfaction and for developing programs and policies that will enhance this satisfaction.

Departments in which there are no communication professionals may have some general notions of their information problems, but may rely on internal communication professionals or contract outside professionals to examine the problems by conducting their own more rigorous analyses. For instance, a software company recently hired OmniCom to determine whether their current training program needed to be changed, or if it would be better to enhance the software itself with better help and tutorial functions. In contrast, simple or obvious projects that don't warrant investigation may also be requested; for example, slides to support an engineer's presentation at a professional conference. In addition, some communication and training departments perform their own analyses and based on them, contract production houses or communication agencies to do the actual production work. For example, a large computer manufacturer recently requested a proposal from OmniCom to produce a series of video programs. Their own training staff had already conducted an extensive needs analysis and had developed specific objectives for the programs.

Unfortunately, problems aren't usually what the client (or you) thinks they are at first glance; in fact, often, the problem hasn't even been identified. When a client shows up at a professional communicator's door, she generally already has an idea of the potential solution depending on her notion of what it is that the communication professional does. For example, if a customer comes to a training department, he probably feels that a particular organizational problem is caused by a lack of skills or knowledge. If another person requests the coordination of a sales meeting featuring a dramatic multi-image program on teamwork, one could predict that she believes there's an underlying motivational problem.

However, it's not the job of a communication professional merely to produce interventions, no more than it's the job of a physician to write out prescriptions for what patients come in thinking that they need, based on what they read in the newspaper the night before or on what their friends are taking. For example, let's say a company that had established an impressive track record in manufacturing and marketing adding machines decided to expand into selling PCs. The company had already established a number of good customers and had a large existing sales staff to serve them; however, after several years of trying to sell PCs, sales figures were far from the established quotas. What could be done? Since the engineering department obviously felt that they had built a quality product, and since the marketing department felt the price was right, nobody questioned the product itself. A relatively easy response from management could be that the sales representatives needed more training on computers—how to use them, how they work, and how to sell them. The marketing department might also feel that it needed more money for advertising and publicity materials. Yet, simply pumping out more information might not get at the root cause.

A good communication analyst would ask the following questions:

1. What are the typical customer objections? Is the product really as good or better than the competition? Do you feel that the sales reps really know what they're talking about and can clearly explain the product?

2. Are sales reps really demonstrating the computers? Perhaps the computers are too difficult to haul around and demonstrate, so they avoid it. Or perhaps they don't feel confident demonstrating them. Perhaps customers don't want to spend the time having a demonstration.

3. Are the sales reps talking to the right people? Their old contacts for selling calculators may be very different from the people who are now in charge of purchasing computers. If they do get to the right people, are they as effective in communicating with them as they were in selling to their old customers?

4. What is the incentive for selling computers versus selling calculators? Perhaps the sales rep makes $50 on each computer sale, but only $2 on each calculator sale. Perhaps it's easier to sell 25 calculators than one PC.

5. Are the sales quotas realistic? Perhaps it just takes longer to sell PCs, which are a newer and more costly product, than it takes to sell traditional inexpensive calculators.

Depending on the analysis performed, the professional communicator may decide that none of his remedies would cure this organizational problem. Perhaps the sales department needs to rethink the commission that representatives make when they sell computers. The human resources department might need to consider hiring a different type of person to sell computers than those who sell calculators. The engineering department may need to redesign the product itself. The executive committee might decide that selling computers wasn't in the company's best interest after all. Yet, there are a number of interventions that might help the situation—depending on the underlying cause or causes:

- training sales reps on how to use and sell computers
- a better, easier-to-use system for demonstrating the new computers, such as a videotape that could be given to potential customers
- a job aid that helps representatives answer questions about the computers
- an integrated advertising and public relations campaign that highlights the new PCs as an integral part of the company's line of office equipment
- a system of letting sales representatives share tips on how to successfully sell the new computers
- a means for tracking progress toward the sales quota, giving recognition to outstanding performers

How would you determine the right causes and solutions in this case? You would need to do more than just talk to the client who requested your help. You would need to examine the facts, and collect and analyze hard data. For example, you could interview a sample of customers who had been approached but who didn't buy a computer. Follow a few sales reps around for a week and find out what they do. Are they really demonstrating or even talking about computers? What happens when they do? Survey the sales staff to see if they feel they need to do a better job selling the new equipment. Examine the differences between those people who are successful in selling computers and those who are not.

It may seem easier just to produce a program or two. Or, it may strike you as unwise to turn down the possibility of developing some intervention if you have the opportunity, especially if your department's or company's profitability is at stake. However, it is unprofessional to do so, just like it would be unprofessional for a physician to just give you a bottle of antibiotic without examining you, or worse yet, if he knew that there was no medicine that would cure you.

When attempting to uncover communication problems, it's imperative to look at the big picture. Regard the entire organization as the client. The particular person who requests the services of a communication professional or department is merely a representative for the organization, and most likely does not have the skills that you have to analyze the root causes and probable solutions to problems. What that person provides you with are the presenting symptoms, which may be incomplete and which may relate only indirectly to the real difficulty.

A very powerful framework in which to place communication analyses is human performance systems. Within this framework, you begin with the assumption that corporate communication systems exist to improve human performance. In conducting communication analyses, the role of the practitioner is to determine which problems and goals can be addressed through which kinds of communication interventions. However, communication must be seen relative to the larger system of human performance. This human performance systems concept was developed by Geary Rummler and Tom Gilbert, two organizational psychologists and training experts. According to Rummler, human performance is affected by

- *performance specification*—Have we clearly told people what we want them to do and, in fact, do we know what we want them to do?
- *task interference*—Are there barriers that prevent people from doing a good job?
- *consequences*—What happens to the performers themselves when they do good or poor jobs?
- *feedback*—Are people given information on how they're doing?
- *knowledge and skills*—Do they know how to accomplish the task?
- *the capacity of individuals*—Do the people have the physical and mental resources to learn and carry out the task? (Rummler, 1988)

Gilbert (1988) advises performance technologists to carefully observe exemplary performers and average performers and to note the differences. He says that there are often only small differences, but they are crucial differences. He also cautions us not to assume that training is the best way to improve performance. The notion that if a little training is good, a lot is better is very dangerous and exceedingly wasteful of time and resources.

Human performance technology, the practice that is based on human performance systems, is "a set of methods and processes for solving problems—or realizing opportunities—related to the performance of people. It may be applied to individuals, small groups or large organizations" (Rosenberg, 1990, p. 46). The primary professional organization whose members develop and practice human performance technology is the National Society for Performance and Instruction (NSPI), a group that is primarily comprised of trainers and educators who realize both the power and the limitation of instruction as a means to make organizations work better. This is a useful perspective in practicing communication analysis and design. (See Figure 4.2.) The model for practicing performance technology involves: (1) conducting environmental scans to determine the current state or actuals; (2) examining the organization's mission and goals (the desired state or optimals); and (3) determining specific performance problems and causes, as well as opportunities for improved performance.

As you can see from Figure 4.2, there are a number of communication-related programs, systems, and strategies that can serve as possible interventions, including education and training, feedback, group process, culture change, employee relations (communications), and information systems. There are also a number of other performance system factors that communicators generally don't control, such as incentives, resources, and personnel selection. However, communication professionals can both recommend and work with other specialists, such as people in line management, compensation, finance, and personnel.

Robert Mager is a well-known author in the area of behaviorally based training design. His book with Peter Pipe, *Analyzing Performance Problems* (1984), and his more recent book, *Making Instruction Work* (1988), provide a very powerful system for sorting out the causes of performance discrepancies in individuals. The flowchart in Figure 4.3 presents the basic model.

The first step in Mager's model for analyzing problems is determining the performance discrepancy: what is the person doing that you don't want her to do or, conversely, what would you like her to do? The performance discrepancy needs to be stated in behavioral terms; that is, terms

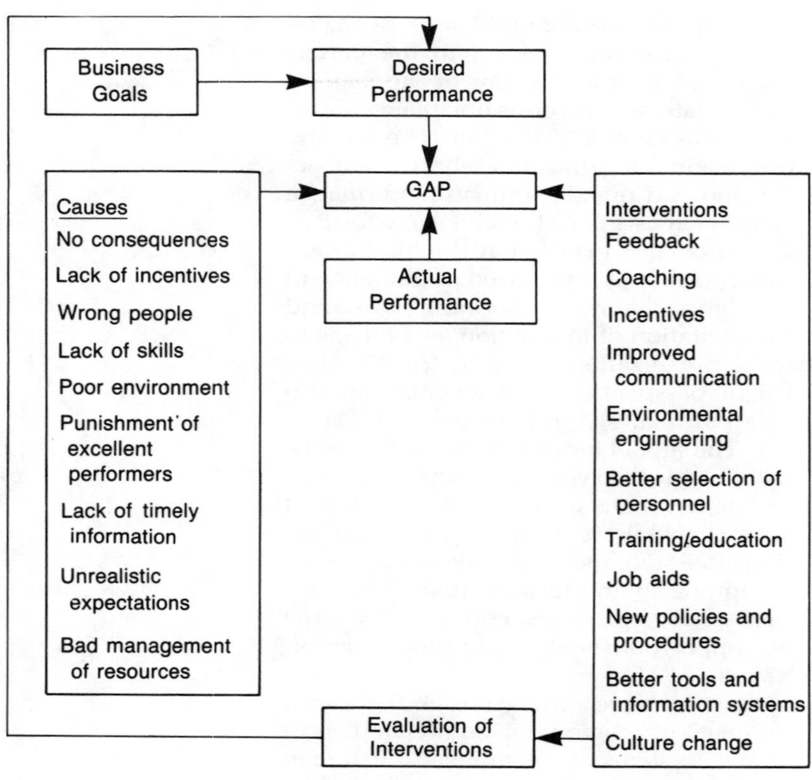

FIGURE 4.2 Performance improvement system. [Adapted from Rosenberg, M. (1990, February). Performance technology: Working the System. *Training,* 43–48.]

that are specific and observable. For example, you can't just say that you'd like a claims processor to pay more attention to his work; instead, you'd say the discrepancy was that his average error rate was 6% and you want that reduced to less than 1%. Then, you need to calculate the cost of that discrepancy and see if it's really important enough to fix.

Once you know what the performance discrepancy and your goals are, you can start figuring out the causes and solutions. Mager says the best way to determine if it's a skill deficiency is to ask yourself, "if their lives depended on it, could they do it?" This gun-to-the-head rule is easy to remember and very effective! In the example of the claims processor, if he really wanted to, he most likely could complete a claim with 100% accuracy, and probably could get his average error rate down to 1%. Thus, according to Mager's model, this is not a training problem; he already knows how to perform the task. Then what could be the cause of the deficiency? Following the flowchart in Figure 4.3, you can see that it could be a problem in motivation or incentive. Perhaps there's no punishment for making mistakes and no reward for being accurate. In fact, there could be a disincentive for taking one's time in being accurate, if performance is measured in terms of claims processed rather than claims accurately processed.

Even if a particular problem is due to a lack of skills, training is not necessarily the answer. It might be easier to simplify the task than to teach someone how to perform the current one. A person may have once known how to do something, but may have gotten rusty, just needing a refresher or some feedback and coaching. A job aid may eliminate the need to teach and to remember facts. Perhaps the individual is not even well-suited to the job. Perhaps hiring someone else would be a better solution, and it would be better to transfer or terminate the current job holder.

Federal Express has one of the most sophisticated and integrated management, communication, training, and incentive systems of any organization today. Larry McMahon, vice president of human resources, says that in the past when there was some problem in performance, people just assumed that there was a training problem. However, when they looked more closely and applied Mager's gun-to-the-head rule, they found that most were problems of discipline. They needed to improve the types of communication and the types of incentives they were using. Today, they employ a number of vehicles to give their employees the tools they need to do a good job; for example, trainers in the classroom; a management force to communicate news, articulate policies, and provide coaching; FXTV to broadcast current news and corporation-wide training programs; interactive video to provide individualized instruction; videotapes sent to employees' homes to encourage family support and appreciation for the organization and its practices; electronic bulletin boards for fast and cheap messaging and announcements; and traditional print materials for long-term reference. Probably the greatest factor in Federal Express's communication success is their pay-for-performance system; that is, employees are evaluated regularly on their work and their knowledge, and the results show up quickly in their paychecks.

McMahon says that FedEx has become an employer of choice so they have the advantage of starting out with good people who feel lucky to have the opportunity to work there. They provide a comprehensive employee orientation that convinces each person that they've made the best career decision of their life. He told me, "If they're good people and you give them the tools, motivation is usually not a problem—it just takes care of itself."

Basically, communication professionals need to sort out a number of different types of problems and related solutions:

- *training problems*—in which it's appropriate to enable people to memorize facts and develop new skills
- *performance support problems*—in which it's appropriate to build tools or job aids that help people to do their jobs without memorizing new information
- *motivation problems*—in which you need to create or point out the incentives for new behavior
- *data problems*—in which the target audience simply needs better access to more facts or more current data
- *communication problems*—in which support is needed to clarify messages or facilitate the sharing of messages among individuals or groups

It should be clear from this discussion that communication analysis and design is a lot like investigative reporting. The communication professional needs to establish a clear, operational definition of actuals and optimals, and to use a variety of sources and techniques for uncovering the reasons for the discrepancies.

Paradigms of Inquiry

Although there are dozens of methods for acquiring data and performing various kinds of analyses, they are all based on one of two philosophical frameworks of viewing reality: objective or subjective. Is the organization a concrete object or is it an idea that exists in its members' shared experiences? Is there one truth that all of us need to strive to understand or is everything a matter of individual interpretation?

Functionalists focus on questions of organizational effectiveness and maintain that organizations can be studied and observed as concrete objects. Their common metaphor for organizations is a machine. *Interpretivists* (sometimes called humanists or constructivists) focus on the social construction of reality and are mainly concerned with revealing symbols and meanings. They view organizations as cultures.

Daniels and Spiker (1991) explain these two views using the following example. Let's say that we want to know whether employees are more satisfied when their managers adopt an open style of communication. A functionalist would observe and classify managers as high or low in communication openness, use a survey to ascertain job satisfaction among employees, and statistically correlate openness with satisfaction. An interpretivist would interview organizational members and ask them to tell stories about and give examples of their

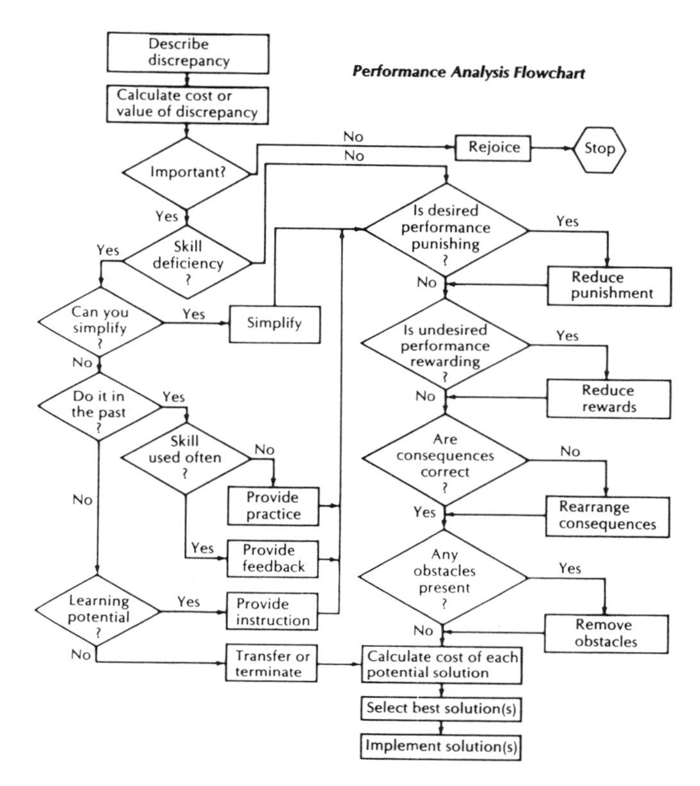

FIGURE 4.3
Performance analysis flowchart. [Reprinted from Mager, R.F. (1988). *Making instruction work*, p. 39. Copyright© 1988 by Lake Publishing Company, Belmont, CA 94002, USA, with permission.]

work, without asking specific questions about openness and satisfaction. Then, the analyst would identify and describe themes in these reports to see how members socially construct their environment, to determine if openness and satisfaction were among those themes, and to explain how they used those concepts to understand their relationships with managers (Daniels & Spiker, 1991).

Communication practitioners employ a number of methodologies to collect and to interpret the data upon which they base their recommendations and designs. Two categories that some writers have used to classify these types of research methods are *quantitative* and *qualitative*. Quantitative methods presuppose an objective reality, generally express observations in terms of numeric values, and use a variety of statistical techniques to test hypotheses and to express relationships among variables. For example, a communication analyst might use a survey to find out how employees rate their managers in terms of their communication practices. This survey might also ask for some demographic and organizational information. Then, the researcher might see if there is a correlation between employees' age, tenure with or division within the company, and the way in which they perceive their managers' styles. Qualitative, or interpretive, research methods assume a social construction of reality and use methods such as interviews and participant observation to gather information. For example, a researcher might spend several weeks observing the daily work of several sales representatives, collecting their correspondence, and audiotaping interviews with them about their work. Based on this large amount of information, the researcher looks for recurring themes in the words and metaphors that people use, and the relationships that emerge. Generally, qualitative or naturalistic paradigms are used to find out what sorts of issues or themes are important to pursue. By not starting out with specific hypotheses or questionnaires, the researcher can determine what is worth investigating. Once these issues or relationships are clarified, quantitative measures can be used to efficiently collect data and express findings.

Most communication professionals conduct research to help them design better systems or programs for a very specific audience and setting. They are not attempting to generalize their findings or results to other organizations. Also, it is generally not possible to set up scientific experiments to test out which kinds of interventions or message designs are best. For example, it might be quite interesting and valuable to find out if employees learn more by watching a videotape than be reading an instruction manual, but within most organizations there is not the time or money to set up an experiment in which you'd have one group use a video and another statistically similar group use the manual, and test them both to see who learned more. Generally, practitioners need to use their best hunch, based on their knowledge of communication theory and other researchers' scientific evidence, to come up with one optimal approach.

Some of the common methods used by professional communicators to get information about communication needs are

- written surveys or tests
- individual interviews
- direct observation
- analysis of data (i.e., production records, reports of accidents, sales figures, etc.)
- focus groups (group interviews)
- reader/viewer response cards
- content analysis of current programs to identify weak or inappropriate points
- telephone surveys

Methods frequently used to evaluate programs or systems once they're in place include

- written tests
- reader/viewer response cards
- before-and-after analysis of data (i.e., production records, reports of accidents, sales figures, prospective customer inquiries, etc.)
- systematic observation of on-the-job performance
- informal feedback from audience or their managers
- patterns and frequency of usage of systems or programs

As we discuss specific communication analysis techniques, you'll see examples of these paradigms of inquiry and of methods for getting information.

Levels of Analysis and Design

Communication professionals work with analysis and design techniques at a variety of levels. These include the following:

1. *the micro level*—the individual or project level, such as developing a single training program or producing a product brochure. This involves what's called *tactical planning*.
2. *the macro level*—looking at the entire organization, such as creating organizational development plans, instituting total quality management programs, or supporting a teleconferencing system. This involves *strategic planning*.
3. *the mega level*—involving society as a whole, beyond the organization, such as responding to environmental conditions by reducing the amount of paper consumed in communication activities or looking at the ways in which a particular industry or profession is portrayed to young people. This involves *proactive strategic planning* (Kaufman, 1991).

Often, communicators' work is stereotyped as dealing with micro level analysis, design, and production, although those with greater experience and influence find themselves working more often on macro- and mega-level problems. While junior staff members work with specific projects, senior communication officers (and Renaissance communicators) spend more time on the bigger issues, like the kinds of communication systems and practices that should be instituted within the organization, as well as within one's entire profession or within the field of communications.

THE MEGA LEVEL—SOCIETAL NEEDS

Neither communication practitioners nor their clients or sponsors operate within a vacuum. Each of us influence and are influenced by our social and economic environments on a daily basis. Some of the broader analysis and design work that communicators perform are

- finding and responding to social and economic issues
- analyzing and improving the status of the organization or the client in their industry or area
- analyzing and improving the status of the communication profession as a whole by developing standard practices and bench marks

In order to make a greater impact, we need to be broadly educated, well-informed, connected to our communities and professions, and continually engaged in scanning the environment. Most senior communication officers are active in civic groups and professional societies and are avid readers of contemporary literature that spans many topics. They use their skills and often their organizations' resources to support global needs and concerns.

The Opinion Research Center (ORC) conducts ongoing surveys on the climate of the American workplace. One of their major research reports, concluded in 1982, found that "all employee groups give their companies generally poor marks for communication, although this is the essence of management." They also found that employees felt their managers weren't willing to listen to their problems or suggestions, and that the company grapevine was their first source of information even though they'd like to get information from their supervisors face-to-face. Major studies such as this one, which were reported to senior managers throughout the country and picked up by the press, set the agenda for new management and communication strategies on a national and even international basis (Ruch & Goodman, 1983).

One major issue for professional communicators is to help their clients decide what kinds of issues the company wants to support. Almost every organization donates money and personnel time to local, regional, or national organizations such as the United Way, educational institutions, and the arts. Of course, not every charity can be supported, so public relations or external communications professionals are often the ones to decide where the money should go. For example, Pet, a large producer of food products was advised by its vice president of communication to concentrate its charitable work in the areas of family, children, and hunger. Many organizations are very active in associations and educational endeavors that support their general industry; for example, most banks have memberships in financially oriented special interest groups, such as the American Banking Association, that provide ongoing training.

Apart from their organizations or clients, many communicators are also involved in their own trade associations, which deal with global issues and professional development. For example, many local chapters of the International Television Association have donated production of televised public service announcements to regional charities, and representatives from the American Society for Training and Development and the NSPI have met with congressional committees studying work force and training issues.

Although we tend to think of corporate communication as an output activity (sending messages out to our employees and customers), it is more appropriately considered a two-way street. Remember that communication systems are the eyes and ears (not just the mouths) of our organizations. We need to set up systems for analyzing trends, conducting continual research on our professions and communities, and evaluating our organizations' roles in the broader sphere of social and economic activity. Some ways in which communication professionals are involved in this kind of environmental scanning include

- reviewing and archiving mass media treatments of topics of interest to the corporation (i.e., new technologies, activities of competitors, new management techniques, etc.)
- taking part in professional and community clubs, associations, and boards
- conducting on-line research using bibliographic search tools covering important speeches, research briefs, government documents and legislation, and court cases
- conducting surveys of customers, community members, and other professionals
- establishing data bases through which information about the organization's environment can be easily accessed and summarized

THE MACRO LEVEL—COMMUNICATION AT THE ORGANIZATIONAL LEVEL

The macro level of communication analysis and design relates to communication systems and policies that exist in a particular organization. This level is responsible for determining general problems or obstacles in internal and external communication and for designing strategies to enhance communication. Obstacles to effective corporate communication can be the result of a lack of communication rules or policies on how information should be disseminated, ineffective networks or outdated technologies, a management philosophy that does not support open communication, too many links in the communication chain, personnel lacking in communication skills, a mismatch of corporate culture or management philosophy with communication strategies, or a lack of knowledge about how best to address an organizations' various audiences' needs and interests.

When analyzing communication at the macro level, examine the following factors:

- *communication structure*—What means are generally used to communicate what kinds of messages with whom?
- *network flexibility*—Who may easily communicate with whom, and which paths and channels can be used?
- *initiation of messages*—Who may initiate and who may seek information in what ways?
- *communication load*—What is the typical number of messages a given person, level of management, or department must send and receive each day?
- *communication encumbrance*—how much of an organization's resources are expended on communication (i.e., memos, meetings, training, newsletters, etc.) ver-

sus the amount spent on processing its main product?

- *communication efficiency*—How quickly, accurately, and inexpensively can a given type of message be disseminated?
- *communication effectiveness*—How accurately can the intended audiences repeat or act on typical messages?
- *communication appropriateness*—Do the channels of communication and typical messages fit the organization's culture and desired goals?

A client of ours who provides software and services to automate billing, insurance reimbursement, and appointment scheduling for large medical practices is undergoing expansion. It started out as the data processing department of a medical center, was then made into a profit center, and now has become a profit-making subsidiary that sells its services back to the parent company, as well as to other clients. This client, having developed a new version of software, plans to expand beyond their firm's region and sell their software through independent representatives and reseller organizations nationwide. The communicaton impacts of this decision are significant. Here are some of the areas that we looked at in our communication analysis:

- How is the company portrayed visually and verbally in its literature, promotional videotape, and customer newsletter? Is the image consistent and appropriate for its intended audience?
- How will resellers be selected, and how will they and the customers they bring in be kept as close to the firm as the company's original regional customers? Will the resellers be able to accurately represent their products and philosophies?
- Will the company be able to support remote customers as well as those nearby? How can this be done cost-efficiently?
- How can the company clone their in-house expertise in training and in configuring systems for use by resellers? By what means can training be standardized now that many different people will be doing the customer training?
- Are there ways to reduce training and documentation by redesigning the software itself or by adding embedded computer-based training and hypertext documentation?
- How can the communication aspects of the software itself, the consultation that the company provides, and the training that they offer be woven into the marketing of the company's products?

The marketing, advertising, public relations, training, documentation, and customer communication policies and tools of this client needed to be studied and designed as a whole system. As we find with many growing companies, it's not simply a matter of doing the same thing more often or with more people. New rules and tools need to be developed.

Developing Communication Objectives

Designing an organization's communication policies and systems can be a daunting task. Usually, it's something that's done from the ground up when a company gets large enough to need a formal structure or when major changes occur. However, the system itself should probably be examined on a regular basis, much like an annual check-up. As we discussed, the way to get started is to develop communication objectives:

1. Develop or ascertain the organization's basic objectives (i.e., profit, growth, diversification, expansion of influence).
2. Prioritize those objectives, and from them develop communication objectives for the long- and short-term.
3. Decide who has a legitimate need to know what kinds of information (both internal and external audiences).
4. Establish a method of determining the cost-benefit ratio for communication interventions.
5. Establish two-way channels of communication for specific types of information and audiences.

Defining Audiences

Organizational messages have a variety of audiences; typically, we think of employees, the general public, and stockholders.

However, there are a number of other audiences who should be aware of important information and have accurate images of our organizations. According to Parkinson and Rowe (1977), these include

- banks
- investors
- stockholders
- customers
- suppliers
- community leaders
- business or trade organizations
- senior executives in other companies
- potential employees
- government leaders
- trade unions
- consumer organizations
- environmental organizations
- teachers
- students
- international political or economic agencies

FIGURE 4.4
Federal Express uses an annual employee survey to receive input on the workplace environment. [Copyright © 1991 by Federal Express Corporation, all rights reserved. Reprinted with permission.]

Frederick W. Smith
Chairman and Chief Executive Officer

INTER-OFFICE MEMORANDUM

2005 Corporate Avenue
Memphis, TN 38132
901 395-3377
U.S. Mail Box 727
Memphis, TN 38194-1841

DATE: March 21, 1991 TO: All Management

FROM: Frederick W. Smith

SUBJECT: **1991 SURVEY FEEDBACK ACTION**

In winning the Malcolm Baldrige Award, our Survey Feedback Action Program was cited as a key strength because it involves our employees in the improvement of our work place and service. Once again, each of us has the opportunity to secure valuable input from our employees by utilizing the SFA process.

Our challenge is to listen to our employees, work with them to develop corrective actions and then follow through with these actions. That is the only way we can assure continuous improvement in the work environment at Federal Express.

FWS:mm
5190

Once the audiences have been identified, other factors must be determined; for example, their current attitude toward the company, their current information sources, their information expectations, their needs, and the company's objectives for the impact of information and communication with them. Combining communication objectives with this kind of audience analysis yields a number of specific action plans.

Taking the Pulse of Existing Communication Systems

As organizations grow and change, it's generally difficult for their leaders to really know if their communication systems are working and are in synch with the company's goals and philosophies. There are a number of ways to do this and just as many underlying goals.

Many organizations conduct regular surveys of their employees to determine their opinions about management communication styles. The intent of this kind of analysis is to reveal the basic management styles in operation, such as bureaucratic (scientific) or humanistic (human relations), and to determine whether the communication system and basic management philosophies are compatible and effective. One way to determine these underlying communication styles is to study individuals' interactions with their managers. Some of the questions that can be asked to determine the nature of managerial communication include the following (Farace, Monge, & Russell, 1977):

- Who initiates interactions?
- How satisfied are you with the implicit communication rules?
- How frequent is your contact with your supervisor?
- Who typically terminates your interactions?
- When you talk with your supervisor, how are outside interruptions handled?
- How are topic changes handled?
- Who selects the topics you discuss?
- How are delays treated?

Many organizations conduct other kinds of analyses to see if employees are generally satisfied with the kind of information they get not only from their supervisors, but from formal information systems. Federal Express, for example, conducts an annual survey that employees now fill out on-line on computer terminals at their local offices (see Figure 4.4).

Communication Audits

The concept of the communication audit was originally developed by organizational development scholar and practitioner George Odiorne, as a means for studying communication flow. An audit consists of standardized questionnaires for gathering data and a set of norms for comparing an organization with other typical ones. Osmo Wiio, a noted organizational communication scholar, developed the *Organizational Communication Development* audit in Finland. The survey has been refined by over 100 scholars in the International Communication Association (ICA) and has been administered to over 5000 employees in 19 organizations (Putnam & Cheney, 1990).

According to Daniels and Spiker (1991), communication audits seek to find out how well employees feel informed about various job aspects:

- how well they're doing their jobs
- job duties
- organizational policies
- pay and benefits
- organizational problems
- new products and services

Using these audits, you can determine the general supportiveness of the organization's communication climate, the quality and accuracy of most communication acts, the nature of managers' relationships with subordinates, and the employees' general satisfaction with the way they receive information.

Auditing in organizational communication has moved into the era of automation. A consulting company in Chicago offers a service called Tele-Survey. Employees in participating organizations can call in from a touch-tone phone and enter their ID numbers (to avoid any employee calling more than once). They then respond to a customized survey by pushing phone buttons and leaving voice messages. The consulting company then offers a written report to management on each survey. Using these kinds of services, even organizations without a communication staff can conduct on going, anonymous surveys of their members.

Information Audits

Similar to the communication audit is the information audit. This audit seeks to find out if the company has the information resources needed to achieve its objectives. It uses the concept of information asset, which includes records, data bases, statistics, and bibliographic material. Some of the areas to be determined are as follows:

- Are the information assets accurate and current?
- Are the assets and equipment used efficiently?
- Does the information system and communication staff operate effectively with all levels of management, using a variety

FIGURE 4.4 (*Continued*)

SURVEY QUESTIONS

1. I feel free to tell my manager what I think.
2. My manager lets me know what's expected of me.
3. Favoritism is not a problem in my workgroup.
4. My manager helps us find ways to do our jobs better.
5. My manager is willing to listen to my concerns.
6. My manager asks for my ideas about things affecting our work.
7. My manager lets me know when I've done a good job.
8. My manager treats me with respect and dignity.
9. My manager keeps me informed about things I need to know.
10. My manager lets me do my job without interfering.
11. My manager's boss gives us the support we need.
12. Upper management (directors and above) lets us know what the company is trying to accomplish.
13. Upper management (directors and above) pays attention to ideas and suggestions from people at my level.
14. I have confidence in the fairness of management.
15. I can be sure of a job as long as I do good work.
16. I am proud to work for Federal Express.
17. Working for Federal Express will probably lead to the kind of future I want.
18. I think Federal Express does a good job for our customers.
19. All things considered, working for Federal Express is a good deal for me.
20. I am paid fairly for the kind of work I do.
21. Our benefit programs seem to meet most of my needs.
22. Most people in my workgroup cooperate with each other to get the job done.
23. There is cooperation between my workgroup and other groups in Federal Express.
24. In my work environment we generally use safe work practices.
25. Rules and procedures do not interfere with how well I am able to do my job.
26. I am able to get the supplies or other resources I need to do my job.
27. I have enough freedom to do my job well.
28. My workgroup is involved in activities to improve service to our group's customers.
29. The concerns identified by my workgroup during last year's SFA feedback session have been satisfactorily addressed.

of media and formats to maximize the impact of decision making?

- What areas will result in better operations and increased profits?

In order to conduct an information audit, communication professionals need to take regular inventories of their information assets. They must evaluate the accuracy and relevance of the information that's being stored and processed, and see how fast and easily it can be retrieved (Meltzer, 1981).

Network Analysis

One of the most important aspects of an organization's communication system is its networks. *Communication networks* are the typical paths that naturally occur as people communicate within an organization (see Figure 4.5). Various kinds of network analysis techniques can be used to see how people use formal and informal communication to meet their professional and social needs. Listed are some of the techniques used to chart out networks:

- *duty studies*—people write down all their communication behaviors during a given day
- *participant observation*—a researcher lives within a group of people and carefully notes how and with whom they communicate

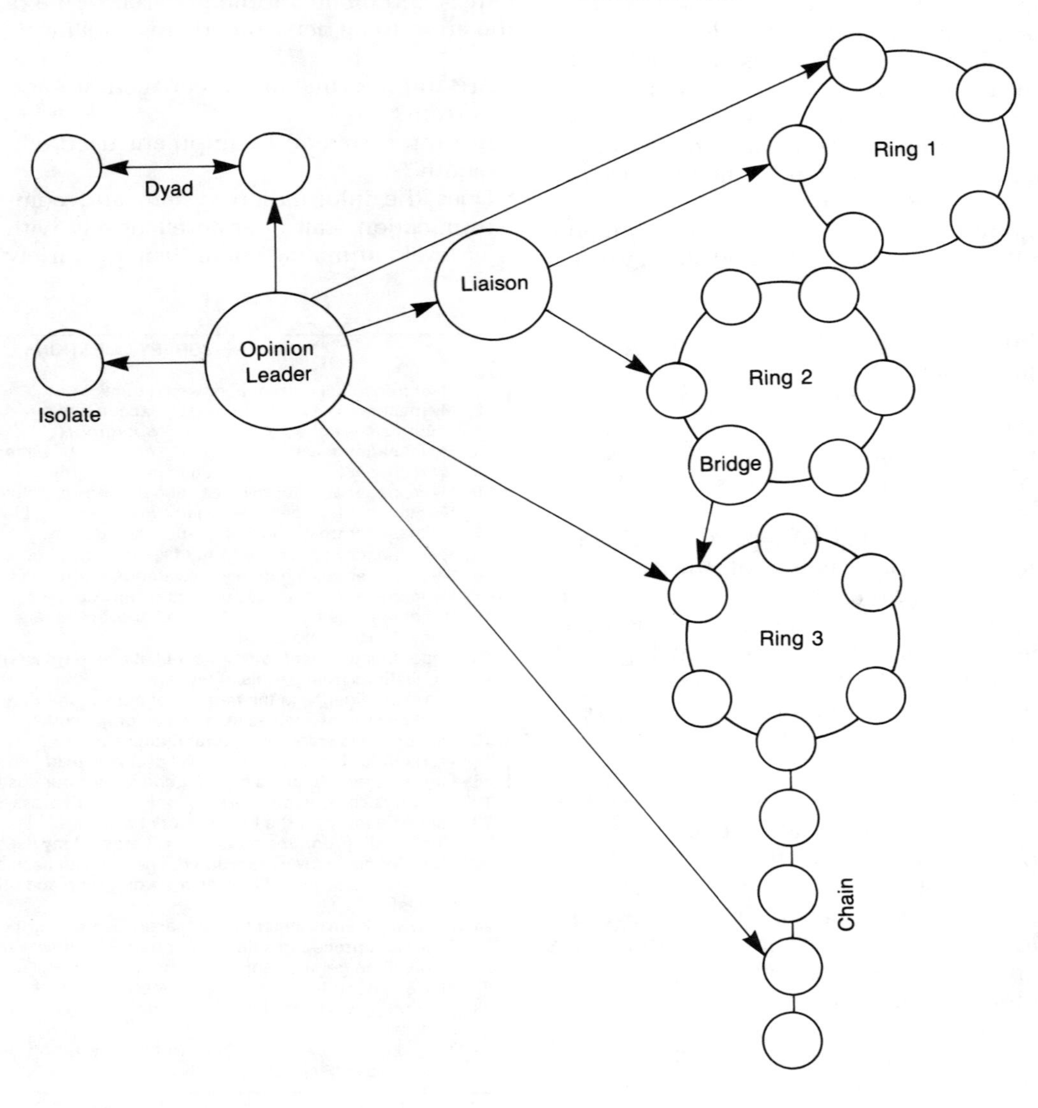

FIGURE 4.5
Typical organizational communication networks.

- *questionnaires and interviews*—subjects are asked questions about their communication patterns
- *small world technique*—a researcher tracks the flow of one given message
- *physical monitoring*—people's phone calls, e-mail messages, and incoming and outgoing mail and memos are tracked and categorized for a period of time

The goal of these techniques is to develop graphic displays of communication roles and channels. For example, the communication analyst can determine which people make up typical groups; this includes discovering the purpose or theme of the group and identifying group members. Within network analysis, there are special names given to particular recurring roles:

- *liaison*—a person who links communication between groups but is not a member of any of the groups
- *bridge*—a person within one or two groups who links the groups
- *gatekeeper*—a person who controls information within and between groups
- *isolate*—a person with few or no communication links

Along with identifying membership in groups and the network patterns within and between groups, a communication analyst also looks for other elements:

- *symmetry*—Is communication typically one-way or two-way?
- *strength*—How often is a particular channel used?
- *reciprocity*—Do individuals agree about the communication links and patterns?

Goldhaber et al. (1979) present the following questions that can be used in performing network analysis:

1. How are messages typically sequenced? based on need?
2. In what direction do messages flow?
3. How structured and rigid is the communication in the organization?
4. To what extent are formal and informal structures subject to critical information overload points? What is the typical ratio of budgeted [scheduled] communication activities versus unbudgeted activities?

Interpretive Methods

Although typical analytical tools, such as questionnaires and interviews, can yield important data, no one would argue that they present a complete or perfectly honest look at communication systems. The very questions that researchers write bias the picture; for example, important areas may be completely ignored or conversely, minor issues may be given undue attention. *Demand effect* is the phenomenon that occurs when people are pressured to respond in a somewhat less than candid manner by the artificial situation of responding to an interview or survey. A person may not want to make trouble for a newsletter editor by saying that she really doesn't find the information in that publication useful. However, when a person is asked how she felt about the new orientation film, she might feel that she'd appear naive if she couldn't come up with some critical comments.

Interpretive, or naturalistic, methods avoid artificial data-gathering situations and instead rely on the researcher's interpretation of existing events. For example, an analyst might look at fantasy themes or picture organizational life as a drama with characters. One way of doing this is *content analysis* of messages, publications, company songs, and graffiti. Another approach is to see how organizational facts are embellished or exaggerated (for instance, a universal story at colleges is about some professor who is said to always fail half his class). In *metaphor analysis*, the researcher records the actual speech of people and isolates metaphors, such as those that relate to activities or objects. A typical metaphor found in some organizations is the one big happy family theme, while others relate to sports stars or military prowess.

Symbols, metaphors, and themes are the building blocks of corporate culture that so many executives are striving to create or change. Looking at corporate life as a series of communication acts with themes, plots, and actors, a communication professional can get a handle on some crucial elements. As Deal and Kennedy (1982) point out, the new symbolic manager is the director or playwright of the corporate drama. Those managers and communication specialists who can shape the performance and build its underlying script can understand and influence the foundation of the organization.

The Communication Flowchart

We've seen that there are formal communication policies and structures, as well as informal networks and themes, that arise within organizations. The communication flowchart is one method that you can use to document and to analyze the ways in which communication should and does occur within an organization. It's a combination of an organizational chart, a network analysis, and a picture of the grapevine. Once you develop one, you can get a clear picture of the ways in which communication is supposed to work in comparison with the way that it actually does work. It's not unlike looking at the grounds of a typical college campus. There are paved walkways where people are supposed to walk and there are beaten-down muddy paths in the grass where people have found it's most efficient to walk.

In order to develop a communication flowchart, you must first make an organizational chart. Add to this chart the typical publics with whom the organization communicates (e.g., stockholders, local citizens, etc.). Then, add lines to show how information is typically provided from top management to employees and to the public. Now, add another series of lines that show how those same messages really get communicated. For example, in a military contracting company where one of my colleagues works, layoffs should first be announced by memos placed on bulletin boards by their communication officer. Then, the information should be read by employees and later sent to the local press by the PR department. However, the way it really happens is that there's a leak to the press, rumors develop about layoffs, employees find out information through phone calls to inside sources, local media interviews with employees, and finally, formal memos and carefully written announcements are provided to the local press. A final set of lines to add is typical paths for horizontal and upward communication. (See Figure 4.6.)

Typically, you'll find that there are few or no formal means of upward communication other than conversations with a successive chain of managers or a token suggestion box. Despite its informality, upward communication is vital and consists of

- status reports
- problems
- suggestions
- feelings about the organization and each other

Downward communication, in which there are typically many formal channels and media, consists of

- instructions
- rationales
- procedures and rules
- feedback on performance
- indoctrination

Horizontal communication, which can also be slow and fraught with many links between people, consists of

- task coordination
- problem solving
- sharing information
- conflict resolution
- building rapport

It's important to discover how both formal and informal communication occur within the organization. Formal communication is officially designed and planned, and it tends to follow the organizational chart. Informal communication tends to be episodic and does not follow specified routes. The *grapevine* is important in network analysis—it is an informal network through which people get information before it is officially announced. Rumors and gossip are generally carried on the grapevine. This pattern tends to be based on people's relationships rather than on their formal places on the organizational chart.

Many people consider the grapevine as the channel that gets used when formal communication fails—as something to be eliminated within organizations. However,

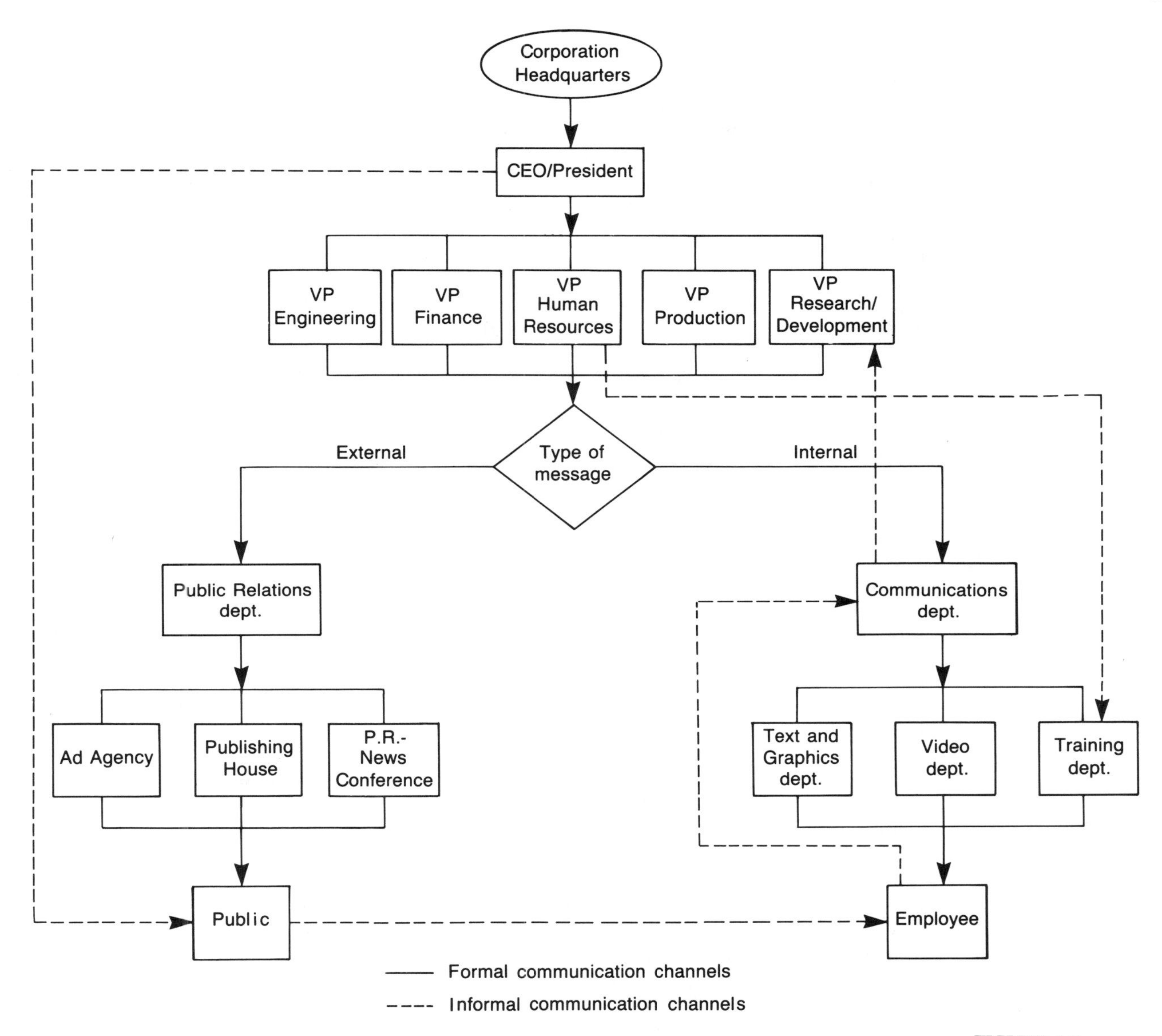

FIGURE 4.6
A communication flowchart. (Based on a model by Michael Dermody.)

we'll probably never do away with it, no matter how good our communication systems. The grapevine is fast, although it is not selective or necessarily accurate. It's most active where there are limited or very bureaucratic lines of communication. Most employees use it, but few prefer it. In fact, most people resent having to hear information first through informal means rather than having their supervisors tell them in person.

Communication Systems Analysis

Every organization has methods or systems that it uses to provide information and training. In a macro-level analysis of them, we look at whether they're effective and efficient. For example, a communication professional might examine whether classroom training was the best way to teach new sales representatives about the company's product line. In doing this, particular courses are not necessarily the object of analysis, but rather the system of training is studied. Generally, the question in systems analysis is whether a particular communication medium or practice is the most cost-effective and powerful method available.

In conducting systems analysis, the first step is to identify the ultimate goals and characteristics of a perfect system. For instance, an ideal training system would be one in which trainees mastered all of the content, opportunities for individual feedback and participation were provided, courses were available when and where needed, and consistency was a guarantee. The next phase is to evaluate the current system against those criteria. Along with that qualitative evaluation comes a cost-accounting of the current method. Using the example of sales training, you might find that the current classroom training was doing a good job of teaching the material and that trainees had some opportunities for individual attention. However, the training was difficult to schedule, varied in effectiveness across different instructors, and involved travel for most trainees. When trainee time, as well as actual instructional costs were computed, a typical one-week course cost $80,000 or $4000 per student. The final stage of systems analysis is to examine alternative communication systems and to evaluate their potential effectiveness and costs. Going back to our training example, communicators might look into media-based training methods that could be used at individuals' work sites as a way to reduce costs; however, the potential for higher production costs, difficulty in updating material, and less opportunity for individual feedback or interaction would need to be examined as well.

A large accounting firm was interested in reducing the costs for a basic course in auditing that they offered to several hundred new hires each year. Management was particularly interested in increasing class size so that fewer sessions would have to be offered. The training department wanted to ensure that such a decision wouldn't result in less effective training, since mistakes would eventually cost the company much more money than a few more training sessions. Evaluators set up an action research study in which several sessions of the same class were offered with different class sizes. They then compared the two groups on measures, including assessments of class assignments, trainer evaluation, and trainee satisfaction with the courses. They found that overall there were no significant differences between the larger and smaller classes, and so recommended increasing class size as it would be an effective way to reduce costs.

THE MICRO LEVEL—INDIVIDUAL INTERVENTIONS

In order to actually work toward societal and organizational goals and to develop solutions to communication problems, we need individual interventions—programs, meetings, and publications. In order to identify specific performance discrepancies and design appropriate responses, analytical techniques similar to those used at the mega- and macro-level are employed, but in a more focused context. Micro-level analysis and design deals with identifying specific performance problems or information needs of a specific audience, and then systematically designing messages and strategies to meet those needs.

Types of Analyses

A number of forms of analysis are used in designing individual programs or interventions. The ones that are used to determine the need for a program and how it might best be structured are collectively called *front-end analysis*. This form consists of needs or performance analysis, cause analysis, task analysis, audience analysis, and communication setting analysis.

Needs Analysis. The first step in designing a communication intervention is to conduct a *needs analysis*, perhaps even more accurately called a performance analysis. "A need is a gap in results, not a deficiency in resources, methods, means, processes, or procedures." It is not simply what people want but rather what is required to meet a goal (Kaufman, 1989, p. 69). Clients or program sponsors may approach a communication agency with a wish list, but this is only the beginning of the process of uncovering actual needs.

Once you or a client has determined that there is some gap in desired performance, you must obtain the details of what comprises optimal performance and contrast it with the actual performance. There are a variety of ways to get this information. It can come from the informed opinions of workers, supervisors, or content experts. It may also come from other sources, such as an examination of sales records, data from secret shoppers, or a focus group.

Unfortunately, many clients are unwilling to pay for or wait for an adequate needs assessment. They may feel that they already know the problem and the solution or that such up-front research is unnecessary. Rossett (1990) advises that it may be better not to call this phase a needs assessment, but rather to use analogies to your industry, such as scoping out work or conducting a study, so that management understands it. Then, measure the impact of the needs assessment, such as if you pared a 4-week course down to 3 weeks or what new information you unearthed. This can prove to your clients that this phase is worthwhile.

Cause Analysis. Once you understand the needs, you must identify why they exist; this can be referred to as a *cause analysis.* The Mager flowchart discussed in the previous section is an excellent framework for this analysis. Simply stated, your job in this phase is to find out whether it's basically a training and information problem or a motivational problem. This information will guide the strategies you use in designing the intervention.

Task Analysis. In order to move your target audience in the appropriate direction, you need to know exactly what expert or model performers actually do. How does an experienced car salesperson really close a sale? What actions and skills are involved in operating a fork-lift truck? What do investors look for when they evaluate stock offerings? *Task analysis* is the set of techniques that are used to determine these things. A task analysis is generally performed by watching one or more expert performers do their job, and then interviewing them about how they do it and how they think about it. In order to provide a contrast, you may also want to observe more average, or even poor, performers to discern the differences. A task analysis can be represented as a flowchart of actions and the logic behind the decision-making processes involved in these actions. For example, Figure 4.7 shows a task analysis for cleaning a spark plug. You can see how even relatively short and simple tasks are really quite difficult to analyze, because no action or decision can be taken for granted.

Audience Analysis. Along with scrutinizing the task, you must also perform an *audience analysis,* which involves specifying exactly who will be the target of this particular communication intervention and finding out as much as you can about them. A good audience analysis not only presents who the people are, but paints a complete picture of their needs, styles, and backgrounds.

For example, one audience for whom I

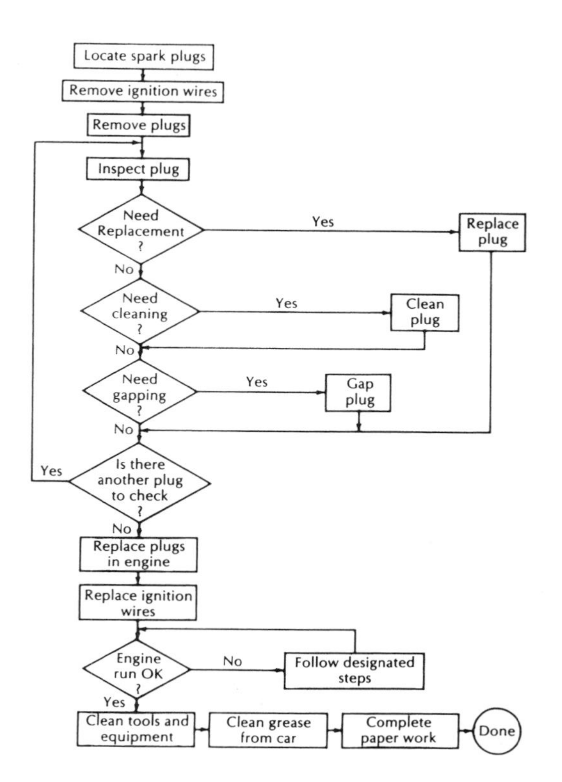

FIGURE 4.7
A task analysis flowchart for cleaning spark plugs. [Reprinted from Mager, R.F. (1988). *Making instruction work,* p. 36. Copyright© 1988 by Lake Publishing Company, Belmont, CA 94002, USA, with permission.]

designed an interactive video was made up of newly hired sales representatives. I found that most of them were young, had a lot of computer experience, were hired directly out of top engineering schools (and, thus, were tired of school and rather cocky about their expertise), and were generally rather competitive. I designed a program that was a parody of a game show that started out by presenting extremely challenging situations, which they were asked to solve by typing responses into the keyboard. My intent was to entertain them, but also to subtly make them realize that they really did have much to learn about the job. This approach was radically different from another technically-oriented interactive video program that I designed for machinists. Most of the machinists had not used computers and had been out of school for many years, and they were rather intimidated by learning to use a new programmable controller.

The following information should be obtained from a prospective audience:

- age and gender
- occupation
- educational background
- work experience
- ethnic, national, and regional culture
- incentives for getting this information or learning a new skill
- aspirations
- needs and desires
- orientation toward the message source
- attitude toward various communication media
- experience with various media or techniques and the content

By determining whether the audience is homogeneous or heterogeneous, you can decide whether there is one approach that is appropriate for all of them, or whether different styles or media are necessary. Increasingly, with the aid of interactive media technologies, different audiences can be effectively addressed in one program—although this one intervention will contain many diverse approaches.

Communication Setting Analysis. This analysis determines when and where the audience will be receiving your message. A program that is aimed at a busy executive will be vastly different if it is to be used during a formal meeting at the workplace, at the home, or in a public location, such as an airport or a store. Listed are some factors to consider when choosing a setting, or when working within a given setting:

- How many people will be simultaneously involved in this intervention (one person reading a leaflet versus 100 people in an auditorium)?
- Is there a fixed and controllable environment in terms of seating, lighting, and media equipment, or are there unknown or variable environments?
- Should the program be used privately, or is it meant for public display or interpersonal interaction?
- What is the general mind-set of your audience in this place (relaxing at home, running to meet a plane at an airport, or actively learning within a classroom environment)?

I spent a day working with representatives of a media production house from Monterrey, Mexico. This company was hired by a large cement company to produce an interactive video program for display in a kiosk at an industrial exhibition center. We worked with them to try to determine what the sponsor's objectives and desired image were, who would be coming to this exhibit hall, and what their mind-sets would be while they were there. The cement company would like to have a program that would explain their various products and projects, as well as show how cement is made. The potential audience in this case are generally people who are attending a conference or trade show at this center and who may take some time to stop by a few standing exhibits of major local companies.

I tried to put myself in this situation: If I were coming to a conference there, why would I want to learn about cement? Would I be in a rush to get to my meetings, or would I perhaps be looking to kill time before it started? What information would I want as a visitor there? Are most people local or from afar? Would people want to buy cement or invest in this company? Why would people spend their time standing at a kiosk to learn about this company? How long could we expect people to interact with this kiosk? Should we put stools in front of the monitor to make it more comfortable for them to use or would this encourage loitering? Should we make the screen small and private or large and public? These decisions are much more crucial than the technical or artistic components of producing the actual program. If nobody comes up to the kiosk, or if they are too tired to stand, or if they are too bored by the company's presentation, most of the program will go unseen.

These various types of analyses lead up to the point where actual program design can begin. Balancing the multitude of audience and environmental factors against the information needs and organizational objectives for the program is the art of clever design.

Designing Interventions

Armed with information about performance discrepancies, the nature of the task, the background of the audience, and the nature of the communication setting, you can start to actually design the message and select the medium. While mega- and macro-level analysis and design processes are considered strategic, this micro-level is sometimes called tactical planning. Most formal communication is carefully designed—whether it be the wording of a speech, the visualization of a video program, the layout of a brochure, or the interactivity of a multimedia program. Although design activities require a high level of creativity, they are not without their particular methods and standards. But before we launch into a lengthy discussion about communication design, a word of caution about this whole process and the philosophy behind communication design.

Perhaps the most important skill in communication design is knowing when not to design. Bassett (1968) believes that there is an important distinction between ritualized communication, or performance, and communication. Performance follows a prescribed pattern, seeks excellence or influence over someone else, cannot tolerate variations in the standard model or pattern, and concludes when an accepted pattern of behavior has been enacted. The signs of communication, he maintains, are quite different; it adapts itself to the unique needs of each participant, there is no competition and no win-lose relationship, error is seen as inevitable and necessary, and conclusion is reached only when understanding is achieved. Interestingly, Bassett (1968) believes that the identifying signs of communication are a lack of planning and control.

> Communication requires openness, randomness, trial and error, and exploration of previously untried paths. . . . Against these criteria, most communication carried on in areas like business—whether it be internal communication among managers and employees or external communication such as advertising, directed toward others—is more accurately described as performance. (Bassett, 1968, p. 75)

We need to think hard about Bassett's points and evaluate whether or not we should design a message on behalf of a client. Will this message become so stereotyped as to become a performance without credibility, spontaneity, or meaning for the intended audience? What kinds of messages should be cleverly crafted and which should spring in an extemporaneous and candid manner from their sources? When we do design communication interventions, how should we remain sensitive to the need for impromptu modifications? These questions have no formula answers, but they are worth pondering before launching into elaborate design projects.

Developing Program Objectives. Just as in mega- and macro-level design, objectives are the foundation for program design. Bloom (1956), developed a taxonomy of objectives that is helpful in broadening our perspectives on this matter. Bloom categorized objectives according to these domains:

- *cognitive*—mental skills, such as defining, analyzing, using rules, and comparing
- *affective*—emotional and aesthetic responses, such as valuing and making decisions according to principles
- *psychomotor*—physical skills, such as hitting a ball, balancing on ice skates or hand-blowing glass

There are increasing levels of complexity for objectives in each of these domains, starting at knowledge and working up through comprehension, application, analysis, synthesis, and evaluation. So, objectives don't simply involve getting the audience to know or recognize some concept or word, they should be written to reach higher levels of cognition as well.

Good objectives specify what the learner must do—actual observable performance. This involves developing an operational definition of what the audience will do if your program succeeds. For example, "the trainee will compare and contrast the Acme widget washer with the Zenus widget washer by orally listing, from memory, at least six features, benefits, and limitations of each one." Notice that the objective doesn't mention knowing about widget washers. You can't see if a person has knowledge, but you can see what he do to demonstrate his knowledge. Also, it specifically lists what is required to demonstrate mastery. In this case, listing six characteristics for each of two washers is considered sufficient. One doesn't have to be able to build a widget washer, but merely recognizing one is not enough.

Well-written objectives also establish specific criteria for acceptable performance. For instance, "upon encountering a fire in the building, all employees will be able to locate the nearest extinguisher, determine if it should be used, call the fire department, and evacuate the building using the closest available exit in 3 minutes, with no direction from another person." In this example, it's crucial that all employees can act not only appropriately, but also quickly and without help from anyone else. In some cases, you might want to list what kinds of job aids or tools are available for the task. For example, "the receptionist will be able to connect a caller with the appropriate party using the company directory and the X25 console within 20 seconds." In addition, include the level of accuracy, if appropriate. For instance, "the trainee will mount the hinges on the compartment door using the standard tool set and diagram 123-abc, within one-sixteenth inch tolerance, so that at least 99.7% of assembled doors pass quality control inspection."

Well-written objectives are specific, observable, and measurable. As you write objectives, ask yourself if a group of other people could watch members of the target audience perform the stated task and agree on whether or not they met the criteria. Use action words, such as "list in writing," "assemble," "draw a diagram of," or "state," rather than ambiguous and invisible concepts, such as "know," "appreciate," "learn," or "understand."

Systems Design. Objectives form the foundation of a set of procedures generally accepted for the design of training, *instructional systems design* (ISD). This methodical and scientific approach to designing programs was begun in the 1940s, and was first widely adopted by the U.S. military in its design of training programs. Since that time, many versions of its basic format, analysis, design, develop, implement, and evaluate (ADDIE), have been generated, and this approach has been adopted for programs other than educational ones (see Figure 4.8). Although there is a great deal of evidence that programs developed using this framework are more effective in meeting their objectives, no specific instructional design model has been shown to be superior to others.

Developing Evaluation Measures. Once objectives are developed, the next step is to mentally jump ahead to the end of the process, which is evaluation. Now is the time to develop methods that you'll use to evaluate the success of the program. In business terms, developing objectives and evaluation measures is your contract with the client—it's the results you promised and the method you'll use to assess whether the results have been attained. In fact, many contractors use objectives and evaluation measures as criteria for their being paid. The contractor doesn't get paid until the target audience demonstrates mastery of the agreed-upon objectives, using the predetermined assessment instrument. The program is revised until this happens. Remember that communicators get paid for developing solutions—not programs.

FIGURE 4.8
The ADDIE model.

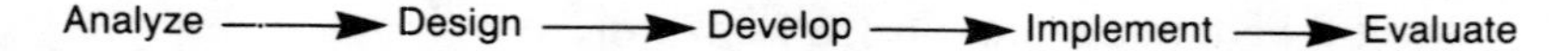

Evaluation measures don't have to be the typical pencil-and-paper tests. They can be a set of criteria against which an experienced person measures someone performing a task. Evaluation can be based on adherence to some stated standard or on achieving a particular goal. Evaluation measures aren't scores on smile sheets—how well an audience liked a particular program. Although communication and training programs don't have to be painful or tedious, their goals are not entertainment, they are bottom-line results.

The next step is to develop a general design or treatment for an intervention. This involves imagining what a program or a product will look like when it's done. Development must consider the following essential elements—who will deliver the message, what medium or media will be used, and what kinds of techniques (i.e., humor, a particular layout or artistic style, or animated diagrams) will be used to make the points and help the audience achieve the intended objectives. Writing treatments is a useful step before beginning to actually write a program, because it allows you to try out one or several general approaches and get the client's feedback and approval of one. (See Figure 4.9.)

New Sponsoring Video - Treatment

This new, 10-12 minute video program is designed to help Amway Distributors in North America sponsor new distributors. It will be brief and fast-paced, utilizing state-of-the-art video techniques and effects, computer graphics, original music, and the services of a "name" talent with mass audience appeal to heighten interest and lend credibility to the lives and motivations of real people as they share with the viewers their reasons for owning an Amway business.

We will see and hear the host at different locations within an ultra-modern airport; the setting itself, as well as what it symbolizes (options, choices, varying directions, excitement, a sense of adventure and progress), subtly underscores our theme of looking toward the future.

Selected distributors, meanwhile, will be videotaped on-location throughout North America, in settings that illustrate and reinforce the values and aspirations they've been able to realize as Amway people. They will be carefully chosen to demonstrate the diversity of individuals and interests found within Amway, as well as to illustrate Amway's compatibility with existing economic and social trends; their comments, woven together by the host, will convey our message: Amway is in tune with the times ... the trends are with us ... the future has already begun. And for those with vision -- and a willingness to work -- Amway offers an opportunity to turn their dreams for the future into reality.

FIGURE 4.9 Treatment for a video program. (Courtesy of Amway Corporation.)

Budgeting. After a treatment has been accepted, a budget is prepared. By this time, you'll know your costs for writing and basic program design and you'll also have a good estimate of production costs. In fact, you may develop several treatments along with the associated costs of each for the client's consideration. A budget includes not only the cost for completing the design and production of a given intervention, but it also should include cost of reproduction and cost of program delivery. For example, if you're designing a sales seminar, the costs include not only the visuals and the handouts, but will involve paying for one or more persons to present the seminar, to rent a space, and to transport the audience to the site.

Writing/Layout. Once a budget has been approved, the final stages of design take place. This can involve writing a speech, script, or text for a printed piece; developing roughs of graphics or photos; flowcharting various paths through an interactive media program; or developing a list of speakers and topics for a conference. Audiovisual scriptwriting includes not just the words that are spoken, but music cues, timing, special effects, exact shots, and visual transitions. (See Figure 4.10.)

Program Outlines and Lesson Plans. The old adage says, "Tell 'em what you're going to say, say it, then tell 'em what you said." This model of introduction, statement, and summary is still a good one. Many communication interventions involve a number of stages, within a given media program and within the larger context of its presentation. For example, a communication professional may be called upon to design an orientation program that consists of an introduction by a trainer, the showing of a videotape, the presentation of several key executives who give speeches, and finally, the distribution of employee manuals.

Learning theory tells us that people retain information better if they are prepared for it, know what is expected of them, have information clearly outlined and presented using appropriate stimuli, and then have an opportunity to review or consolidate the information. Figure 4.11 presents an adaptation of what Gagné (1970) has called the events of instruction, which can guide you in developing any kind of informational presentation (see Figure 4.11).

FIGURE 4.10
Script for a multi-image program. (Courtesy of Q1 Productions.)

BAYER SELECT
OPENING MODULE
Job # 2009
July 16, 1992
Hansen/Pearce

SECOND DRAFT SCRIPT - 6/8/92

VIDEO:	AUDIO:
The screen is black...	NARRATOR *(over moody, mysterious music)*: Once in a generation ... a change occurs.
The blackness turns to dark storm clouds, rolling and colliding across the sky.	*(Music builds.)* A change so powerful...so fundamental...that the world is never the same again.
A ray of light appears through a chink in the clouds, opening up to reveal an extreme close-up of a human eye.	Incredibly, such changes are set in motion by an act of thought...an idea...*(Music hits dramatic chord as eye is revealed)*...A vision -- that provides a new understanding of the world around us. And every great vision creates new challenges and opportunities that ripple across history like waves across the sea.
Dissolve to footage of the surface of the ocean, an endless expanse of waves.	*(Music segues into sound effects of the crashing of waves.)* History abounds with such visions, and the changes they made. Like the one that is being commemorated this year, all over the world.
The projection areas are filled with images suggesting the Age of Discovery, in the 15th and 16th Centuries: engravings of ships, caravans, etc., portraits of contemporary rules, etc., concluding with a map of the world as understood at the time.	*(Appropriate music and sound effects suggesting Age of Discovery)* For centuries, trade between Europe and Asia depended on long, difficult routes. Further progress was barred

not because of geography, or technology, but simply because of old ways of thinking.

The 16th Century map animates to become a globe. New shipping routes appear, across the Atlantic.

Finally, this barrier to progress was broken by a new vision that revealed the true nature of the earth as a globe.

Dissolve to footage of replicas of Columbus' three ships at sail.

With the voyage of Christopher Columbus, the Age of Exploration began.

Clouds appear, as the images seem to fade back into the mosts of time.

Today, a strong vision still has the power to change the course of history.

Suddenly, the image areas are filled, one by one, with images suggesting graffiti-covered cement walls, like the Berlin Wall.

Dissolve to actual footage of the Berlin Wall being broken through and pulled down, showing the excitement and joy of the people.

Close-ups of people in cheering crowds.

Laser beams begin to dance around the room.

(Music is suspenseful, tense.) Vision can change history itself. And such changes are so powerful that they happen once in a generation.

The laser beams converge into one bright beam that focuses on the rear wall of the set. The rear wall starts to glow as though with white heat --

(Music builds in a dramatic crescendo.) Only once in a generation are the conditions just right. Only once in a generation is everything in place, waiting for the power to be unleashed. Only once in

(continued)

FIGURE 4.10
(*Continued*)

	a generation does such a time come. And for Sterling Health, that time is -
Suddenly the rear wall breaks apart, revealing the Sterling Health logo on the large, circular video screen.	--now!
Sterling Health products are depicted in an animated montage "behind" the logo.	*(Music is celebratory, triumphant.)* The Sterling vision of greater success through equity management has revealed exciting new opportunities. Now Sterling makes that vision real.
Dissolve to a dramatic, animated reveal of the "Once In A Generation!" meeting logo.	Get ready to take a giant step into the future. Now is the time to launch a wave of change -- an innovation so powerful that it can only be described as coming -- once in a generation!
The logo holds for a moment. Then, with a final dramatic display of laser light, the "light bridge" is illuminated. The first speaker appears out of the vortex-like entrance, crosses the stage to the lectern, and begins the meeting.	*(Music hits climax as meeting theme logo is revealed, then continues until the speaker reaches the lecturn.)*

Evaluation of Interventions

To round out our discussion of design models, we need to look at the evaluation of programs and interventions. Although evaluation is the most neglected step in program design, it may be the most crucial in terms of the long-term success of communication systems and practitioners. Two factors make evaluation more powerful. These are linking the evaluation of interventions to organizational objectives and the ongoing evaluation of interventions.

FIGURE 4.11
Events of instruction. [Adapted from Gagné, R.M. (1970). *The conditions of learning,* 2nd ed. New York: Holt, Rinehart and Winston.]

Gain audience attention

Inform the audience of the objectives of the message

Stimulate audience recall of related information

Present your message

Provide guidance in how to learn the new concepts

Elicit some performance or action from the audience to see if the audience understood the information

Provide feedback and coaching on audience performance

Evaluate the final outcomes of the message

Enhance retention of the material and transfer of training back to the workplace by providing follow-up, materials, and job aids

Too often, evaluation is merely a set of test items presented after a class is given or a program is shown. However, an organization's success doesn't depend on how well people answer questions on a piece of paper 5 minutes after they've been told something. Rather, you need to look for long-term and on-the-job performance change. For example, if you develop a sales training program, you might want to track certain statistics:

- ratio of new accounts to old accounts
- turnover of sales personnel
- percentage of calls resulting in an order

- size of average sale and overall sales volume
- items per order
- accuracy of systems specified by sales force
- repeat business
- average number of calls per day
- customer satisfaction data

If you develop an interactive employee news kiosk that features individual department's progress toward goals, you might want to track the following information:

- number of interactions per day
- number of news item contributions
- employee climate survey data
- number of applications for jobs posted
- use of company resources mentioned
- increased output
- decreased rejection rates and waste
- decreased tardiness and absenteeism

Keeping systematic records of analysis, design, and evaluation activities that occur within a given project makes it easier for someone else to step into the process, if necessary, and provide hard evidence of communication professionals' contributions to the organization.

Philosophical Paradigms and Design Strategies

The general ADDIE framework of instructional systems design springs from the *objectivist,* or functionalist, tradition that there is one true reality, external to the knower, which can be known and explained. In this tradition, the role of instruction is to map concepts or things to learners. The newer, competing theory of *constructivism,* or interpretivism, hold that reality is determined by individuals who build symbols through their interpretations of experiences. (See the previous discussion on modes of inquiry and the section on Objectivism in Chapter 2 for a more thorough discussion.) If we were to adopt a constructivist viewpoint, there would be several changes in our orientation and practice with regard to designing and evaluating interventions.

Movement from Objectives to Goals. The goal of instruction in the objectivist tradition is to specify a narrow set of operationalized objectives that each learner is expected to master. The constructivist viewpoint would hold that because individuals interpret reality differently, the outcomes of instruction should vary. Abandoning objectives, general goals would instead set a direction for learners. Goals would never be completely achieved; rather, they would exist as a state toward which we would continually strive.

Creation of Diverse Enabling Environments. Since learners vary in their interpretations and experiences, instructional designers would create "mental construction 'tool kits' embedded in relevant learning environments that facilitate knowledge construction by learners. . . . These tools, and the environments containing them, should not only accommodate but also promote multiple interpretations of reality" (Jonassen, 1991, p. 12). There would be no one right way to present instruction, or any one correct content or reality to learn. Furthermore, the means for human development should be primarily constructed in the real world, rather than in artificial training environments.

Movement from Target Audiences to Collaborators. As we have seen from the development of communication theory, we generally no longer think in terms of senders who deliver messages to passive receivers. Communication is considered as a transactional process, involving active information-processing and feedback on the part of all parties. In the constructivist paradigm, knowledge is not transmitted or delivered, it is constructed by the learner; therefore, the involved parties are active partners in the implementation of the intervention. Many effective communicators have realized that "Commercials that attempt to tell the listener something are inherently not as effective as those that attach to something that is already in him. We are not concerned with getting things across to people as much as out of people" (Schwartz, 1984, p. 96). As McLuhan (1964) stated, the audience becomes a work force instead of a target.

DESIGN CHALLENGES IN TODAY'S ORGANIZATIONS

It would be nice if clients could just provide you with a clear description of what they'd like their intended audience to do, and leave you with a generous budget and few creative constraints. However, such cases are rare. Although experienced designers and producers can tell you hundreds of war stories, each featuring different problems, they generally boil down to a few fundamental categories.

There is rarely one right way to do a task or approach a problem. Different people—even master performers—will have different ideas. The communication professional must resolve these differences somehow, both politically, and in terms of presenting a complete picture to the audience. For example, at one large bank for which we produced instructional programs, there were five different ways to count money, depending on the region. Demonstrating only one way in an interactive video program would be incomplete and misleading. No one way was correct.

Content changes rapidly. Often, products, organizational structures, or policies change several times during a design and production cycle. The same bank that had five different ways to count money went through the same number of rapid management changes that many organizations face today. We couldn't even complete the editing of an orientation program before the content and personnel were outdated.

Often, nothing will have been documented on a proposed program's content. It's not a matter of simply reading a book or policy manual and making a creative production based on it. The designer needs to help clients create concepts and policies. Sometimes you'll need to develop training or sales materials on a product that doesn't even exist yet, or you'll be expected to develop a motivational message supporting new management philosophies that haven't been clearly articulated. This can involve brainstorming, consensus, and even organizational development activities before design can even begin. I can vividly remember producing a computer-based tutorial on a new graphics program to which our client had bought out the rights and would begin marketing under their own name. The only problem was that there was little documentation. We were given a trunkful of computer hardware and software, a small pamphlet of instructions, and wishes for good luck as we found out by trial and error how the software worked. Once, I accidently pressed a key combination and a grid appeared on the screen. We had to first find out how the program worked before we created a tutorial for it.

Content experts may not be adept at explaining what they know. They may lack verbal, written, or interpersonal skills; may not have any incentive to help you develop a program; or may just not be able to explain what they do or how they do it. A given concept or skill may be so natural to them that they can't articulate the steps or decision-making processes that they use in performing a task, or their backgrounds may be so different from yours that communication is difficult. I've had the opportunity to work with leading medical researchers and experienced engineers, writing programs for which I had essentially no background. Being a fast learner is essential.

As we've seen in Chapter 3, the very nature of organizational life is changing. Professional communicators have moved from communicating stable and agreed-upon content by having homogeneous audiences memorize deterministic terms and procedures to attempting to provide fluctuating and debatable information to heterogeneous audiences through abstract models for problem solving. No wonder communication design is a difficult job!

The Communication Analysis, Design, and Development Model

Although traditional communication and instructional design models provide some helpful guidance in terms of how to approach the analysis and design of interventions, they are lacking in many respects when we try to apply them to the real world. These models are biased in terms of the potential solution; that is, they are directed toward developing persuasion, instruction, changes in management or environmental structures, or the use of some particular type of medium. However, problems don't present themselves that way;

therefore, communication professionals need a system of analysis that is not prejudiced in favor of a category of intervention and that incorporates a variety of solutions. These models also spring from the objectivist tradition that there is some correct content out there somewhere, and that the designer's job is merely to package it attractively so that it can be assimilated efficiently by the target audience. They neglect the universal constraints of time, budget, and client preferences. Designers don't start out with a blank slate, they operate within a complex political and economic environment. Finally, these models fail to include making a business case for interventions. They just assume that if a client wants it and there's enough money in the budget, it should be done.

To address the limitations of existing models, OmniCom has developed and applied the *Communication Analysis, Design, and Development* (ComADD) *Model* to communication projects. This model began life as *The Participatory Design Model,* a process that, through research and practice, proved useful in developing a series of instructional videotapes produced on Southern and Eastern European-American ethnicity in modern American life, which were funded by the U.S. Office of Education in the 1970s (Gayeski, 1981). This model later became the foundation for my doctoral dissertation and for guiding our firm's subsequent development of communication interventions in complex corporate and educational environments (Gayeski, 1983).

Several concepts are fundamental to the ComADD Model and are infused in each of its processes. These are depicted in the oval shapes near the top of the model. (See Figure 4.12.)

1. *Which step?*—This principle acknowledges that not all projects begin at the beginning, and that although there is a default top-to-bottom flow of the process, there may well be some skipping around or recursion back to previous steps. The processes that have already occurred and those that have been skipped should be identified in order to determine where to start in the design model. In actual practice, you may find that some decisions and deliberations have already taken place. Although this is not necessarily ideal, it is realistic. For example, it may already be known that there are conflicting opinions and practices regarding the subject with which you're dealing or that sponsors of an intervention may be looking for a way to use the new satellite video network.

2. *Diversity/commonality*—This principle acknowledges that there are few singularly correct approaches or facts. Various SMEs may well have conflicting opinions. Instead of just getting input from one person and finding out the other existing opinions later, or trying to force consensus, you should always look for areas of diversity and areas of commonality. Don't consider conflict or disagreement to be a failure; rather, acknowledge it, respond to it, and make sure your potential solution incorporates a diversity of views.

There are also several terms that are used in a specific way in the ComADD model:

- *solution*—the one or several interventions that are being developed to solve the problem(s). The solution might be an article in a news magazine, a training program, an electronic performance support system, or a redesign of work areas.
- *constituency*—those groups of people who will somehow be affected by this project, including its design, development, and implementation. This includes every person who has requested the intervention (your client), has input to the content (managers, SMEs, regulators, etc.), will be involved in the development of the solution (the training department, contract graphic artists, the marketing department, the organization or department for whom the solution is being developed, etc.) and those who will use it (trainers and trainees, presenters, readers, etc.).
- *developers or development team*—these are the actual people who will work on designing and developing the intervention. They represent the constituency.
- *developees*—those groups or individuals whose performance the client wants to change. They may be all the people in an organization who read the company newsletter, customers who need to learn to use the client's new product, trainees, etc.

FIGURE 4.12
ComADD Model, [Copyright © 1992 by OmniCom Associates, all rights reserved.]

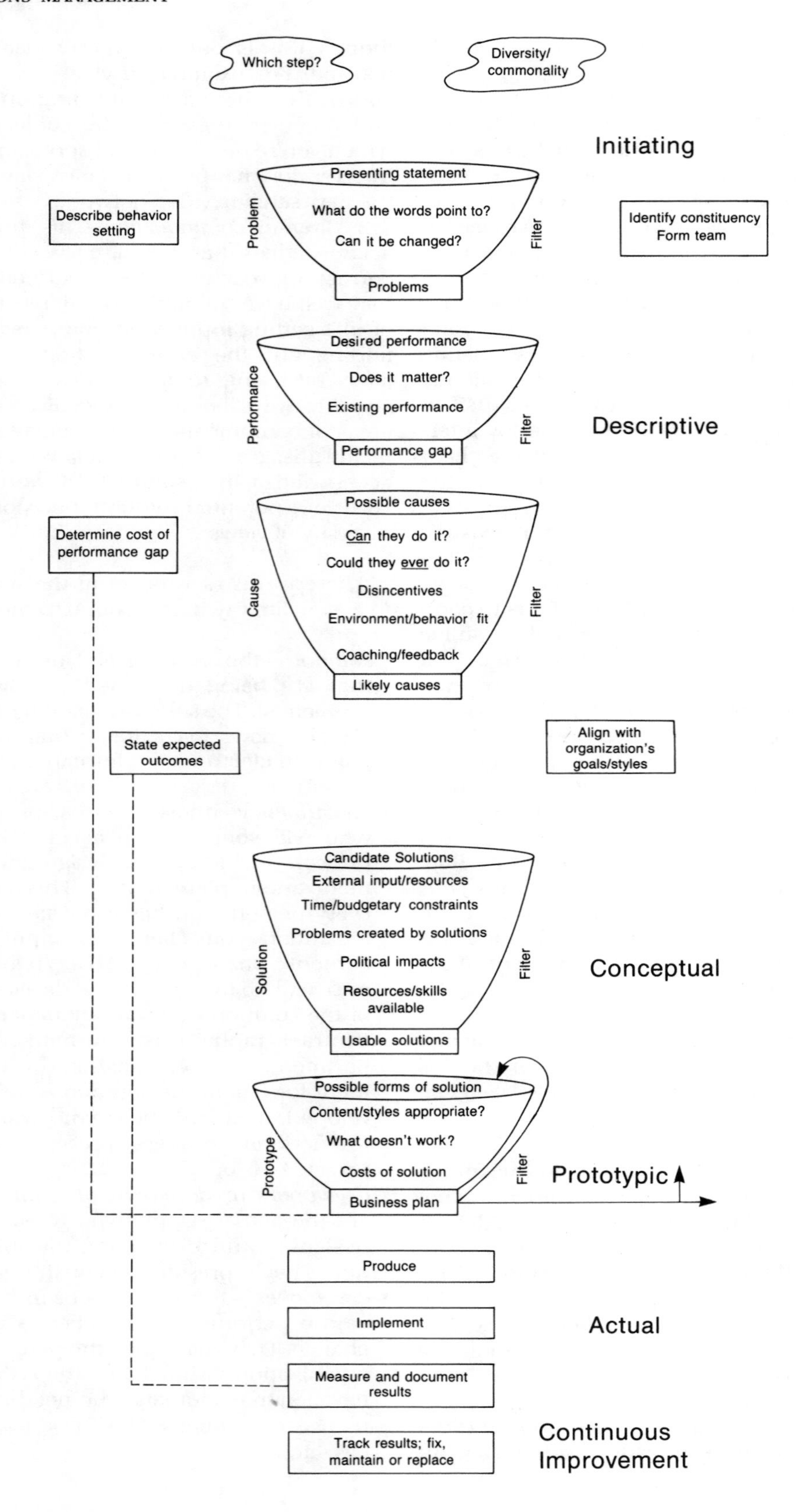

The model consists of six phases: *initiating* (identifying legitimate problems and forming a development team); *descriptive* (defining performance gaps, their costs, and potential causes; forming expectations and goals), *conceptual* (generating and assessing possible solutions and formulating recommended solutions); *prototypic* (developing and trying out draft versions of solutions, and from that developing a business plan for the project); *actual* (the production, implementation, and results of the solution); and *continuous improvement* (the ongoing tracking of results and maintenance of the solution). It relies on filters through which problem statements, potential goals, possible causes, candidate solutions, and likely formats and strategies for interventions are sorted out to determine the most realistic and potentially most cost-effective methods. Using it, you can determine which client complaints probably can't be fixed (at least by you), which expectations are realistic, and whether the cost of the potential solutions is, in fact, less than the cost of the existing problem.

The Problem Filter. Clients arrive with a *presenting statement*—what they think the problem is or what they want done. It might be something like, "The tellers need to increase their efficiency" or "I want a 10-minute videotape on customer service." In order to respond to this, you first need to find out what they're talking about. What do the words point to, in terms of observable and concrete things or behavior? Once you see it, can it be changed? For example, what does the client mean by efficiency, and is it possible or realistic to expect tellers to work more quickly? If nothing can be done about a complaint, it's not a problem—it's a fact. For example, if there's a lot of turnover among waitstaff in your restaurant because the college students who make up most of your staff keep graduating and leaving the area, there is probably not much you can do about it. Once you have a set of problems that are operationalized, you can move on to the next set of procedures.

Describe the Behavior Setting. Where are the identified problems taking place, and what are the various aspects of that environment? For example, if your goal is to reduce injuries on the shop floor, what are the components of that setting, including the equipment, the workers, the workers' attitudes and behaviors, ergonomic aspects of the work setting, the worker's attire and safety equipment, and signs. Perhaps there are different environments for various developees.

Identify Constituency and Form Team. Determine your constituency—for whom or about whom are you developing the project, and what groups will be involved in the design, production, and implementation. Next, form a development team of constituency representatives. Be sure to include both *subject-matter experts* (SMEs), those who understand the content, and *process experts*, those who understand how the content is actually applied on the job and can identify the common performance problems associated with this content. Take, for example, the development of a product knowledge videotape for the sales force. Your development team might include several engineers who developed the product, a first-line supervisor who observes the mistakes that employees make when trying to sell the product, an instructional designer, a scriptwriter, a video producer/editor, the client who is the product sales manager, a trainer who will use the tape within a classroom environment, an experienced sales rep who has worked with similar products, and a brand-new sales rep who has never been trained in such products.

The Performance Filter. In this process, you'll perform a needs analysis and documentation of performance discrepancies. First, state the *desired performance* in observable and operational terms. For each behavior or factor, ask,"does it really matter?" For example, a vice president of a bank may not like the kind of clothes the tellers are wearing, but unless it can be determined that their appearance makes a difference in terms of bottom-line costs to the bank, this problem should not be addressed. Now, determine exactly what your developees or target audience isn't doing that they should be doing (or vice versa). The difference between ideal (but realistic!) performance and existing performance is the *performance gap*.

Determine Cost of Performance Gap. How much in real dollars is the performance gap costing the organization? Is it driving customers away, wasting time or materials, causing injuries, reducing the work flow (and, therefore, necessitating the hiring of more employees), or failing to sell more product to an existing market? This step is crucial because unless you know how much the problem is costing, you can never determine if your solution is cost-effective!

The Cause Filter. To sort out possible causes, first determine if the performance gap is due to a deficiency in skills or information (could they really perform in the desired manner if they had to?); a lack of practice or just forgetting something (could they ever do it?); a lack of incentives or the presence of disincentives; less than optimal fit between the work environment and the desired behavior; the need for feedback and coaching; or a combination of these factors. For example, if most of your sponsors' claims representatives process about two claims per hour, while exemplary reps process an average of five an hour, there could be a number of causes. Perhaps the work environment (keyboards, terminals, lighting, or background noise) differs; maybe they have no incentive to work faster; perhaps they have not been trained sufficiently in dealing with customers or the computer system; or maybe the ones who are stellar performers are shunned by the group for being rate-breakers!

State Expected Outcomes. It is important to take into account previous team work in order to determine possible causes and to develop specific expectations or goals for the program. These are demonstrable behaviors or end results (i.e., $25,000 in donations, an increase in production quality so that no more than one item in 1000 does not meet inspection standards, or a 25% decrease in turnover) that would be expected if the intervention were to be considered totally successful. It's important to recognize (1) that the evaluation of the solution will not be limited to looking at predetermined, narrow objectives, and (2) that these expectations will probably never be met, but are goals toward which you will measure objective progress. These expectations are directly applied to the construction of assessment measures.

Align with Organization's Goals and Styles. Consider the overall goals and styles of the organization, and align the expectations with them. For example, my husband worked in Swaziland developing curriculum materials. An overall goal, as articulated by the king, was to make agriculture seem like an attractive career option, since too many young people were leaving the farms to work in the city. So, in each of the books and instructional materials, farming was depicted in a positive way, as a technical, interesting, and rewarding pursuit and as an example of the application of engineering, math, and science. Likewise, in an organization undergoing a cultural change, teamwork and innovative decision making by line employees might be featured in articles or training materials, even if the primary topic is quite different. Take into account the corporate culture or style. In some organizations, for example, it's quite usual to poke fun at the executives or use them in humorous media productions. In other organizations, this would never be tolerated.

The Solution Filter. Now that we have defined the possible causes for the problem, we can generate and filter candidate solutions. The first screen is to access external input and existing resources in order to identify which models, research, or materials might be applied. For instance, consider what research or case studies might point to the best solution. Also, check to see what sort of existing training materials or job aids may already exist within your organization or that could be purchased. The second filter is time and budgetary constraints. Some solutions that seem ideal may not be realistic in terms of the time frame or financial situation. Next, examine what kinds of problems may be created by implementing the solutions themselves; for example, implementing a new incentive system may create temporary performance problems by increasing competition and the learning curve necessary to get the sales force to understand and implement the new policies.

The next filter is political impacts. What are the consequences of various solutions in terms of individuals or departments? For example, creating a new interactive system might lend needed visibility to the video manager who is trying for a promotion or choosing an imbedded help system might put money into the hands of the MIS department. Some interventions may have a bad history with the organization. One of my clients had such a bad first experience with interactive video that it will be years until those words can be uttered again in the training department. Finally, look at the in-house resources and skills available to undertake various interventions. If the company has some staff members experienced in human-computer interface design, has in-house programmers, and has access to the source code of the software that's causing trouble, it's quite feasible to develop a better front-end to make it more usable. If this isn't the case, but expertise exists in computer-based training, than that solution would emerge as more feasible. From these filters emerge one or several usable solutions.

The Prototype Filter. In this phase, a rough, working model of the solution, or prototype, is developed starting from several possible forms of the solution. This might include rough drafts of a job aid, a script and storyboard for a video, or some sample screens from a multimedia program. As the prototype is evaluated by the development team and used or reviewed by other representative developees, it is checked to see if its content and styles are appropriate to the problem and the organizational situation, and more importantly, against the expected outcomes to see what does not work. Next, the prototype is filtered by the costs of the solution as currently designed, keeping in mind the costs of the problem and the resources available to the project. The prototype is iteratively refined until it meets these criteria, and out of this process comes a business plan. The business plan describes, in detail, problems and associated costs, the approach to the solution, the look and feel of materials to be developed, the expectations, and the cost of developing and implementing the solution. Here, the most important question is—is the cost of the solution less than the cost of the problem? If not, another prototype solution might be developed that would be cost-effective. Alternatively, the team might decide to reframe the problem or expected outcomes, or to live with the problem.

Produce, Implement, Measure, and Document Results. Once a business plan has been approved, the next steps are to actually develop the solution, implement it, and measure and document its results. Those results are then compared to the expected outcomes developed in the cause phase. However, it is important not to limit the examination of results to the list of expectations. Other unanticipated positive and negative outcomes may have resulted and these should be noted.

Track Results—Fix, Maintain, or Replace. Solutions to communication problems are not complete until a system for long-term tracking of results is in place. Portions of a solution may become a part of the everyday environment, such as changes in the design of work areas or the use of a job aid. In this case, it's necessary to see if the results of these continue to improve preformance and are still appropriate in terms of content and style. Even in the case of a one-time solution, (i.e., holding a meeting), it is crucial to see if changes in performance hold up over time or if improvements are only temporary. Using data from this tracking, the solution can be continuously improved by fixing it, maintaining it, or even replacing it.

The ComADD Model provides an overview of the program development process at the macro level. With this larger view in mind, let's now address the micro level of actually interviewing SMEs.

Working with Subject-Matter Experts

It is rare that the communication professional is an expert on the topic of a program being developed. We work with SMEs to learn the content ourselves and then translate it for the audience. This is a time-consuming process that requires much communication skill. SMEs are not just experts on the theory of some topic; the best informants can often be proficient performers of some activity.

In most cases, the problem solving knowledge of SMEs has become automatic or tacit through extensive use. Some researchers even doubt whether much of this knowledge and skill is really accessible. Have you ever tried to explain how you solved a problem in one of your areas of competency?

A favorite exercise of mine when teaching message design is to divide my class into two groups: those who drive stick-shift cars and those who can only drive cars with automatic transmissions. In each class, the automatic-only drivers have at least been a passenger in a standard transmission car and, thus, have some idea of how it's done. However, their knowledge is usually full of misconceptions. By interviewing the class SMEs, the automatic-only drivers can develop an explicit step-by-step procedure for driving with a standard transmission. Unfortunately, this is often an excruciating process—they usually only have a few procedures and processes down after 3 hours! Try it yourself. How do you know when to shift? How do you know what gear you're in? How do you know when to downshift? How do you explain the timing and feel of letting the clutch out?

The process of interviewing SMEs and bottling their knowledge is a high-level skill called *knowledge elicitation.* Since a communication program designer is often a relative novice in the subject or domain of the elicited knowledge, there is a substantial error potential in the message transmission process. If you recall the communication models reviewed in Chapter 2, you'll remember that communication is improved when there's a great deal of similarity between the message transmitter and the message receiver. The disparities in backgrounds between communication professionals and their SMEs can cause a number of specific problems.

One major problem stems from oversimplification of the SME's knowledge. This occurs for two reasons. First, the developer will try to look for regularities and structure in the domain. This tendency may lead the developer to adopt an incorrect view of the domain; for instance, a designer may be looking for simple rules or categories that don't exist in real life. The second related problem is that SMEs have a tendency to produce simplified descriptions, or translations, of their expertise when they speak to novices. While these simplified accounts are adequate in the SME's social circles, they do not contain enough detail to inform detailed training or to be used for informational purposes.

Two common examples of oversimplification involve using inappropriate analogies and common connotations. An analogy may be too simple, or we may erroneously try to extend attributes of the familiar situation to the unfamiliar one. An example of this is the misconception that videotape, like the somewhat similar media of audiotape and film, can simply be physically cut and spliced to edit it. Common connotation errors occur when everyday terms are used in technical ways, but their technical meaning is not understood by the audience. For example, most people think they understand the word "cure," but pharmaceutical sales representatives must recognize that physicians who follow the World Health Organization definition of this term may mean something very specific and quite different (Gayeski, Wood, & Ford, 1992).

The most common methods for knowledge elicitation are personal interviews and observations of actual work situations, although sometimes the program designer also examines books, pamphlets, manuals, or data. All of these methods are tremendously labor-intensive and, therefore, expensive processes. In addition, many, if not most, projects require the input of multiple SMEs. The debriefing process is also complicated by the fact that there are few tools that help the designer to document and to organize the information. When learning about a new knowledge domain from an SME, a developer is bombarded with a great deal of new and unfamiliar information that is generally not well-organized by the SME (who, after all, lacks instructional expertise!).

Some researchers and practitioners have tried to enhance the process of SME interviewing by borrowing techniques from the disciplines of anthropology and ethnography. Researchers in those disciplines face a similar problem when they are attempting to elicit cultural information from informants, or cultural experts in unfamiliar societies. For example, one technique often used by ethnographers to elicit informants' natural descriptions of terms and concepts is to ask how terms, tools, and concepts are used rather than what they are or what they mean. For example, you might ask, "How is the word 'mullion' used when you talk about styles of windows?" Questions should prompt SMEs to provide real-life situations, as well as common mistakes or misconceptions, which occur when concepts are being learned or applied. You might also try to construct if-then rules, which represent the logic you see emerge from your observation of the SME's decision-making processes. When documenting processes, another useful technique involves video-taping or audio-taping the SME while he is engaged in a task, and then reviewing this record with the SME later. This procedure frees the SME from trying to talk about procedures while also attempting to perform them.

Software Tools for Program Design

Many areas of endeavor have been aided greatly by computer-based tools. Consider, for example, the benefits afforded to architects by computer-aided design, the ubiquitous use of spreadsheets for financial analysis, and the application of simulation systems in areas such as ecology and city planning. However, communication and instructional design have not benefitted by similar programs, at least not on a widespread basis. In general, programs that communication professionals have used are limited to the production side, including word processors, desktop publishing software, and graphics programs. Slowly, this is changing.

OmniCom developed a software tool to make the process of interviewing content experts more efficient and effective. Following the ComADD Model, we generally employ a number of SMEs on projects; however, we wanted to reduce the time and the expense involved in personally interviewing each one. We developed the *Content Expert Interviewer (CEI)* as a tool for doing the initial debriefing of SMEs in a structured fashion. This allows us to send out a diskette that interviews the SME, and then provides a structured report of the SME's ideas (see Figure 4.13).

Other tools also help the brainstorming process. There are programs that help marketing specialists develop names for products and programs that help organize one's thoughts while developing a speech or paper. In addition, there are various software tools that help to create vivid, creative words and images.

At which question would you like to start the review/revision?

1. Topic
2. Who is the Audience?
3. Audience traits
4. Identify performance problems
5. State information points about each problem
6. Common mistakes/misconceptions
7. Methods of imparting content
8. Prerequisite concept or skills
9. Content people might be able to skip
10. Pretest questions
11. Questions to evaluate mastery
12. Probable responses to mastery questions
13. What should the viewer be told next?
14. Menu or viewer options

0. Return to the main menu

Your selection:?

Interview with Diane M. Gayeski, Ph.D.
Regarding: NEMA motor types
5/5/92
As Recorded By The Content Expert Interviewer
Copyright (c) 1988 Omnicom Associates

THE TOPIC OF THE PROGRAM:
How to select the correct NEMA type AC motor for typical materials handling applications

THE AUDIENCE:
WHO THEY ARE:
newly hired sales reps
CHARACTERISTICS:
Have M.E. or E.E.; little sales experience; motivated by commissions, interested in technology and sports

FIGURE 4.13
A menu screen and the first part of the interview output generated by the Content Expert Interviewer (CEI). (Courtesy of OmniCom Associates.)

John Hamm, a speechwriter for the Illinois Department of Children and Family Services, uses IdeaFisher computer data base to help him develop his speeches, two of which, in fact, have won regional awards from the IABC. One of his speeches was the keynote address at a conference called "Windows of Opportunity in Social Work." Searching through Idea-Fisher computer data base, he found 44 types of windows and wrote his speech around the five windows of opportunity in the social work community: "the bay window of research projects, the stained glass window of public/private sector cooperation, the kitchen window of social service in the homes of troubled families, the computer window of technology and the cashier window of legislative financial support" (Heger, 1991, p. 20).

Rapid Prototyping

Our organizations are moving from slow, carefully preengineered, assembly-line modes of production to rapid, custom-crafted, just-in-time paradigms. So, too, is the process of designing and of producing communication interventions. Since the production of print and video materials was once so expensive and was generally done by outside experts, programs had to be correctly designed the first time around. This is changing. Now, we have desktop publishing, desktop video, and desktop multimedia systems that can be used for design, production, and sometimes, even dissemination. A new mode of communication design and production called *rapid prototyping* is emerging (See Figure 4.14).

FIGURE 4.14 **Assembly-line versus rapid prototype model of communication project development.**

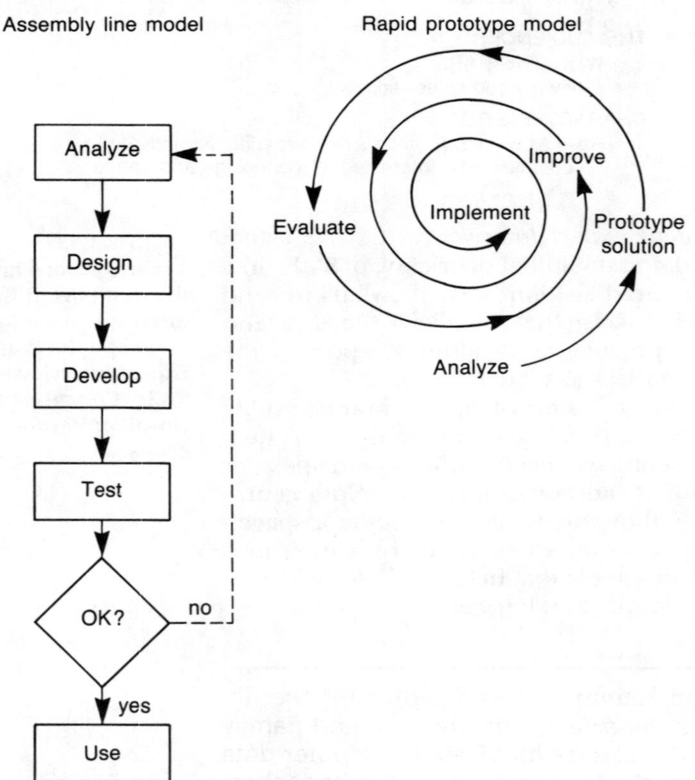

In rapid prototyping, instead of trying to specify just one set of words, images, or layout up-front, a number of quick and inexpensive models are created. This gives clients and SMEs an opportunity to review rough ideas of the way a final program might look and feel. This is especially necessary when using new interactive technologies that are not easily described on paper and that may be unfamiliar to potential clients and audiences. Many new technologies are making it easy to create rapid prototypes; for example, desktop publishing and computer graphics software can take the same text and create different layouts by simply applying different style sheets. Animation can be simulated using inexpensive PC-based software. Video can be shot with inexpensive camcorders and rough edits made with in-house equipment before actual productions are executed and edited.

Many multimedia producers use inexpensive software programs, such as HyperCard for the Macintosh and ToolBook or LinkWay for MS-DOS computers, to rough out interactive programs. No programming is involved. A designer simply creates text and buttons on the screen, and specifies the action of each object. When special routines are needed, a designer can simply lay out the screen and add a note that provides instructions for the programmer. Within a few hours, a client can review the program's look and feel.

Rapid prototyping provides the bridge between communication design and communication production. Although these two processes were considered different specialties practiced by different professionals in the past, this model is changing. The Renaissance communicator will be a message designer and a message producer.

REFERENCES/ SUGGESTED READINGS

Basset, G.A. (1968). *The new face of communication.* New York: American Management Association.

Be ready in case crisis hits. (1991, November). *IABC Communication World, 8*(11), 13.

Bloom, B.S. (1956). *Taxonomy of educational objectives.* New York: David McKay.

D'Aprix, R. (1977). *The believable corporation.* New York: AMACOM.

Daniels, T.D., & Spiker, B.K. (1991). *Perspectives on organizational communication.* Dubuque, IA: Wm. C. Brown.

Deal, T.E., & Kennedy, A. (1982). *Corporate cultures: The rites and rituals of corporate life.* Reading, MA: Addison-Wesley.

Emmanuel, M. (1981). Auditing communication practices. In C. Reuss & D. Silvis (Eds.), *Inside organizational communication* (pp. 49–61). New York: Longman.

Farace, R., Monge, P., & Russell, H. (1977). *Communicating and organizing.* Reading, MA: Addison-Wesley.

Gagné, R.M. (1970). *The conditions of learning,* 2nd ed. New York: Holt, Rinehart and Winston.

Gayeski, D. (1981, June). When the audience becomes the producer: A model for participatory media design. *Educational Technology,* 11–14.

Gayeski, D. (1983). *Corporate and instructional video.* Englewood Cliffs, NJ: Prentice Hall.

Gayeski, D., Wood, L., & Ford, J. (1992, August). Eliciting and organizing information from subject-matter experts. *Training & Development,* 55–62.

Gilbert, T. (1988). The 10 most important lessons I've learned about productivity. In G. Dixon (Ed.), *What works at work: Lessons from the masters* (pp. 17–21). Minneapolis, MN: Lakewood Publications.

Goldhaber, G., Dennis, H., Richetto, G., & Wiio, O. (1979). *Information strategies: New pathways to corporate power.* Englewood Cliffs, NJ: Prentice Hall.

Haynes, G. (1981). Organizing and budgeting techniques. In C. Reuss & D. Silvis (Eds.), *Inside organizational communication* (pp. 62–76). New York: Longman.

Heger, J. (1991, November). Whiz . . . bang . . . Eureka! *IABC Communication World, 8*(2), 18–21.

Jonassen, D.H. (1991). Objectivism versus constructivism: Do we need a new philosophical paradigm? *Educational Technology Research & Development, 39*(3), 5–14.

Kauffman, R. (1989, September). More than questionnaires. *Training & Development Journal,* 26–27.

Kauffman, R. (1991). Trainers, performance technologists, and environmentalists. *Performance Improvement Quarterly,* (4)2, 69–76.

Mager, R.F. (1962). *Preparing instructional objectives.* Palo Alto, CA: Fearon Publishers.

Mager, R.F. (1988). *Making instruction work: Or skillbloomers.* Belmont, CA: Lake Publishing Company.

Mager, R.F., & Pipe, P. (1984). *Analyzing performance problems,* 2nd ed. Belmont, CA: Lake Publishing Company.

McLuhan, M. (1964). *Understanding media: The extensions of man.* New York: McGraw-Hill.

Meltzer, M. (1981). *Information: The ultimate management resource.* New York: AMACOM.

Odiorne, B. (1965). *Management by objectives.* New York: Pitman.

Parkinson, C.N., & Rowe, N. (1977). *Communicate: Parkinsons' formula for business survival.* London: Prentice Hall International.

Putnam, L., & Cheney, G. (1990). Organizational communication: Historical development and future directions. In S. Corman, S. Banks, C. Bantz, and M. Mayer (Eds.), *Foundations of organizational communication: A reader* (pp. 44–61). White Plains, NY: Longman.

Rosenberg, M. (1990, February). Performance technology: Working the system. *Training,* 43–48.

Rossett, A. (1990, March). Overcoming obstacles to needs assessment. *Training*, 36–41.

Ruch, R.S., & Goodman, R. (1983). *Image at the top: Crisis and renaissance in corporate leadership.* New York: The Free Press.

Rummler, G. (1988). The 10 most important lessons I've learned about human performance systems. In G. Dixon (Ed.), *What works at work: Lessons from the masters* (pp. 2–13). Minneapolis, MN: Lakewood Publications.

Schwartz, B. (1984). *Psychology of learning and behavior*, 2nd ed. New York: Norton.

5 Communication Interventions: Rules and Tools

Now that we have reviewed the ways in which communication should occur within organizations, as well as some guidelines for planning communication in specific contexts, we can take a look at the production aspects of the job. A *communication intervention* is a focused attempt to solve a documented problem or to take advantage of an opportunity for improving organizational performance by creating and implementing a communication program, process, or policy. A *program* is an individual presentation, such as a training videotape, a multimedia expert system, or a meeting. A *process* is a function or system that will become an ongoing part of the organizational environment and contribute to performance improvement, such as an electronic mail (e-mail) system, a video teleconferencing set-up, a committee structure, or an ongoing series of management meetings. A *policy* is a set of rules for communication, such as guidelines to avoid the use of sexist language in written and oral discourse, or a plan of action for crisis communication.

As communication professionals assist their clients in moving toward their goals, they create rules and tools to improve the quality and the efficiency of sharing messages. These need to be aligned with societal needs and customs, as well as with corporate visions and values. They must also be consistent across interventions and departments. For example, interventions should take into account

- generally accepted needs and values of society
- the corporate culture
- differences within and between audience groups and corporate locations
- documented performance gaps
- organizational plans and goals
- popular tastes in graphic design and writing style
- corporate identity in terms of graphic standards, trademarks, and slogans

MEGA-LEVEL INTERVENTIONS

Similar to analysis and design, interventions can be considered to be mega-, macro-, or micro-level in terms of their orientation and their scope. Mega-level interventions address communication systems that encompass more than one organization. Their audience is the general public or a large sector of people with some common characteristic, such as persons who are members of the same profession. Of course, many of the mega-level systems might be defined as mass communications: newspapers, television networks, and commercial films. Although it goes without saying that these pervasive tools are powerful influences in modern society, their primary focus is on the individual rather than on the organization. Their general purpose is to make money by providing entertainment. However, there are other forms of public communication systems that are aimed primarily, or at least partially, at organizations.

Professional communicators may produce interventions that are generic; for example, they may create instructional films on how to manage meetings, or they may create information utilities and electronic bulletin boards that are available to the public. Although these interventions may be used within individual organizations to address specific goals, their aim is much broader.

CompuServe calls itself an information utility. Started in the early 1980s, this telecommunications network consists of powerful computers and massive data bases housed in Columbus, Ohio. Subscribers to CompuServe enter the system via their computer modems and have access to thousands of newspaper and magazine articles, share ideas within special-interest group forums, check on the weather and airline schedules, and send other subscribers electronic messages or files. Many users are individuals for whom on-line communication is a hobby. They can talk with electronic pen pals about mutual interests, such as sports or computer games. Others use it as a research tool; for example, students can find encyclopedia articles to help with term papers, and consultants, lawyers, and physicians can gain access to current information and ask questions of colleagues. In addition, organizations use it as a means to communicate with their employees. For example, several airlines have specific forums to which only their employees can gain access. Pilots and cabin crews use the system to place electronic bids for the schedules they'd like, and the software automatically compiles these bids, creating individuals' assignments based on their preferences and seniority.

Individual organizations often create interventions that are aimed at the public at large. For instance, most utility companies provide films and brochures on energy conservation and safety measures regarding gas and electric lines (see Figure 5.1). Automobile manufacturers sponsor safe driving films. Ben & Jerry's Homemade has a traveling road show to promote the merits of solar energy.

MACRO-LEVEL INTERVENTIONS—RULES

Macro-level communication interventions target overall communication systems in individual organizations. These interventions are aimed at providing what we might call the three A's in communication:

- adequacy
- accuracy
- accessibility

> The major qualitative elements of organizational communication are the degrees of openness, timeliness, distortion, and reciprocality. The more complex the tasks of the organization, the more open, timely, and distortion-free should be the communication. To create these conditions, organizations must train employees, especially managers, in communication skills and must establish a culture (norms) which encourages this type of communication. (Tichy, 1983, p. 91)

It is at this macro level of communication interventions that an information-supportive climate can be created. It is a

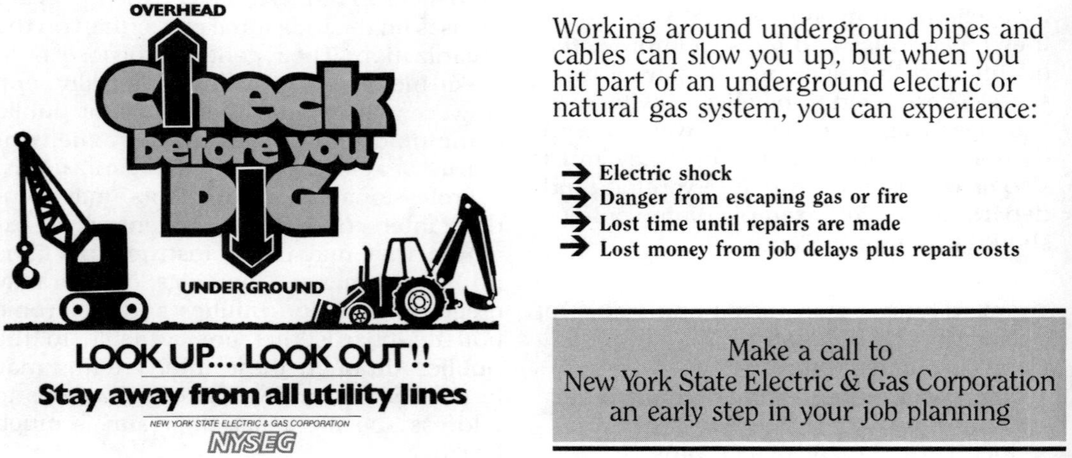

OVERHEAD

Check before you DIG

UNDERGROUND

LOOK UP....LOOK OUT!!

Stay away from all utility lines

NEW YORK STATE ELECTRIC & GAS CORPORATION

NYSEG

Working around underground pipes and cables can slow you up, but when you hit part of an underground electric or natural gas system, you can experience:

→ Electric shock
→ Danger from escaping gas or fire
→ Lost time until repairs are made
→ Lost money from job delays plus repair costs

Make a call to
New York State Electric & Gas Corporation
an early step in your job planning

FIGURE 5.1
A public safety brochure produced by a utility is typical of mega-level communication interventions. (Courtesy of New York State Electric & Gas Corporation.)

heady task indeed to develop communication policies and strategies for an organization. Based on what people in democratic societies and popular management theories hold as values, it would seem that an ideal system need only support open communication and uniformly accessible information. Although these goals are, without a doubt, noble, they may obscure other elements that characterize optimal communication systems.

Based on data generated by analytical tools and design specifications, the judgment may be made that it's necessary to restructure an organization's communication system. Or, in other cases, communication professionals find themselves in charge of developing overall communication processes and channels for a new or emerging company. In either of these situations, comprehensive and sensitive communication planning is necessary.

Developing Strategies and Policies

The first step in developing a communication program is to establish an overall policy. That policy should state the organization's general beliefs and values about communication:

- Better communication will encourage productivity.
- This organization will maintain downward, upward, and lateral systems of communication.
- Our employees should be told promptly and by their immediate supervisors about any major organizational events that may impact them.

In addition, communication policies must be aligned with overall organizational values and goals. For example, if an organization says its values include continuous improvement and teamwork, more specific communication policies must reflect those concepts. Often, communication professionals will be actively involved in helping top management articulate those values and goals. Some examples of slogans that reflect core values are those of GE—"Progress is our most important product"—and of DuPont—"Better things through better living through chemistry."

Communication philosophies can be articulated by writing statements such as "Our organization rewards loyal employees;" "This is an interesting and challenging place to work;" "Our management team is composed of personable individuals dedicated to leadership within our industry" (Haynes, 1981, p. 64). These statements can then shape the underlying themes and messages of the company, including its annual report, the way it conducts its meetings, employee publications, and its use of technologies.

We may find we need to bring organizational policies and practices in line with it's stated philosophies. Existing rules may give mixed or conflicting messages. For example, we ask employees to become personally and totally involved in doing their jobs, to put aside family and social activities, and to take work home. However, most companies have strict rules regarding employee's use of tools or supplies for personal projects. We say we want innovation and truth, but punish mistakes and information that is counter to what's politically acceptable or comfortable. We call for total quality and zero defects in products, but do nothing about incomplete or inaccurate messages. Astute communicators will seek out any such opposing messages and bring them to the table. The goal is to develop honest and clear principles that everyone can live by.

The Bank of Montreal welcomed a new CEO in 1990, and he and his executive management team set out to create a new culture within the bank. They articulated a series of values and goals, along with specific business strategies (see Figure 5.2). Part of the new effort was an increased focus on training—the bank is building a new $40 million training and education center.

As I write this book, I'm involved in a project in which OmniCom will develop standards and strategies for training at the bank, aligning them with the overall corporate goals and philosophies. The project started out with a focus on standardizing the look and feel of their training materials. It quickly grew to encompass the design and development procedures used to create the materials, an examination of the basic business of the training department, and recommendations to leverage the work that is being done by the department's colleagues in public affairs and policies/procedures. Our focus is on ensuring that consistent messages are being sent by the style, tone, and design of all of the bank's print and audiovisual programs, and that along with teaching specific skills and concepts, all training materials reinforce the bank's values of service, valuing diversity, and teamwork.

Developing Standards

Organizational strategies and values, along with the particular history and culture of the company, form the basis of its image. Image is "a managed perception on the part of the customer of the way the company does business." It's said we manage the image by managing thousands of "moments of truth" (Albrecht & Zemke, 1990, p. 61). Both customers and employees may find it difficult to relate to an organization that has no image, or worse yet, has several conflicting images. Once one is developed, every aspect of communication, including advertising, marketing, letterhead, training manuals, brochures, business cards, samples, news releases, trade show displays, newsletters, and videos, must be coordinated and consistent. (See Figure 5.3.)

Emerson Electric has enjoyed more than 30 years of consistently increasing earnings. Charles Knight, CEO of the firm, believes that one of its core principles is open communication. Division presidents and plant managers meet regularly with employees to discuss the specifics of the business, methods of cost-control, and an analysis of the competition. As a measure of the success of communication, Emerson claims that every employee can answer four questions:

"1. What cost reduction are you currently working on?

2. Who is the 'enemy' (who is your competition)?

3. Have you met with your management in the past six months?

4. Do you understand the economics of your job?"

Knight says communication is a two-way street. Emerson spends a lot of time conducting and analyzing opinion surveys of every employee, and Knight personally reviews a summary from every plant. "When I spot a downward trend, I immediately ask the managers in charge for an explanation. It has to be a good one" (Knight, 1992, pp. 60–61).

Although openness may be desirable in many cases, in this era, when information is even more valuable than products, organizations need to create safeguards for their intellectual properties. Policies and practices need to ensure that proprietary information is kept secure and that individuals' privacy is protected. Managers also need the freedom to speak openly and debate issues without their every comment becoming public knowledge. Even though individuals should be free to speak their own minds, there must also be official company

FIGURE 5.2
The Bank of Montreal articulates its values and visions through this brochure. (Courtesy of Bank of Montreal Public Affairs Dept.)

VALUES

Our values are what each of our stakeholders can expect of us, and what we, in turn, can expect of them.

Our customers come first ...

- They will be served at all times by friendly, caring, responsive and well-trained employees. Our service will be characterized by fairness and respect, professionalism, integrity and ethical conduct.
- We will expect from customers a reasonable return for our services, for them to honour any commitments to the Bank, and to reciprocate the respect and courtesy which we extend to them.

Employee satisfaction leads to customer satisfaction ...

The bank is committed to attracting, developing and retaining the best people.

- Employees can expect us to demonstrate respect for the individual, to be a fair and equitable employer, and to provide the opportunity for a full and challenging career. We'll offer competitive compensation and benefits programs, and will recognize and reward personal achievement as well as teamwork. And we'll provide opportunities for continuous professional development through training and through mobility within the Bank.

GOALS

Our overall objective is to outperform the competition, achieving consistently superior results through continuous improvement in all we do. We will:

- generate a superior return on investment for common shareholders,
- achieve superior levels of customer satisfaction and protect the safety of the funds they entrust to the Bank,
- achieve distinctiveness in the degree of competence, commitment, and cost-effectiveness of the Bank's team of employees, and
- achieve a superior reputation for ethical conduct and corporate citizenship in the communities we serve.

statements and spokespersons who are responsible for speaking for the organization. Departments and teams should be encouraged to formulate their own communication styles and methods, yet, there must also be organizational standards for the look and feel of publications and media. We need to set standards for assessing and for reducing data to that which is absolutely essential. Finally, a good organizational communication design includes methods for accepting and soliciting information from the outside, and managing the storage and the flow of that information internally.

Once a communication policy and philosophy is in place, more specific standards are developed:

- requirements for use of logo, company colors, trademarks, and other graphic design elements (See Figure 5.4.)
- policies to ensure equal treatment of all employees, regardless of gender, age, and race
- benchmarks for employee satisfaction with management communication

Developing Ongoing Interventions and Systems

Along with specific standards, ongoing interventions are also developed:

- a formal program of communication with all employees
- a regular printed piece (i.e., a newsletter)
- regular supervisory meetings
- a formal orientation program
- annual employee surveys

In developing interventions, it is essential that professional communicators seek input from a broad base of not only their clients and sponsors, but also their target audiences.

In this issue ...

Fast Facts

1981		1991
3.0	millions of Class 6-8 trucks in U.S.	2.8
5.0 mpg	average fuel economy of Class 8 trucks	6.5 mpg
536,000	average trade-in mileage for Class 8 trucks	757,000

Cooperation Yields Big-time Payoffs

Espar heaters are now written into West Virginia's standards for school buses - creating a potential market of almost 3,600 heaters. Inroads like this don't come overnight, but this case study points out how cooperation and long-term relationships can really pay off -- so take note!

Several years ago, Espar's Steve Cross and Eric Jessiman called on the West Virginia Pupil Transportation Office to see what Espar could do to get its products in their buses. The answer: install four heaters and have them tested for two years. After that, Espar had to supply a certification from an independent testing lab that the heaters met all federal minimum standards. The most important of these is the Federal 301 standard which mandates that in accident simulations, no matter what happens to the heater, not more than 5 ounces of fuel would be spilled in the first 5 minutes which could contribute to a potential fire. Espar passed the test easily (note - some of the competition can't meet this standard without extensive modifications to the fuel system).

FIGURE 5.3
A consistent style is evident in these internal and external communication materials. (Courtesy of Espar Products.)

Allstate Insurance Company employs a systematic approach to corporate communication, ensuring that top executives as well as the broad base of employees are involved. Each year, the Allstate Research and Planning Center conducts a survey of 5000 randomly selected employees. The results of this survey are presented to top management and to the communication executives, then communication objectives are set. Editorial boards, consisting of decision makers from each of the company's business units, decide what stories will go into which publications and then the editor of each publication is responsible for developing them. One person on each board is responsible for securing approvals of articles from her business unit. As a result of this process, employee communication interventions are closely tied with organizational goals.

Allstate's publications contain only substantive, informative articles. There are no announcements of birthdays, anniversaries, weddings or other nonbusiness related per-

(continued)

FIGURE 5.3
(*Continued*)

sonal information. Instead, the publications cover news about how the company operates, what contributions individuals or sections have made, how the business environment is changing and how Allstate is responding. (Putting a communication program, 1989, pp. 9–10).

Along with creating standards and ongoing communication vehicles, it's essential to establish objectives, key indicators, and critical success factors. These will form the basis on which the success of communication interventions can be measured. Objectives should be based on the company's broad objectives and future plans, management style, business profile, legal requirements, and reputation (Parkinson & Rowe, 1977). These key indicators may be measured by communication audits, assessments of training, informal feedback, and public opinion surveys.

Kaiser Permanente is the largest private health service organization in the world. They use communication audits to help them plan their communication strategies and interventions. Based on focus groups, interviews with top managers, and questionnaires, they discovered the following:

- employees used the grapevine, regional magazines, and immediate supervisors as information sources, but wanted more contact with top management
- employees liked the company publications, but didn't want them to be too slick or expensive
- employees preferred quick, short, timely news stories to long features
- supervisors liked the quarterly video magazine, but were not showing it at staff meetings as intended

Based on this feedback, Kaiser's communication department switched to desktop publishing and a newsier content for magazines, revamped the video program, placed more emphasis on inexpensive weekly news sheets, and focused on employee appreciation activities and meetings. These changes reduced communication expenses and improved the effectiveness of communication strategies (Tapping into employees' views, 1989).

Intertwined with the development of communication objectives is the establishment of roles, departments, and functions for communication-related activities. For example, some organizations have separate departments for internal communications, external communications, information systems, and training; in other organizations, these functions may be combined under one umbrella. Moreover, communication professionals and the organization's top management must decide what falls into the realm of communication-related de-

FIGURE 5.3
(*Continued*)

partments, and what is the job of every manager. For instance, the communication department may provide fact sheets and presentation support for divisional meetings, but it is probably the responsibility of divisional managers to see that they're used. Following this line of reasoning, it may become the job of the communication or training department to provide courses or coaching on communication itself for line managers, or even every employee in an organization, so that the company's communication objectives can be met.

At the macro level, organizations must develop systems for

- training (internal and, perhaps, customer training)
- public information/public relations
- information resources (i.e., MIS, environ-

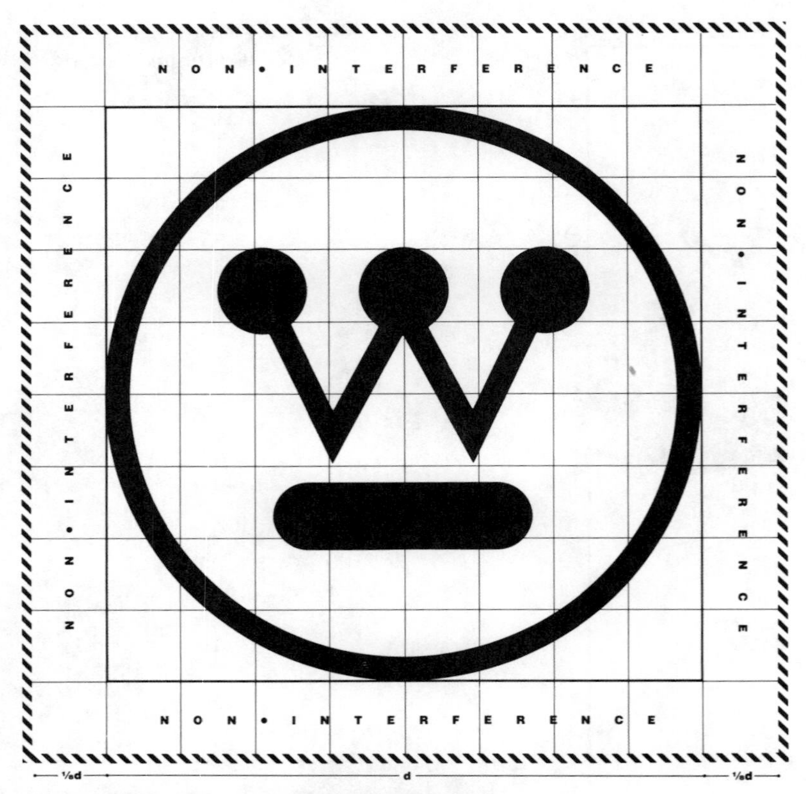

To avoid crowding the symbol, a non-interference zone, illustrated above, has been created to isolate it. The zone is based on one-eighth the diameter of the standard Circle W symbol, and shows the minimum space allowed between the symbol and any other element.

The zone does not mean that the symbol should appear in a box. It means that once the symbol has been applied or reversed out of the background, it must not be nearer to any other element than the outer edge of this imaginary zone. If the nearby element is another trademark, there are other considerations. Because closely placed trademarks suggest Corporate endorsement, these must be reviewed by Corporate Graphic Design. The general rule is that other trademarks should not appear closer than four times the diameter of the Circle W.

FIGURE 5.4
Many organizations have specific standards for the use of logos and for the layout characteristics of print materials. (Courtesy of Westinghouse Electric Corporation.)

mental scanning, research, documentation, etc.)
- marketing and advertising
- employee information and internal communication
- coordination with key external bodies (i.e., investors, lenders, government regulators, professional associations, etc.)

A prime example of a communication system is a crisis communication plan (see Figure 5.5). Every organization needs to have one of these in place, since a company's reputation is often made or destroyed by the way it handles emergencies. A classic example of an excellent response to a crisis is the Tylenol case; the opposite is generally true about Exxon's handling of the Valdez oil spill. Even more important, the safety of an organization's members and the surrounding community need to be ensured by a system of coordinated and accurate communication. The communication department must develop a list of key contacts, a plan for notifying the public and employees about any crises, and must name a spokesperson. Lou Williams, president of L.C. Williams & Associates, a communication consulting firm, offers this advice on dealing with crises:

- appoint a company spokesperson
- make an immediate disclosure of the facts—don't say, "no comment"
- make sure you notify family first
- develop a communication network so that everybody can get information
- set up a press room and have information easily available
- be accessible, offer immediate assistance to victims
- follow-up (Be ready, 1991)

Barbara Charmichael handled the silicone breast implant crisis at Dow Corning, starting in October 1990. She says:

It's my job to be an advisor, a trip-wire on issues, and to make sure that Dow has a 360-degree line of sight in making a decision. One of the first steps I took when I started this job was to set up a database to monitor all media coverage—both here and overseas. . . . By February 1991, after this extensive monitoring, one of the company's biggest problems became very clear to me: Public perception was that Dow Corning was being secretive. . . . With the data I had accumulated I was able to go into the executive team and say, 'Listen, we've got an issue of corporate reputation on our hands.' This led to the company putting into the public domain every shred of research results it had on implants. (Sobkowski, 1992, p. 67)

Another common system in organizations is Robert's Rules of Order. This traditional set of guidelines for running a meeting is widely practiced as a way to ensure equal access to input and voting, to increase the efficiency of meetings, and to eliminate personal attacks and sidetracking. Even in situations in which a formal system like Robert's Rules are not used or appropriate, other conventions are useful. For example, in labor disputes, a neutral third party enforces a rule that every time someone speaks, he must begin by summarizing the points made by the previous speaker. "In other words, no one is permitted to make his own point until he demonstrates that he has been listening to the messages from the other side of the table. The effect of this simple prescription can be dramatic. For the first time, the disputants are required to receive the messages their adversaries are transmitting" (Bassett, 1968, p. 126).

Planned Parenthood of Tompkins County, New York, has an explicit rule for staff meetings that promotes a cooperative and appreciative atmosphere for its employees. The first item on the agenda each month is "thank-yous." The entire staff spends about a half hour during which managers publicly thank their supervisors, staff, and peers, and express their appreciation for extraordinary effort (a pattern of performance that is not at all uncommon at this agency!). This sets the tone for the organization and makes clear its goals and values. Cassandra George, director of development, credits practices like this for the agency's ability to retain and motivate staff, even under conditions of low pay, long hours, and much public controversy about their mission.

In an era marked by charges of ethnic discrimination and sexual harassment, many companies have developed formal standards of behavior toward women and minorities—and obviously, many of these are grounded in communication practices. Many organizations offer courses in managing the diverse work force and eliminating gender bias in the workplace. In addition, internal communication departments often develop formal rules for the ways in which they will attempt to eliminate bias. For example, when I am invited to speak at conferences, the organizations who sponsor them generally send out a list of guidelines for speakers that include such recommendations. (See Figure 5.6.)

Typography

The graphic design system for employe publications features two type sizes for text material. Here are the ways we recommend using them:

- The text is set flush left, ragged right, to a measure of 15 picas (one column).
- As much as possible, lines of text should not be set less than 13 picas.
- Words should not be letter spaced.
- The first line of the first paragraph of an article is set flush left. For subsequent paragraphs, the first line is indented one pica (12 points). No more than the regular two point leading separates paragraphs.
- For major stories the lead paragraph (or paragraphs) can be set in 14 point two columns wide to call attention to them, leaving 12 points after last line before returning to our normal type size and measure.

The recommended type styles are Helvetica or Times Roman.

The specifications for the typesetter are:

Body Copy

• Size	10 point
• Leading	2 point
• Word space	4 to em
• Paragraph Indentation (except for the first paragraph of each article)	12 point
• Maximum line length	15 picas
• Minimum line length	13 picas

Lead Paragraph (when used):

• Size	14 point
• Leading	2 point
• Word space	4 to em
• Paragraph Indention	14 point
• First Paragraph	Flush Left
• All Succeeding	14 point
• Maximum line length	31 picas
• Minimum line length	28 picas

FIGURE 5.4 (*Continued*)

MACRO-LEVEL INTERVENTIONS—TOOLS

In order to implement practices and interventions, an infrastructure of technologies must be built. For example, most organizations have e-mail or voice messaging systems. Many dispersed companies have established private video networks consisting of video playback units in each office. Others are quickly moving to teleconferencing and satellite business TV channels.

Although they are generally part of the information systems and data processing department, management information systems (MISs) are also an integral part of the communications infrastructure. A typical MIS consists of an operating subsystem (everyday data/business applications, such as payroll and inventory); a reporting subsystem (that may contain data not in the operating subsystem, such as competitor

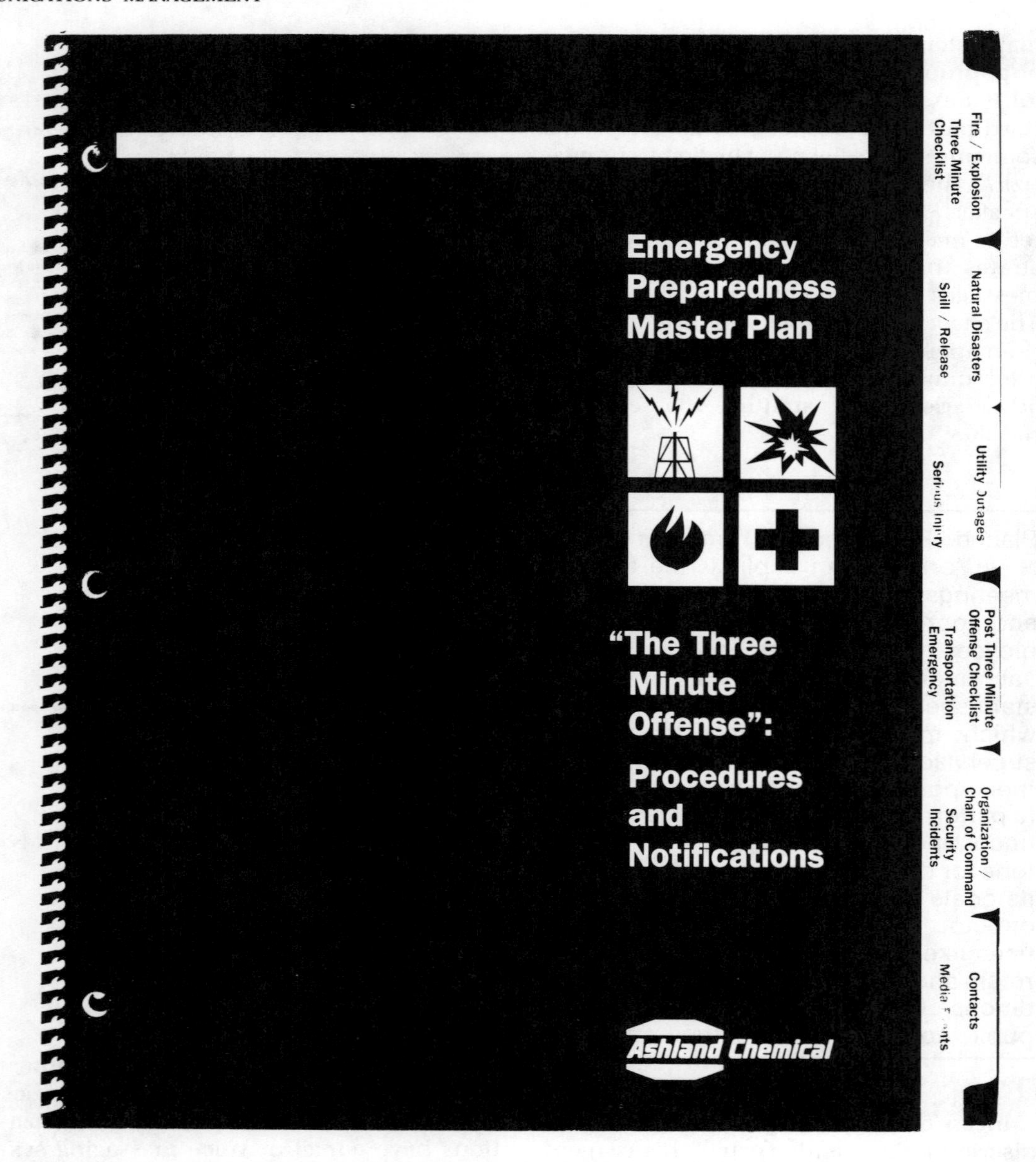

FIGURE 5.5 Communication professionals generally prepare crisis plans such as this one, which includes specific communication protocols. (Courtesy of Ashland Chemical Public Relations Dept.)

information); and a decision-making subsystem (mathematical models and simulations used to make decisions) (Meltzer, 1981). A typical information system might include

- *individual facts*—such as sales of grapefruit in the Main Street store during February
- *summarized and sorted data*—such as sales of grapefruit in all stores, from highest to lowest, during the winter months
- *graphically displayed data*—such as a bar chart displaying sales of grapefruit in February for the past 10 years
- *causal explanations*—hypotheses about why sales of grapefruit were so high this year in the Northeast
- *predictions*—graphs or figures based on some model, such as what would happen if the chain ran a special on grapefruit next year

Executive information systems are designed to summarize data in an easy-to-understand format for decision making, control, and communication. For example, it might include e-mail and presentation graphics for communicating information, data bases for obtaining and summarizing

Emergency Preparedness Master Plan | **Three Minute Offense**

"The Three Minute Offense": Procedures and Notifications

This section provides you with the opportunity to develop a plan for the early minutes of various types of emergencies, specific to your site. It should include the information you need to have at your fingertips in the event of an emergency. Each organization should establish procedures mobilizing emergency response personnel and provide resources for these emergencies:

Fire / Explosion
Three Minute Checklist
Natural Disasters
Spill / Release
Utility Outages
Serious Injury
Post Three Minute Offense Checklist
Transportation Emergency
Organization / Chain of Command
Security Incidents
Contacts
Media Events

information, and decision support tools, such as spreadsheets and models for analyzing information and generating what-if scenarios. According to Alter (1992), executives need several types of information:

- *comfort information*—general state of the business
- *warnings*—indications of things that require action or changes in plans
- *key indicators*—measurements of key variables that indicate organizational performance (i.e., output per hour, rework rate)
- *situational information*—data on an important project or problem
- *gossip*—informal information
- *external information*—competitor information and general economic indicators

There are two categories of communication channels: *direct* (face-to-face) and *mediated* (using technology); interventions or campaigns generally include both. Traditionally, research has shown that person-to-person communication is preferred in the workplace—employees have reported

1. Use gender-neutral terms: instead of chairman, use chair; instead of firemen, use firefighters; instead of man-hours, use staff time. It can be helpful to use the plural: instead of "a manager should always return his or her phone messages within a day," say "managers should always return their phone messages. . . ." When it is necessary to use a pronoun, vary the use of "he" and "she" among sentences or examples. The constructions "he/she" and "his/her" are awkward.

2. Make sure you do not stereotype particular genders nationalities, or ethnic groups. For example, do not always assume that physicians are men and nurses are women. Use names that represent a wide range of national origin: e.g., "Dr. Rao and his research assistant, Luis, . . ." Ensure that photographs and illustrations depict both men and women as well as a variety of ages and ethnic backgrounds, within a range of roles. Of course, you cannot include a black woman in a picture of your management team if such a person does not work on the team. However, you should try to represent both the *status quo* as well as the image that a variety of people *could* be included in such a scene. Be sure that the images do not consistently portray white men as superior; i.e., a man standing over a woman as she types.

3. Refer to individuals in a respectful and appropriate manner and make sure that persons of equal status are addressed equally. For example, when introducing two professors, one an older man and one a younger woman, refer to them both as "Dr." or "Professor," or use their first and last names; e.g., "Loretta Mangiola and Sam Goldman." Do not call them "Dr. Goldman and Loretta." Do not refer to women as "ladies" or "girls" or "females."

4. Avoid jokes that perpetuate negative stereotypes of religion, race, nationality, or gender. It is acceptable to make fun of yourself, and it is often funny to refer to groups of people who are familiar to the audience, but be sure to represent a wide variety of backgrounds. For example, you may make fun of another branch office of your company (if this is an "in" joke), or the residents of a town whose football team just beat the local favorites, or members of a given profession (like lawyers—*especially* if the audience is primarily composed of lawyers!).

FIGURE 5.6
Guidelines for using inclusive language.

that they like to get information from their direct supervisors and have the opportunity to talk things over. However, this preference may change over time as younger workers and new technologies make other faster, more responsive means of communication, such as e-mail, available throughout the organization. No matter what the means of communication, many researchers have found a clear contingency relationship between communication satisfaction and communication structure—this model can be called a *distance-direction paradigm.* Goldhaber (1983) indicates that his research found that employees were most satisfied with information sources in their immediate job environment and dissatisfied with sources far away. Employees also tended to be more satisfied with the role of sender than with the role of receiver, especially when they were able to share information with their immediate work group. They were least satisfied in the role of far-receiver, when they were most often in the position of being told information from sources who were not in their immediate group. Organizational members seem to blame distant sources for most communication failures.

> There is a clear analogy with the two-factor theory of motivation of Herzberg, Mauzner, and Snyderman. Communication with distant sources seems to relate to the "motivation" factor, and communication with close sources to the "hygiene" factor. Thus, organization members tend, in general, to be satisfied with communication with the immediate social environment and dissatisfied with the distant environment. The latter needs improvement if true motivation is to take place. (Goldhaber, 1983, p. 65)

Face-to-face communication is often impossible in today's dispersed organizations, and it can be very inefficient and expensive. One of the primary skills of professional communicators is choosing the channel(s) used to provide information and facilitate coordination of work.

Are new media potentially undesirable because they are too able to substitute for face-to-face conversations, or according to the uses-and-gratifications research, are they desirable because they promote future interactions with other people? This is a question often hotly debated among communication scholars and practitioners. Most applications have found that mediated communication does not supplant direct communication. For example, in a study of an interactive cable TV system in Rockford, Illinois, teachers said that the interactive system increased the diversity of new ideas within groups, but the "increase was regulated by the structure of interpersonal communication within groups" (White, 1986, p. 117). Companies who are using teleconferencing and e-mail may eliminate the need for some travel, but meetings and conversations have certainly not been reduced.

Dialogues

Much of the communication that occurs in organizations can be categorized as dialogues or as two people talking face-to-face. These include

- *interviews*—to screen job candidates, to gain information, to solve problems
- *performance appraisal*—regular formal

meetings in which an employee's performance is reviewed and formally assessed
- *counseling/coaching*—sessions in which a mentor, supervisor, or employee development professional provides personal or professional advice
- *work coordination*—informal meetings with a supervisor or peer to share information and to plan work

Although most of these dialogues happen quite naturally and are not the responsibility of professional communicators, a general system for certain types of dialogues should be developed and articulated formally. For example, most organizations have ground rules for interviewing job candidates, which ensure conformity to government guidelines for equal opportunity, and protect the rights of other candidates and the privileges of the screening committee. Another communication activity, *performance appraisals,* are similarly formalized in most organizations. There are standard forms and rating sheets to be filed, and training is generally offered for supervisors in this very important task of evaluating their subordinates' performance.

Meetings

Everybody seems to hate meetings, but many people spend most of their work days in one kind of meeting or another. Dislike them or not, for most people, their preferred information source is their manager. This has been documented in research studies since the 1940s and, more recently, in company surveys done by companies such as Federal Express. In today's environment, which stresses employee participation and teamwork, meetings seem inevitable.

Similar to dialogues, most meetings happen without any assistance from professional communicators. However, many organizations provide written meeting planners for managers who are unsure of what to say or how to say it. These guidelines can present sample agendas, ice breakers, answers to common questions, or visual aids.

Meetings take many forms. They may consist of

- informal discussions
- brainstorming for creative problem solving
- decision making or voting
- quality circles to discuss ways in which to improve organizational products and processes
- focus groups to gain input and reactions
- symposia and conferences to share research and to learn new concepts
- yearly sales meetings or product kick-offs
- press conferences
- formal reports, announcements, and speeches

Larger and longer meetings, such as product launches or an annual stockholder's meeting, are generally coordinated by the corporate communications department. Indeed, many departments have, or will contract for the service of, a professional meeting planner who can coordinate everything from agendas to facilities to audiovisual support to food service. In many companies, executives' speeches are written by the communications department.

Other kinds of meetings, such as focus groups and press conferences, are integral to the formal communication systems in organizations. Focus groups are often a primary means of garnering a target audience's reactions to a proposed product or service. For example, before a product is marketed or a commercial is aired, it may be presented to one or several small groups made up of people typical of potential customers; their reactions are recorded (sometimes on videotape). Once products are ready, they are often announced to the public in a press conference. This may be done at corporate headquarters or during a professional conference. Here, the new product can be demonstrated, key executives can provide information and answer questions, and press kits can be provided so that it's easy for material to be run in appropriate mass media.

Professional communicators may be asked to facilitate other meetings and to suggest strategies for problem solving or brainstorming. There are a number of group process techniques that can be used to manage collaboration, to enhance the generation of novel ideas, and to ensure parity among group members. Some group process techniques include

1. *synectics*—a method for tackling difficult problems in a creative way by having

participants generate metaphors and unusual perspectives. Participants take a client's stated problem, generate how-to statements for the problem, and try to develop solutions. The client then identifies things he likes about each proposed solution, and asks in a positive way, for more information on what he'd like to see next.

2. *nominal group technique*—a method for generating ideas by having each participant silently generate and then offer ideas in a round-robbin session (with no criticisms allowed), clarify ideas, and then rank items based on their preferences

3. *delphi technique*—qualified experts respond to general questions individually and in print, the results are tabulated, and several more rounds of the questions are presented to seek consensus. This spans a long time period, and is generally designed to provide information that would be the basis of subsequent meetings.

Many group processes are significantly aided by mediated support, such as computer-based voting and tabulation systems, large-screen display of information as it is generated, and remote conferencing technologies. Systems that help to maintain a history of the session, ranging from simple notepads and flip charts to word processing and hypertext software, can make these processes more efficient. Participants can actually see what they're creating and react to it more easily and rapidly.

The Interactive Management group process technique was used by a midwestern Native American tribe council working on implementing a self-sufficiency plan.

During the first day of meeting, participants generated a master list of ranked obstacles using the nominal group technique. Interpretive Structural Modeling (ISM) was then used to establish the priority and relationship of all the obstacles to each other in terms of relative priority. The PC and computer projection system were used to display each pair of obstacles for the participants to discuss and to jointly decide on the relative priority. The pair decisions were transmitted using a modem to a mainframe computer, which used a transitive inferential algorithm to minimize the number of pair judgments required by the group and calculate overall obstacle relationships. The second day began with a graphic display of the computed results of the ISM, showing participants' judgments in the form of an obstacle map.

This structural map showed obstacle priorities and relationships. The map provided a comprehensive picture clarifying what was previously a complex and confusing plethora of issues. The obstacle map seemed to infuse the group with energy and a desire to continue.

At the end of 2 days, the participants had developed an integrative picture of problems facing their efforts to become self-sufficient and several solutions for each of 14 key problems. Virtually every participant expressed a strong sense of achievement and satisfaction with what they had accomplished (Jensen & Chilberg, 1991, pp. 400–401).

Mediated Systems

Although most communication in organizations is direct person-to-person, most formal communication these days is mediated. The array of media available to store and to display data is overwhelming, and one important role for professional communicators is staying on top of the kinds of technologies that can help them fulfill their duties more effectively. The power of media is evident from everyday experience with television, radio, publications, computers, and telephones. It is hard to imagine how organizational life would be without pervasive use of one or more of these technologies.

Although media can be very effective tools, they are not without their limitations. Remember that the singular of media is medium, which is something that goes between two other entities. Technically, all communication is mediated, whether by our bodies, our voices, writing on a sheet of paper, or by a touch-screen interactive digital video program. While we can't ignore the efficiencies and capacities afforded by them, we also can't be dazzled to the point of neglecting the quality of the message itself. Unfortunately, too many business communicators think that merely using some flashy new technology will persuade or instruct, that some boring or ill-formed data will undergo some sort of transubstantiation when a new channel is used.

> Some of the worst offenders, of course, are advertising agencies whose staff people and account executives think that to bedazzle the client is commensurate with demonstrating creative thinking and imaginative solutions to complex marketing, public opinion, or attitude problems. . . . But to assume advertising techniques are appropriate tools for communicating management's messages to the whole corporation, or even to some of its important constituencies, is a case of a method being applied in an inappropriate setting for the wrong reasons. (Ruch & Goodman, 1983, p. 92)

Although media selection is generally considered to be one step in designing individual interventions, purchasing and implementing media systems is a macro-level process. For example, it is unlikely that interactive video will be chosen for an individual training project if interactive systems have not been procured as a part of a long-term strategy for employee instruction. One can only choose between using the regular postal system and e-mail if an e-mail system is in place. Since technological support for communication is generally an expensive undertaking, it's important that a careful business plan be formulated first. Consider these issues:

1. What are the problems that a new medium could potentially solve better (more efficiently/effectively) than current systems?
2. What are the total costs of purchasing, implementing, and maintaining a new system, including the costs of training people how to use it? (For example, a PC network may cost $80,000, but the requisite software, installation, and training may cost three times that, not to mention yearly maintenance and updating costs!)
3. Do members of the organization have the skills to design and to develop programs for the new medium productively, and if not, how can those skills be acquired? For example, an organization may make a good case for live satellite business TV broadcasts, but if the in-house video crew has no skills in directing fast-paced, live, multi-site programs, they must either acquire those new skills or the organization must hire free-lancers.
4. How might the new medium change the nature of communication and with it, the corporate culture? For example, voice mail may make the organization sound more impersonal than having office professionals answer the phone, and electronic bulletin boards may encourage more communication across organizational lines, regardless of any existing formal hierarchy.
5. Who will be responsible for buying and maintaining the new system, and who will have access to it? For example, if the training department buys desktop publishing equipment out of its budget, will professionals in corporate communications or marketing have access to it? How will costs be controlled?
6. What kind of standards in terms of content and format should be established for the use of the new system? Desktop publishing, for example, can enable departments to produce their own literature; if this happens, what kind of standards should be maintained so that their material meets basic corporate standards for its style of writing, accuracy, and visual image?
7. How long is the system likely to work, and how quickly will it become outdated? What means will be provided for updating?
8. Is this system an industry standard (i.e., VHS videotape), or is this type of medium not yet standardized (i.e., interactive multimedia)? This will impact how programs can be shared with other organizations, and will determine if off-the-shelf programming can be bought.
9. What systems will this replace (if any) and what will happen to those older communication vehicles? For example, will desktop video replace the TV studio and its technical support staff?
10. Will this new system spawn ideas for even more messages and programs, and if so, how will the organization handle the increased information load and costs?

Kay asks, "What does a medium ask you to become in order to use it?" (Schrage, 1990). He says that print requires a rational reader; television, a passive observer; the telephone, a conversationalist. Eddy (1992) adds that the new interactive media "demand continuous choice and navigation" (p. 56). As we put media systems into place, we must consider the current skills and proclivities of the users, as well as the ways in which we want to shape their behaviors and develop their expertise.

Media Matrix

The Media Matrix (Figure 5.7) categorizes media by *bandwidth*—the type of modalities that can be transmitted or reproduced—and *applications*—the purpose of the medium. The bandwidth includes text only (memos or e-mail); graphics (presentation support media, such as slides or displays); audio (phone conferencing or audio tapes); video (full-motion images that can be displayed on videotape or in satellite conferencing); and synthetic images (computer-generated images that can be controlled vicariously by users). There are three main categories of applications: (1) to inform (such as memos, bulletin boards, or corporate video news programs); (2) to instruct (systems such as computer-based training or manuals); and (3) to automate or reduce the human judgment factor in a job (such as print job aids or expert systems).

Obviously, there is a lot of overlap between different kinds of systems. Business TV obviously can display text, audio, graphics, and full-motion video; hypertext can be used as a reference tool or an instructional system; and similarly, interactive video can be used for point-of-purchase information or tourist kiosks, as well as for teaching hard-core technical skills. This matrix is a representation of the palette of tools available to modern-day communication practitioners, and keeping up with the continual addition of new technologies to this array can be a full-time job.

Print/Desktop Publishing

Print media are the traditional tools of formal communication in organizations.

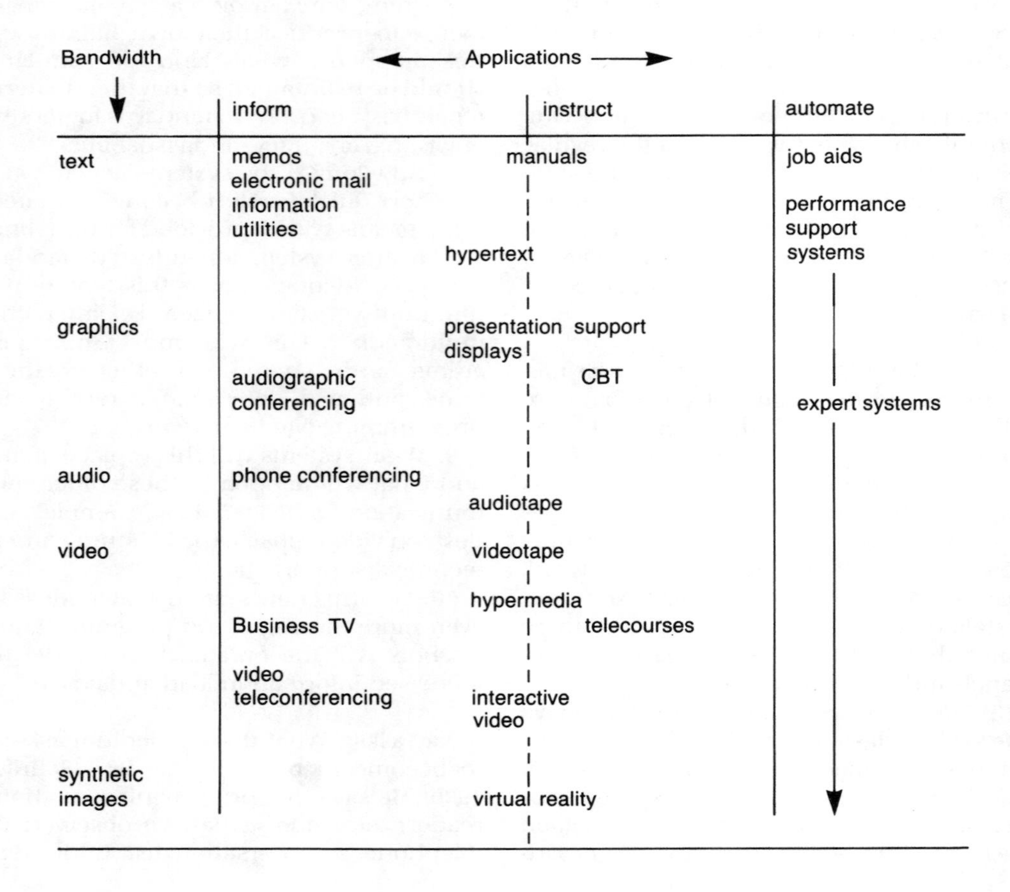

FIGURE 5.7
Media Matrix.
[Adapted from Gayeski, D. (1992, May). Making sense of multimedia. *Educational Technology*, 9–13.]

Memos, brochures, newsletters, and manuals are so ubiquitous that we don't even notice them in the environment. Their prevalence doesn't, however, mean that they're generally done well.

Many print pieces are written by people whose only education in this area was their freshman English course, and they don't realize the different styles of writing and layout that are called for in various publications. Journalists may do well in developing employee magazines, but are likely to write poorly designed instructors' guides or computer documentation. Instructional designers might not have the knack of writing promotional pieces or press releases.

The mainstay of many corporate communication departments is the employee newspaper. A recent review of an employee magazine published by a large Canadian manufacturer in the mid-80s found that it contained the usual articles on personal news (birth, deaths, weddings), company social events and sports teams, new offices, promotions, safety messages, and an announcement of a ban on smoking in the workplace. This probably describes most current newsletters. However, this magazine was put out not in the 1980s, but in the 1880s! Newsmagazines have been around since the fifteenth century and, unfortunately, not much has changed since then (Johansen, 1990). To be considered as elements of the business strategy, newsmagazines must take on major issues; provide timely, complete, and sometimes controversial information; and stay ahead of the grapevine.

Most people recognize that typical house organs are superficial and never honestly discuss critical issues. Bassett (1968) recommends that company failures be analyzed in company newsletters so that others can learn from past mistakes. He says that the underlying message of most house organs is "management's unwillingness to communicate with its employees about issues of genuine importance" (Bassett, 1968, p. 153). Given this, we can expect employees to be no less cautious in communicating with managers.

Other typical print pieces include annual reports (which are required of all publicly held companies), documentation and instructions for product use, brochures, and manuals. Managing these media include

- establishing a schedule and budget for the creation and update of each piece on a yearly and monthly basis
- formulating means for generating the content from outside sources, internal reporters, free-lance journalists, and company SMEs,
- developing standards for writing styles and layout
- generating a distribution scheme

Most organizations have guidelines that relate to printed materials: these include how the logo must be reproduced (in style, color, position, etc.); what kind of stationery is used by whom and under what circumstances; which style guides are used in publications (such as *The Chicago Manual of Style*); and the general corporate tone for various kinds of pieces. For example, it is virtually unthinkable for most corporations to have a humorous annual report, but it would be against Ben and Jerry's Homemade, not to use humor. Other stylistic elements include corporate colors, the use of certain company slogans or icons (i.e., cartoon characters), and the reading level at which different kinds of publications are to be written.

The use of computers to generate print materials has radically changed this enterprise since about 1988. Beginning with electronic typesetting and blossoming into full-blown desktop publishing systems, computer-based print layout and production has made it possible for organizations to turn out professional-looking documents in-house. Software programs, such as Ventura Publisher and Pagemaker, turn PCs into typesetting devices, and laser printers can output close to typset-quality masters.

Desktop publishing systems generally consist of these components:

- a personal computer, such as a Macintosh or an IBM-compatible, with a color monitor and mouse
- a laser printer
- a scanner—to scan images and turn them into computer files that can be imported and manipulated
- a word processor for entering and editing text

- one or more graphics programs for creating pictures and business graphs
- a page layout program to integrate text and graphics using specified templates (i.e., a three-column newsletter or a two-column phone listing)

Optional elements include

- a still or motion video camera for capturing live images and importing them into a document
- a CD-ROM player for accessing high-density CD-ROMs, which may have clip art, stock photographs, or publishing tools such as a thesaurus, dictionary of quotes, or zip code finder
- a data-base program to file information (i.e., listings of employees' office and home locations), which can then be imported into a document
- style-checking tools, such as grammar checkers or style guides
- clip art, in hard copy or on floppy diskettes
- predeveloped software templates for different publication layouts
- a modem for downloading text or graphic files from internal or external sources, or for shipping files to printers
- a graphics tablet for ease of drawing sketches
- a full-page monitor—an extra tall monitor that can show an entire page or facing pages of layout
- software fonts—either provided on a cartridge that is inserted into a laser printer or on diskette; these expand the number of typefaces that can be used in publications

Desktop publishing has decentralized formal communication within organizations more than any other technology. While these systems have made it easier for communication departments to do their jobs quickly and cheaply, it has also eliminated some of their functions, since individual managers and their secretaries can now create their own documents. Deciding on who should be allowed to create their own materials and on what kinds of checkpoints should be instituted before materials are sent out have become major issues for communication departments. While the management styles that espouse empowerment, two-way communication, and local decision making certainly favor decentralizing print production, many organizations have found themselves swimming in poorly produced newsletters, or having to defend inaccurate brochures or embarrassingly bad writing.

Desktop publishing can free professional communicators from many menial and time-consuming tasks and can also liberate budgets for other purposes. Standardization can rather easily be incorporated by generating style sheets for different kinds of publications. This makes it virtually impossible for novices to create really poor documents and builds a consistent look for publications. Computer files for logos and other corporate symbols can be provided to anyone who can use them. Finally, software style guides can be preset to the established organizational tone and reading level for various categories of documents, eliminating the need for inexperienced writers to guess at what's appropriate. (See Figure 5.8.)

On-Line Communications/E-Mail

Gradually replacing the ubiquitous memo and the annoying game of phone tag are electronic versions of messages. E-mail uses a host computer to run messaging software and store mail, and remote terminals or PCs to gain access to it. Most large organizations have their own e-mail systems running on their mainframe computer. Many people also tap into public mail systems, such as those available on CompuServe and Bitnet. Another form of e-mail is voice mail, a computer/phone system, similar to an answering machine, that allows callers to leave messages.

Companies have found that e-mail does more than avoid paper and phone tag. The Beneficial Corporation uses an extensive MIS, which connects over 1200 employees, to provide e-mail and to transmit data. All messaging between managers is done through the system rather than on paper. The chairman/CEO who established this system says, "I can communicate with or yank figures from any manager anywhere. We can make a decision now in a quarter to half the time it took before." He also predicted, as have executives in DuPont and PPG, a flattening out of the organization as e-mail allows managers to deal with a greater number of subordinates more effectively (Bowen, 1989, p. 271).

Westinghouse Corporate Internal Communication

Taking the Plunge into annual reports

Each year, as the winter freeze thaws and the trees begin to blossom, another rite of spring occurs: publicly-held companies file their annual reports, disclosing the financial "picture" of their organization's past year.

In the Securities Act of 1934, the Securities and Exchange Commission (SEC) required that publicly-held companies which raise capital by selling securities file annual and other periodic reports. By doing so, these companies provide potential and current investors, employees, customers and suppliers with up-to-date and accurate information.

Unfortunately, many would-be readers of annual reports are confused by page after page of financial data. As a result, they draw their conclusions from just half the story — the words and pictures that typically fill the front section. For those of you for whom this is true, this article attempts to guide you through a typical annual report, suggesting financial measures that you should consider when evaluating a company.

The back door

Business author and commentator Jane Bryant Quinn recommends that readers jump into an annual report from the back: "We don't want to be surprised at the end of this story. . . so start with the report of the certified public accountant — a third-party auditor who will tell you if the company's report conforms with 'generally accepted accounting principles.'" In a sense, the independent auditor is like a book critic or reviewer — now you know what to expect from the rest of the book.

Most annual reports have two sections: words and numbers. The "words" describe the organization's opportunities, plans for the future and community involvement. The section usually begins with a letter from the chairman to shareholders, summarizing the company's strategies and growth prospects. This is your glimpse into the opinions of top management — its assessment of the company's past performance and future expectations. From this letter, you can often gain insight into how a company's executives promote the organization to the public, and a sense of its personality and style.

Next, the numbers

The second section of the report is the financial review, which provides the "numbers" that illustrate the relative "health" of the company. The review begins with management's discussion and analysis, which summarizes the company's financial performance and outlook from the viewpoint of top management.

Two portions of the annual report that are especially helpful in assessing the company's stability and growth are the balance sheet and the statement of income. Let's take a closer look at how each provides insight into a company's financial picture.

A balancing act

The **consolidated balance sheet** summarizes the condition of the company's assets, liabilities and

FIGURE 5.8
This news bulletin tells employees how to read a typical annual report. (Courtesy of Westinghouse Electric Corporation.)

E-mail systems allow messages to be sent instantly to one recipient or a group of recipients who are also a part of the system. When a person signs on to his account, the arrival of any new mail will be indicated. Those messages can then be deleted, saved, copied onto the recipient's own disc, or printed out. E-mail systems send messages that are private and sequential. A person can only read his own messages and, generally, two people are not actually on-line at the same time—rather, they exchange messages by turn. Other forms of on-line communication include bulletin boards, computer systems that allow users to post and read messages, and computer conferencing systems, which allow groups to exchange information sequentially or simultaneously. Related trains of thought, or threads, can be searched so that a user may look back over what has been said about the topic "free trade" over the past 2 months.

Quite a bit of research has been done on the effects of e-mail and of conferencing systems on group communication. Generally, people find it easy to use and find that it increases their efficiency. Surprisingly, there is an increase in the emotional and controversial content of messages using these media. *Flaming* is a term that describes outbursts of criticism or anger, which are not uncommon among electronic workgroups attempting to solve problems or discuss issues. People tend to express their emotions candidly and send more negative social messages while using these forms of communication. These systems produce less agreement, reduce social cues and the effect of status and appearance, and are more task-oriented, but are less effective in managing conflict. Users have even developed special typewriter art symbols to express the kinds of nonverbal facial cues that they would use in direct communication (see Figure 5.9).

Some of the major factors that influence the acceptance of computer conferencing and e-mail include

- the number of people, particularly of high status, already on the system
- one's anticipated level of use
- the relative priority of kinds of tasks to be performed on system
- ease of access to the system
- ease of use of the software and hardware
- perceived need to communicate quickly

Companies such as the Insurance Company of British Columbia (ICBC) and AT&T have on-line newsletters; ICBC's is weekly, and AT&T's is updated daily and available to over 120,000 employees worldwide. Unlike print newsletters, e-mail equivalents allow readers to reply immediately—and they do. AT&T reports that it receives as many as 50 letters each day (McGoon, 1992, pp. 16–17). Other organizations use a combination of e-mail and desktop publishing. Westinghouse's corporate communication department provides stories of general interest and sends them by e-mail to their regional offices and subsidiaries. Newsletter editors in those locations can download the stories and import them into their own desktop-published newsletters.

Digital Equipment Corporation (DEC), a large manufacturer of computer equipment, found that its employees were frustrated with the company's apparent inability to provide them with news directly and promptly before they found out about situations through local mass media. DEC turned to its own technology—computers—to provide quick and tailored information. Using their videotext system, they created LIVEWIRE. It uses a menu of topics that allow over 70,000 DEC employees throughout the world to learn about what's happening and to contribute memos and news items. In just 1 month in 1988, the system was accessed more than 175,000 times. Said one employee, "LIVEWIRE gives me the ability to get a substantial amount of current information about the company without leaving my desk" (Communicating on-screen, 1989, p. 44).

FIGURE 5.9
E-mail affect codes.

:-)	smile
:-(	frown
;-)	wink
|:-O	surprise (note raised eyebrows)

When I call organizations these days, there's about a 50% chance that my call will be answered by a voice mail system. Although some people are annoyed by the seeming impersonality of this, voice mail accomplishes one important goal—it eliminates phone tag. One study by AT&T found that only 9% of most business calls were completed on the first attempt. Moreover, voice mail allows users to screen calls and return them when they are ready to respond—they can be better prepared and less distracted and can let the emotional content of a message sink in and diffuse before a reply is necessary. Another survey found that unnecessary or unproductive phone calls cost at least a month out of every executive's year (Clampitt, 1992). And Rice (1984) found that the true cost of each successful call by typical managers and salespersons was $16.24.

The difference between voice mail and an answering machine is that a computer records the incoming call and controls what is done with it. Simple systems may just record calls and allow the recipient to play back the messages; however, most voice mail systems do much more. Many voice mail systems have a calendar function that allows a message to be sent some time in the future and a broadcast feature that allows messages to be sent to a predefined distribution list. Users can send themselves messages as reminders, and communications professionals can send out information to an entire group of people, which will surface when it's most likely to be effective rather than too early or too late.

Some techniques can help make voice mail more productive:

- Recorded outgoing messages should be explicit and updated each day, or even more frequently. For example, don't just say, "I can't take your call now . . . ;" rather, say, "I am on vacation until May 15 and you may leave a message with my secretary by dialing 3456 . . . "
- Incoming messages should also be explicit. Don't just say, "This is Sam Shmoe, please call me . . . ;" rather, say, "This is Sam Shmoe. Please send me any literature or catalogs you have concerning computer graphics software for the Macintosh . . . "
- Help recipients process and prioritize their messages by indicating the subject and the action they need to take. For example, "Hello Judy, this is Diane and I have something relating to the Rapse project that requires your immediate attention," or "Hi, Jake—this is just an update and there's no need to call me back on this one . . . " This way, if someone has 54 messages waiting for him when he calls in, he can skip over those that aren't urgent and prepare himself for the information on those calls that do require a rapid response.
- Systems managers should consider giving out special mail boxes to specified callers; when the callers call, a customized message can be given to them, and all calls from them can be accessed as a group. For example, a key client or associate may have her own mail box; when she calls and gets the voice mail system, she can enter her code. Then, she will hear a private message, such as, "Hello Allison. We've come up with a price for the workshop and I'd like you to call Pat at Acme and tell them it will be $2500; if they seem to balk, tell them I can talk with them on Friday."

Similar structures can be used to facilitate e-mail. A prototype system, called the Information Lens, distinguishes between common types of messages (i.e., a request for action or a meeting announcement) and displays a template that must be filled in with the required information. Based on that information, which can include items such as the date or the urgency of the request, the system decides how to classify and how to display the message. A similar commercial system called Coordinator requires that users classify their messages into fixed types of conversations, such as a request or an offer. If, for example, someone makes a request, she must assign a respond-by date. The system then requires that the recipient respond by an acknowledge, promise, or decline, and maintains a record of all open, or uncompleted, action items. In this way, Coordinator ensures that people are clear in their requests and that they are aware of their obligations (Alter, 1992).

Increasingly, corporations are making use of powerful conduits, which can carry voice, data, and images. *ISDN* (integrated service digital network) is a standard for digital phone lines. It can provide transmission of up to 1.5M bits per second, or multiple channels for linking computers, local area networks, and phone systems. Using this, it's possible to transmit both voice and computer or fax data during the same call, so that people can work on the same computer application while they talk.

Display Media

Although they don't have the glamour of video or interactive systems, display media such as bulletin boards, posters, and trade show booths are important tools. These low-cost, easy-to-use media are often overlooked. I remember a situation in which colleagues of ours were asked to specify a video system for employee communications in a manufacturing plant. They wound up telling managers that there was no appropriate place for employees to watch videos and that, instead, they should put up bulletin boards.

Displays are often badly used; most bulletin boards are full of scraps of advertising and old news. Many posters and graphics designed to promote a new theme or product never seem to be placed in a strategic position, and there is no system for retiring old campaigns and replacing them with new ones. For example, when we went into several branches of a major bank to produce a video, we ended up spending several hours taking down old posters, straightening out little boxes of leaflets, and removing

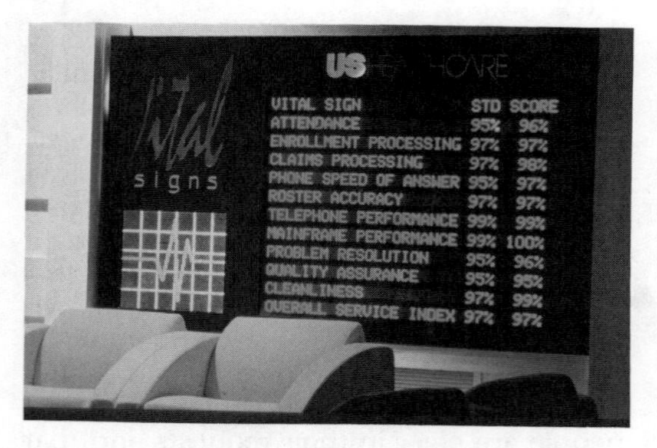

FIGURE 5.10
The Vital Signs electronic bulletin board displays employees' progress toward daily goals. (Courtesy of U.S. Healthcare. Photo by Stephen Barth.)

stickers from the teller windows. As these media are produced, it's important to put into place some strategy for their use. A person at each location should be given the responsibility for maintaining these media, and policies should be established as to who can post what kinds of announcements, ads, and posters.

Display systems can turn an entire environment into a communication system. For example, U.S. Healthcare, a large HMO, created a new customer service building that prominently features computer-generated displays with up-to-date information on the company. A large panel, called Vital Signs, shows daily progress in meeting important goals, such as timeliness, attendance, and cleanliness. However, not all display systems need to be high-tech. Preston Trucking started charting and posting data, such as number of shipments per driver and miles driven without breakdown. Large graphs of this information are prominently displayed in a conference room at headquarters and at all their terminals. A bulletin board near the cafeteria also tracks Preston's stock prices. Systems like these make expectations clear and allow employees to get the big picture quickly and easily. (See Figure 5.10.)

Trade show booths are an important part of many organizations' advertising campaigns. Although they may be 20 feet long, most of them are designed to collapse and fit into rugged suitcases for shipment. Many of them rely on rather plain backgrounds that can be customized for a particular show or product launch. For example, we designed a booth for Espar, a company that markets small diesel-powered heaters for various vehicles. Their booth has simple gray panels, lighting, and their logo. They can velcro large graphics, which depict the particular market addressed by each show (i.e., trucking, marine, off-road construction vehicles, school busses), onto the panels. (See Figure 5.11.)

FIGURE 5.11
Trade show booths and collateral sales materials are an important part of many organizations' marketing efforts. (Courtesy of Espar Products.)

Presentation Support Media

Some media, such as the telephone and voice mail, merely act as channels for personal messages between people who cannot or choose not to meet face-to-face. Other media, such as videotapes or brochures, replace interpersonal communication. And still another category of media is designed to enhance live presentations and meetings. These include

- slides
- overhead transparencies
- flipcharts/writing boards
- computer-generated visuals
- interactive facilitation software

Some of these media are used to visualize a presenter's information, giving the audience something relevant to focus on and helping them tune in to the important aspects of a presentation. For example, graphs can illustrate financial trends, and word slides can aid students with note-taking in a lecture. These types of media are generally created beforehand and are controlled by the presenter. Other media, such as flip charts and writing surfaces, are used to jot down important words, especially if they are to be copied down or if the spelling is ambiguous. These devices are also used to record responses from the audience so that they can be remembered and referred to at a later time. Still other devices are used to help facilitate meetings. For example, interactive response systems and brainstorming software help to generate and to record ideas.

Today, visual support for sales and planning sessions is almost a requirement. For example, Emerson Electric uses a standard three-screen system for management planning meetings. Each screen shows a different level of detail and position within the entire presentation; thus, at any given time, the audience knows not only the immediate details, but where they are in the overall plan. In the past, the visuals were displayed by way of an elaborate system that changed transparencies loaded into drums on each of three projectors. Currently, the charts and graphs are displayed on a Macintosh-based computer system. This allows the presenter to zoom in on details of the charts and to easily go back and forth between images.

Many meeting rooms now contain group decision support software and hardware. As a presenter or facilitator leads a group discussion, participants use individual keypads linked to the facilitator's computer and projector to register their opinions or votes on a topic. This form of voting allows people to remain anonymous but still express their viewpoints, and it promotes a fuller and more open deliberation. Other software systems allow the leader to display slides, computer graphics, videotape scenes, and videodisc segments or frames. Still other systems make it easy for the group to brainstorm—the facilitator can quickly enter words or phrases into the computer system. These can then be expanded or contracted to show more or less detail, and groups of items can easily be edited and moved around. At the end of the meeting, a clean and clear laser-printed copy of the outcomes is available to all participants. Furthermore, charts and graphs can be readily created from the input for inclusion into subsequent reports. (See Figure 5.12.)

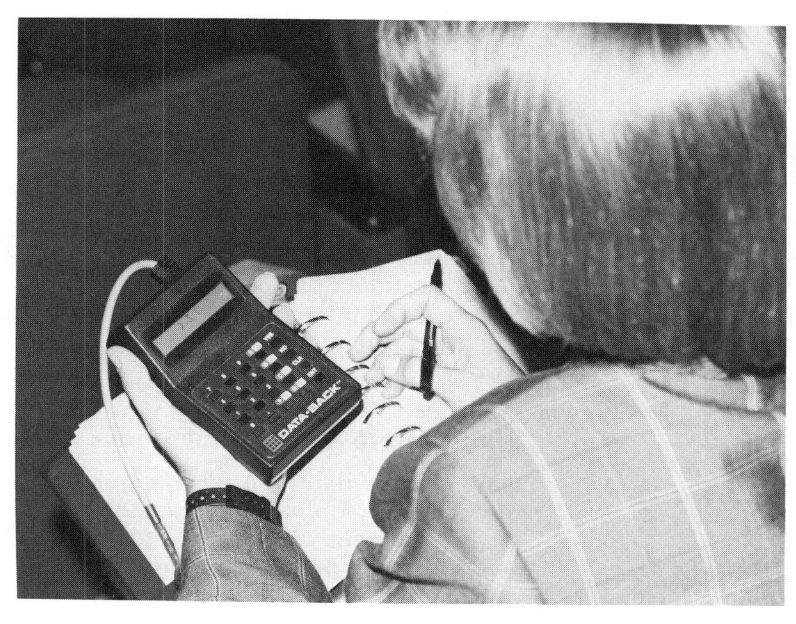

FIGURE 5.12 Participant response systems are powerful ways to get feedback on ideas generated in planning sessions and to test a group's understanding of material presented in a training course. (Courtesy of Dow Chemical.)

Creating simple presentation support materials is now quite easy. Software tools, such as Harvard Graphics or Powerpoint, enable anyone with basic computer skills to create attractive graphs and charts, which can be output to slides, to overhead transparencies, or straight to a computer monitor or projection system. It is a trend to display images directly from the computer using large monitors or an LCD panel, which fits on top of an overhead projector. This eliminates the need to develop film or copy transparencies and allows the presenter to modify material right up to the time of the presentation. Most of the current graphics programs have default templates that produce attractive and readable graphics; however, some guidelines are useful, such as guidelines for producing projected visuals:

- Use a colored background; light letters on a dark background are easiest to read.
- Limit colors to three or four, and use them to highlight words or indicate subordination of ideas. For example, titles can be in yellow, major bullet points in white, and footnotes in light blue. Make sure there's sufficient contrast between foreground and background colors so that visuals can be easily read, even if the room is not completely darkened.
- Use no more than six words per line and no more than about 10 lines per visual.
- Use simple typefaces and both upper and lower case letters. All caps are difficult to read.
- Use short phrases rather than long sentences.
- Use visuals to emphasize key points.
- Use consistent backgrounds, typefaces,

and colors throughout the presentation. This can be done easily by creating templates.

- When presenting a list, display one item at a time so that the audience won't get ahead of what you're saying. You may gradually reveal items while displaying previously discussed items in a more muted color.
- Visuals should present just a summary; if more detail is needed, use a handout.
- In general, a horizontal format is easier to project.

Stand-Alone Audiovisual Media

Film, video, and multi-image programs represent a more sophisticated category of media, the content and design of which are generally carefully planned and executed. Unlike print, e-mail systems, or presentation support media, they are generally produced by professionals rather than end-users. Of the three media just mentioned, video is by far the most popular. Due to its ease of use and low cost, video has surpassed film as the most popular medium for public relations and training programs. With the increasing availability of high-power large-screen projection systems, video is also displacing the awkward, expensive, and nonstandard multiprojector slide and tape programs generally known as *multi-image*.

Video is the most popular medium in use for training today and, needless to say, VCRs are available in most offices and homes. The technology to play back prerecorded tapes has been mastered by almost everyone, so there is little resistance to its use. Moreover, simple camcorders have made it possible for amateurs to record practice sessions, document events, or gather images for later analysis. Many organizations have established their own in-house video production units, although corporate downsizing trends and the increasing availability of outside video production houses has curtailed the growth of large internal video departments.

Video's ability to reproduce sound, motion, and color and to rapidly play back or pause images makes it a powerful medium for persuasion and instruction. Some of the popular uses for video within organizations include

- corporate news programs
- interviews with executives
- documentation of important events or speeches
- internal and external marketing pieces
- hard skills training (i.e., how to use a machine or safety procedures)
- soft skills training (i.e., customer relations or management communication)
- recording and play back of training events (i.e., presentations, role-plays, or psychomotor activities) for analysis and feedback

Video, however, is not inexpensive or quick to produce. A typical program costs $1500 per finished minute to produce. Factors that escalate costs are elaborate graphics or animation, multiple locations, professional actors, special effects, or rapid pacing of edits. Once produced, tapes can be duplicated rather inexpensively; for example, a 30-minute VHS tape in quantities of 100 costs under $10 (including the tape stock and duplication service). (See Figure 5.13.)

Desktop video is a generic term for video programs that are recorded and edited on small format equipment, including graphics and special effects generated by a personal computer. These systems potentially allow end-users and managers to create their own professional-looking video messages. This set of technologies is closely allied with interactive multimedia (see the next section for a discussion of this class of media).

Film is still used for high-quality image programs, which are destined to be shown to large groups. Its superior resolution, especially when displayed on a large screen and the ability to carefully manipulate lighting, colors, and textures within scenes still make it an attractive, although expensive, medium. Not many organizations produce or sponsor films these days. The most common use of film is the form of rented generic training programs.

Multi-image programs are designed for high-impact, emotional presentations (see Figure 5.14). Often using 36 or more slide and film projectors, these fast-paced programs display overlapping or adjacent images on a large screen, accompanied by music. These programs are typically used for product roll-outs and motivational introductions to meetings. Since this medium

FIGURE 5.13 Excerpts from a video news program script. (Courtesy of Metropolitan Life Insurance Co.)

MARVIN O/C	(MARVIN) COMING UP ON THIS SECOND EDITION OF MET LIFE NEWS FOR 1989.....
V/O VIDEO	MET LIFE PREPARES FOR A CHANGE AT THE TOP AS PRESIDENT AND C-E-O JOHN CREEDON ANNOUNCES HIS RETIREMENT.
CLAIRE V/O	
V/O VIDEO	(CLAIRE) THE FUTURE IS NOW, AS EXPERT SYSTEMS COME ON LINE. EXPERT HELP FOR MANY MET LIFE EMPLOYEES IS AS CLOSE AS THEIR COMPUTER KEYBOARDS.
MARVIN V/O	
V/O VIDEO	(MARVIN) AND A PRESCRIPTION FOR HEALTHY LIVING... MAMMOGRAPHY. A SURE STEP FOR EARLY DETECT-
MARVIN O/C	ION THAT SAVES LIVES. THESE STORIES AND MORE WHEN MET LIFE NEWS CONTINUES.
MARVIN O/C JOHN CREEDON GRAPHIC	
	(MARVIN) MET LIFE PRESIDENT AND C-E-O JOHN CREEDON HAS ANNOUNCED HIS RETIREMENT, EFFECTIVE

(continued)

FIGURE 5.13
(*Continued*)

	SEPTEMBER FIRST. MR. CREEDON'S LONG AND DISTINGUISHED CAREER WITH MET LIFE BEGAN IN THE MAILROOM FORTY-SEVEN YEARS AGO. SINCE THAT TIME, THE COMPANY HAS GROWN IN SIZE AND STATURE, AND JOHN CREEDON HAS PLAYED A LARGE ROLE IN SHAPING BOTH ITS PAST AND FUTURE GROWTH. SINCE HIS 1983 APPOINTMENT AS PRESIDENT AND C-E-O, JOHN CREEDON HAS OVERSEEN MET'S
MARVIN O/C JOHN CREEDON GRAPHIC	DIVERSIFICATION INTO THE FINANCIAL SERVICES MARKETPLACE. AND ITS EXPANSION OVERSEAS.
CLAIRE O/C CREEDON GRAPHIC	(CLAIRE) MARVIN, EXCEPT FOR A BRIEF TIME DURING WORLD WAR TWO, JOHN CREEDON HAS BEEN A MET LIFE EMPLOYEE SINCE 1942, WHEN HE STARTED IN THE GROUP DEPART-MENT. AND AT THE CORPORATE DAY PRESENTATIONS IN
HAWAII CONFERENCE GRAPHIC	HAWAII RECENTLY, HE SHARED SOME THOUGHTS ABOUT MET LIFE TODAY, AND TOMORROW.
	SOT CREEDON AT HAWAII CONFERENCE.

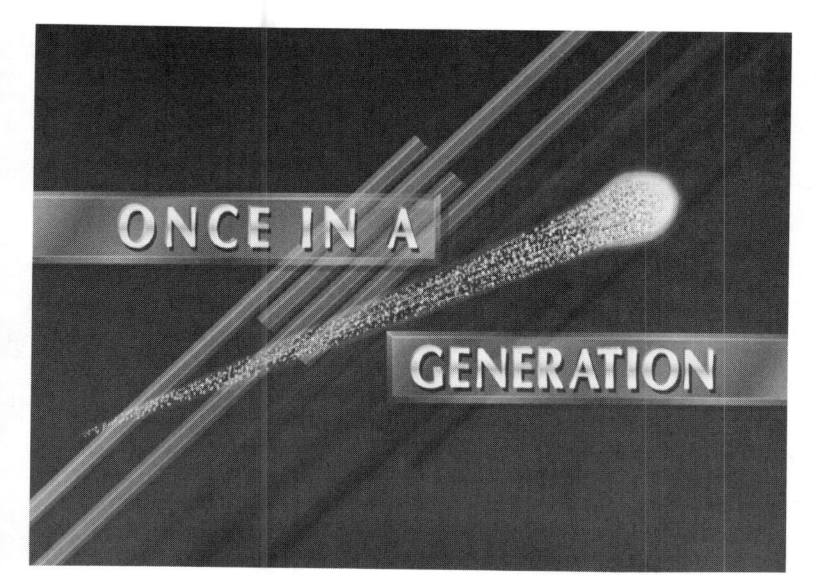

FIGURE 5.14
A multi-image corporate theatre program. (Courtesy of Q1 Productions.)

A

Q1 productions INC

OPENING MODULE
SCREEN 3

STERLING HEALTH

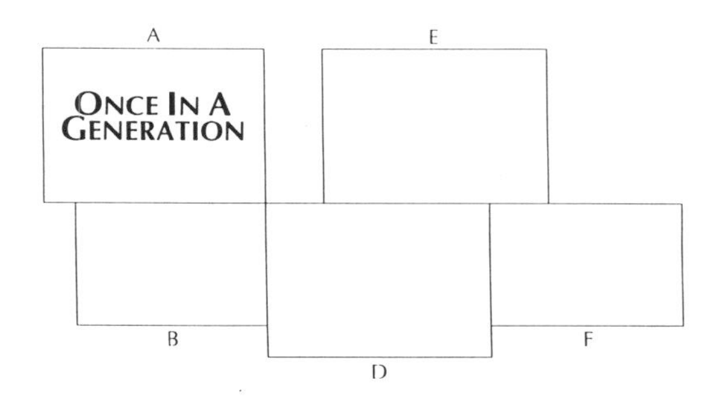

1 FADE IN WHITE TITLE ON BLACK

B

OPENING MODULE
SCREEN 3

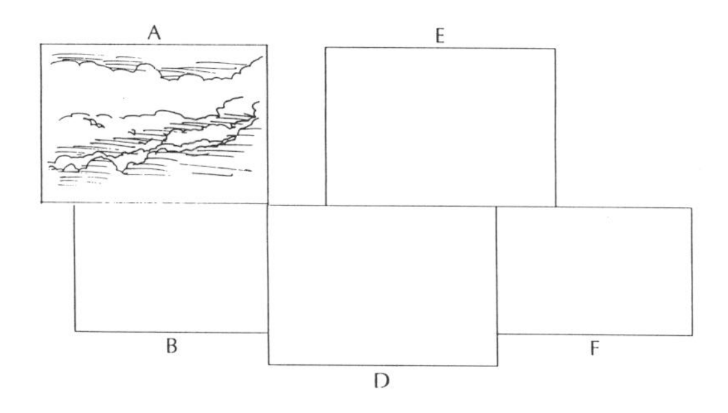

2 DISSOLVE THRU TO OMINOUS DARK STORM CLOUDS

C

(*continued*)

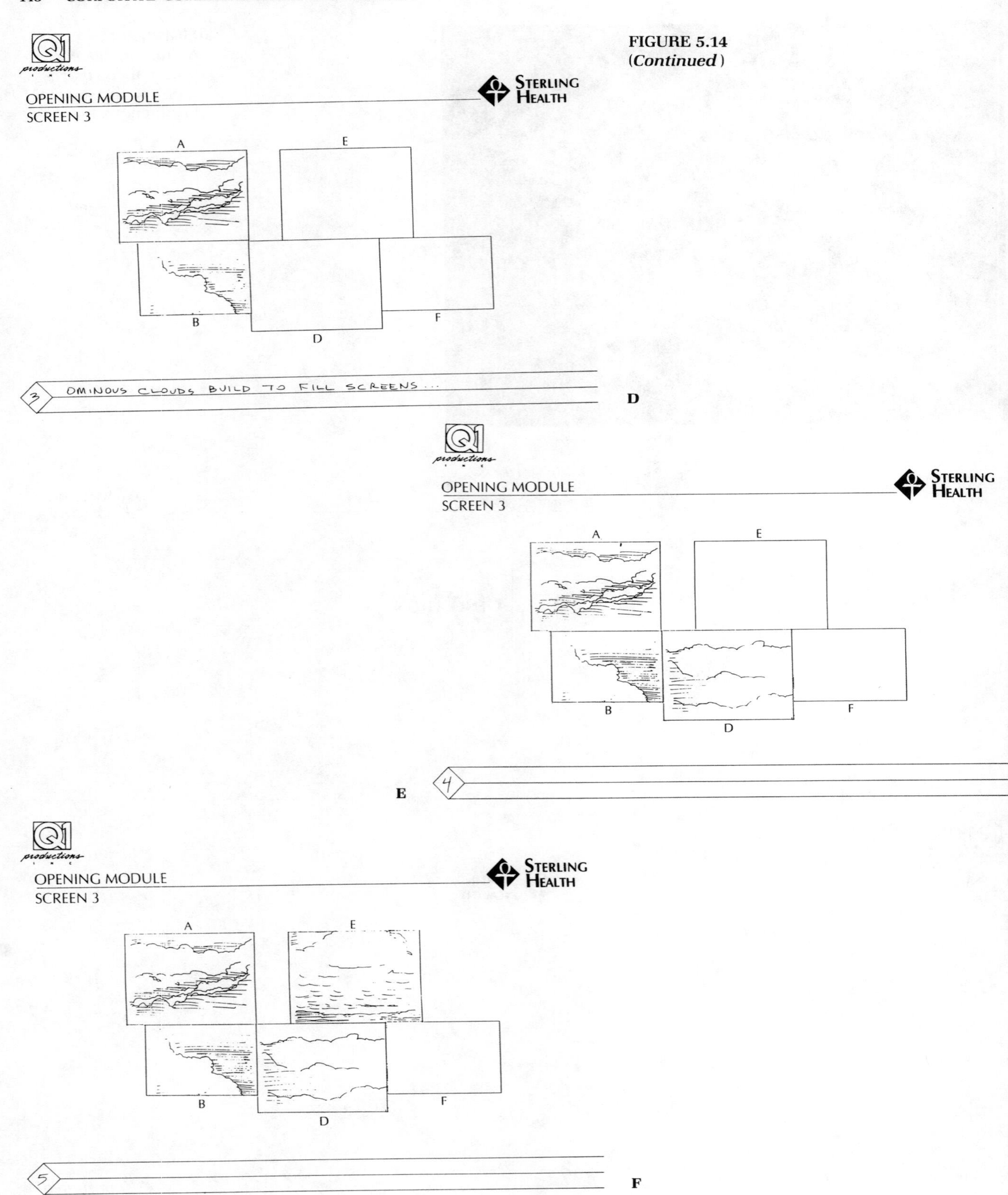

FIGURE 5.14
(*Continued*)

OPENING MODULE
SCREEN 3

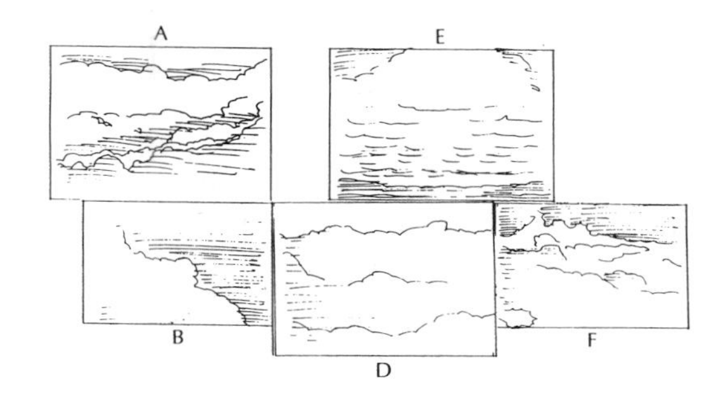

6

G

OPENING MODULE
SCREEN 3

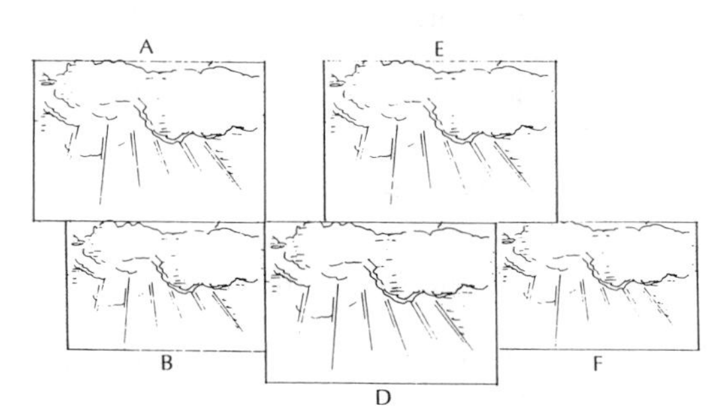

7 DISSOLVE TO A RAY OF LIGHT BREAKING THRU CLOUDS / BLUE SKY. IN REPEAT 5X

H

OPENING MODULE
SCREEN 3

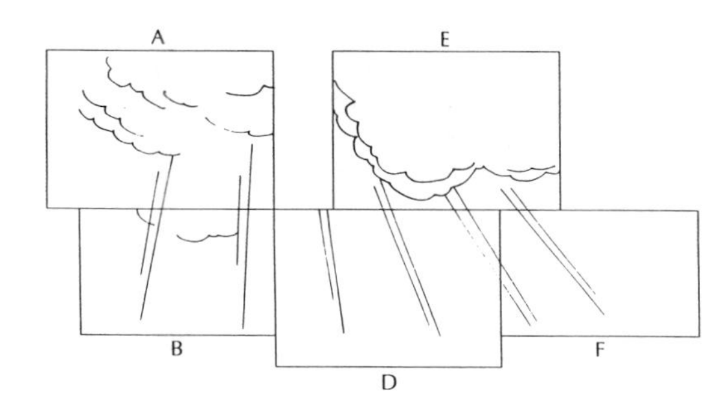

8 PREVIOUS IMAGE NOW SPANS ALL 5 SCREENS

I

(*continued*)

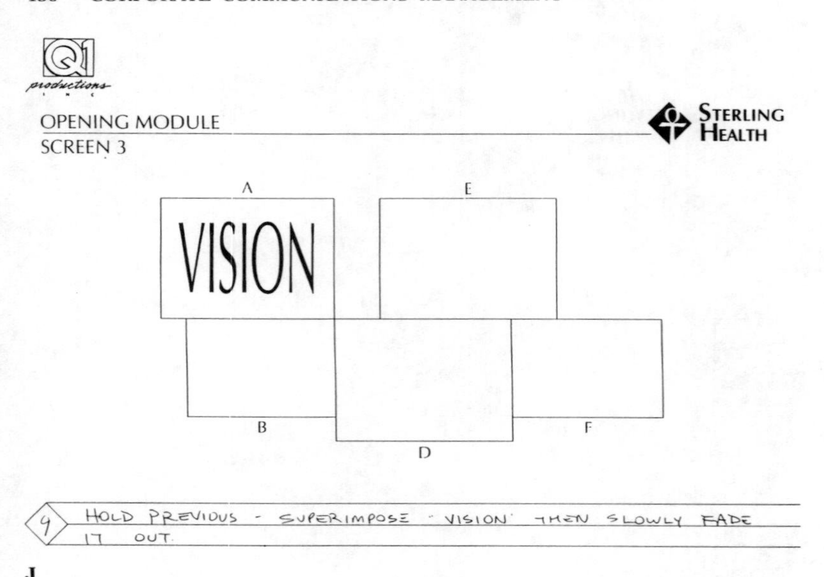

J

FIGURE 5.14 (*Continued*)

calls for precise alignment of quite a bit of expensive equipment, as well as a good measure of know-how on the part of the person playing back such presentations, it is generally limited to one-time, large-group, high-budget applications.

Interactive Multimedia

Multimedia is "a class of computer-driven interactive communication systems that create, store, transmit, and retrieve textual, graphic, and auditory networks of information" (Gayeski, 1992, p. 1). Variously known as computer screen shows, interactive video, computer-based training, interactive kiosks, and computer simulations, this category of media is rapidly emerging as a major format for corporate communication. Not only do these powerful platforms support presentations that reproduce text, graphics, animation, sound, and even live motion video, but these presentations are also interactive in that their sequence and content depend on user input. On top of that, the systems use powerful personal computers, which are already in many homes and workplaces, as their base, eliminating the need for extraneous hardware.

Interactive multimedia got their start in the late 1960s and early 1970s in the form of mainframe computer-based training. Learners could sit down at a terminal, read many screens of information, respond to questions, and immediately be given feedback about the correctness of their responses. As PCs emerged in the early 1980s, this form of self-instructional program migrated to the new platforms, which were cheaper and could display graphics and color. The open architecture of PCs allowed them to be connected to other devices (i.e., videodisc players), so that instead of just reading text off a screen, users could see live motion video that would vary depending on their responses. The 1990s will see even more miniaturization and integration of these systems. Instead of having separate videodisc or videotape players to reproduce audio and video, these signals are being digitized and stored on internal optical discs, such as CD-ROMs, or even on a hard drive.

The potential benefits of multimedia are matched only by the confusion currently surrounding it. Each week, another incompatible hardware system or software tool is announced. It seems that as soon as a staff learns to use a given system, it is "de-released" and another improved, but incompatible, system replaces it. (See Figure 5.15 for a current glossary of terms.) The design considerations, given so many options of *branching* (moving to different sections of a program given different user responses or choices) and display characteristics, can be overwhelming—not to mention learning how to connect the hardware and use the software necessary to generate multimedia programs.

Most applications for interactive multimedia are in the training arena. Since instruction can be personalized and self-paced, questions can be asked of the learner and immediate feedback can be presented. *Response files*—computer files of trainees' actual responses to questions—can be maintained and immediately uplinked to central data bases so that the accomplishments of individuals can be recorded for later personnel and salary decisions. Interactive video programs allow the trainee to vicariously interact with situations that are rare, costly, or dangerous. Such examples are medical education, operation of equipment in hazardous situations, or learning to use complex equipment, such as nuclear submarines. In addition, interactive media can be available when and where the end-users are located.

Authoring tool (also called authoring system or language)—A software program that allows nonprogrammers to create multimedia programs by selecting options from menus and laying out screens.
CBT—Computer-based training: tutorials and simulations consisting of text and possibly graphics, which provide training by means of a mainframe or personal computer. Also known as CAI (computer-assisted instruction), CAL (computer-assisted learning), or CBI (computer-based instruction).
CD-I—Compact disc-interactive: a CD-ROM based self-contained system that is attached to a regular television set and displays interactive stills and motion clips. It is controlled by a simple handset. This standard has been developed and is being promoted primarily by Phillips and Sony.
CD-ROM—Compact disc-read only memory: a small optical disc capable of storing and playing back digital data. In storage capacity it is roughly equivalent to 500 floppy disks.
CDTV—Commodore dynamic total vision: an interactive multimedia player, incorporating a CD-ROM player and an Amiga computer engine, which can display programs on an RGB or television monitor and is controlled by an infrared remote control.
Digital video—Video that is turned into a series of bits and bytes and that can be stored and played back in a computer-readable format.
DVI—Digital video interactive: a standard for compression and decompression of digitized video and audio that is stored on a CD-ROM, digital tape, or large capacity computer hard drive, enabled by specialized cards, which are inserted into a personal computer that also controls the flow of the program. This format was developed by GE and RCA and is now sold through Intel and IBM.
Hypermedia—A classification of software programs that consist of networks of related text, graphics, audio files, and/or video clips through which users navigate using icons or search strategies.
Hypertext—A classification of software programs that consist of networks of related text files through which users navigate using icons or search strategies.
IVD—Interactive videodisc: analog optical discs capable of storing and playing back 54,000 still frames or 30 minutes of motion video and two channels of audio by means of a videodisc player that can be controlled by a remote control or an external computer. This format has been the mainstay of interactive video for the past decade.
MPC—Multimedia personal computer: a trademark for software and hardware systems, which conform to the MPC trade association standards and include support for CD-ROM, digitized audio, and high resolution graphics in a Windows environment.
Quick Time—Apple Computer's multimedia technology that supports the storage and distribution of motion video, stills, and audio over local area networks.
Ultimedia—An IBM trademark for their multimedia hardware systems composed of personal computers with CD-ROM players, and optionally DVI digital video capabilities.
Virtual reality—The display and control of synthetic scenes by means of a computer and peripherals that sense a user's movements via datagloves, helmets, or joysticks. These systems allow users to vicariously interact within "virtual worlds."

FIGURE 5.15 Multimedia terms. [Adapted from Gayeski, D. (1992, May). Making sense of multimedia. *Educational Technology,* 9–13.]

Federal Express maintains interactive video systems in each branch office through which employees can upgrade their knowledge and skills. The major portion of the video is stored on a videodisc, but updates to the training are made by way of new CD-ROMs that arrive every 4 to 6 weeks. Larry McMahon, vice president of human resources, said that they made about 17,000 changes in the curriculum in 1991, in order to keep the material current. They have found that the time taken for training with multimedia has been compressed about 40% as compared to classroom delivery of the same material. The success of the system is part of a large company philosophy—their pay-for-performance system. When a trainee completes a module and gets a passing score on the final assessment, he receives an increase in his next paycheck. Site administrators are trained to maintain the interactive systems so that they're always in operation. (See Figure 5.16.)

Increasingly, multimedia is also being used for marketing and employee information. Amway produced an interactive videodisc on some of its major products for an internal trade show they put on each year. The program quizzes users on important product features and benefits, and allows them to score points and win prizes for

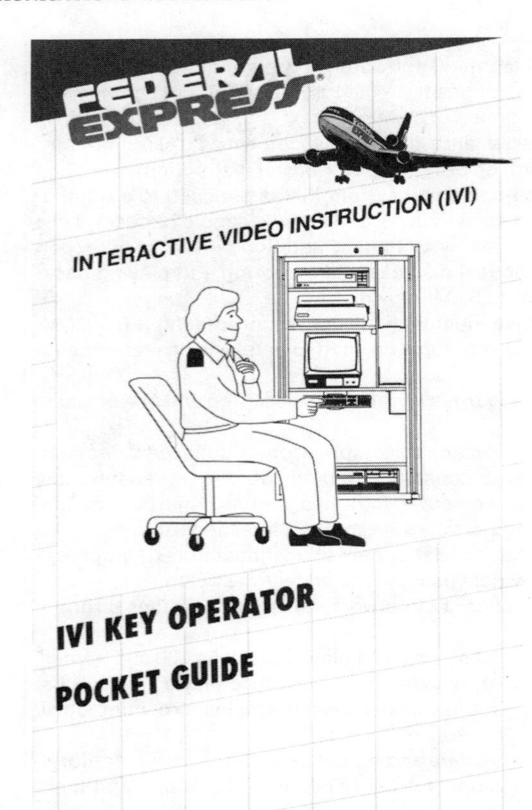

FIGURE 5.16
Once media hardware is placed in the field, there must be a system in place to maintain and update it, such as the use of key operators at each Federal Express site. (Courtesy of Federal Express Corporation.)

FIGURE 5.17
At Amway's 1990 internal trade show, one booth featured interactive video systems, which quizzed users on Amway products, offering the chance to win prizes based on proficiency. (Courtesy of Amway Corporation.)

demonstrating mastery of the content. Other companies also use kiosks in malls or other public areas to advertise their products and services (see Figure 5.17). Another example is the Levi Strauss Jeans Screen kiosks found in their retail stores.

Virtual reality, sometimes called Cyberspace, is the next frontier in multimedia. It is defined as a multisensory experience generated by computers and audio or video input that mimics reality. Participants can control what the computer generates and what they sense by ordinary body movements, such as moving a hand or head. For example, AutoDesk uses 3-D goggles with built-in LCD screens to view and to control a synthesized office. One of the common control mechanisms for virtual reality is the DataGlove, which users wear to manipulate objects—somewhat like PowerGlove for Nintendo entertainment systems (Fritz, 1991). These technologies are useful in the corporate setting to train people to operate devices that don't yet exist, or to put them vicariously in dangerous situations.

Teleconferencing/Business TV

Another form of interactive video is *teleconferencing*—"real-time multi-point communication using voice with still frames or motion video by means of phone lines or satellite broadcasting. This may include multiple channels for two-way phone conversations and transmission of additional computer or graphic data" (Gayeski, 1992, p. 2). Basically, teleconferencing is the use of picturephones, which were demonstrated many decades ago. This technology is now a part of many corporate conference rooms and, as it becomes incorporated into PC systems, it will be found on many office desks.

Teleconferencing is most often used to allow small groups of people at two different locations to collaborate or discuss ideas with the support of live motion video, as well as fax, still-frame graphic, or live computer displays. The addition of the video channel enables people to show things or demonstrate processes to each other and increases social presence as compared to a conventional phone conference. Increasingly, this technology is also used for training and coaching.

Miles, Inc., a vitamin manufacturer, uses a remote viewing system (RVS) to communicate with its print shop, design firm, and advertising agency. The RVS consists of a PC, two video monitors, a tablet with an electric grease pencil, a modem, a voice channel, and a video compression board. Two of these complete systems cost about $30,000 in 1991. The system can capture an image from desktop publishing systems or scanners and transmit it over phone lines in about 90 seconds. Collaborators can look at files and electronically zoom in, pan, point, or mark them up. Once the parties have completed their review and editing, a color copying device, which is similar to a fax machine, sends signed documents for final approval. The system saves time, travel, money, and confusion (Stevens, 1991).

Strictly speaking, *audiographic conferencing* uses conventional phone systems with the addition of another line that displays still frames of graphics or prints out material that a person writes on a special board. *Videoconferencing* includes live, motion video. It is possible to pump video signals over phone lines because of compression schemes and equipment such as the *video codec* (coder/decoder).

When ARCO underwent a major reorganization and hired a new president, employees were understandably anxious. The manager of employee communications at ARCO created PrimeTime, a series of video teleconferences between two ARCO sites (at one of which President Lod Cook was actually present). Volunteers were sought from all employee levels except managerial. These representatives asked candid questions of the president on behalf of their colleagues. This intervention capitalized on Cook's personable style and, in follow-up surveys, employees indicated that PrimeTime increased their confidence in Cook's leadership and in the future of ARCO. Many of the volunteer interviewers prepared extensive reports to share with their peers. Finally, Cook set a model for the style of open communication that he wanted to characterize the new ARCO (Two-way communication, 1989).

Business TV (BTV) uses satellite transmission of video signals from one, or a few, central points, or *uplink sites,* to a number of receive sites. The receive sites can watch full-motion video and respond by phoning or faxing comments or questions, much like a talk show. A number of large, dispersed organizations are using BTV networks to broadcast regular news updates and training courses to their field offices. Other companies who may not have a regular need for this technology can lease uplink equipment and have field staff watch a program in rented hotel conference rooms, where satellite receiving dishes are commonly available.

The use of AppleTV, a seven-site dedicated digital teleconferencing network, results in an annual savings of $10 million in travel and time. FXTV was the brainchild of Federal Express CEO Fred Smith, who wanted a quick way to get information out to his worldwide employees. Their system includes downlinks at each FedEx office. If employees are not there to watch the daily broadcasts of information each morning, the broadcast automatically turns a VCR at each site into the record mode and records the program for later viewing. FedEx fields a Leadership Institute, a monthly broadcast on management training, including such topics as sexual harassment and the creation of a drug-free work environment. The broadcasts are supplemented with a precourse text, live phone-ins, and are followed up with on-line individualized computer-based testing. The results of the tests are automatically fed back to headquarters and become a part of each employee's record and performance assessment.

Performance Support Systems

As work becomes more complex, as procedures change more quickly, and as people in organizations become drowned in a sea of manuals, audiovisual programs, and memos, there is a need to provide condensed information when and where it is needed. While organizations have used checklists, posters, and other printed re-

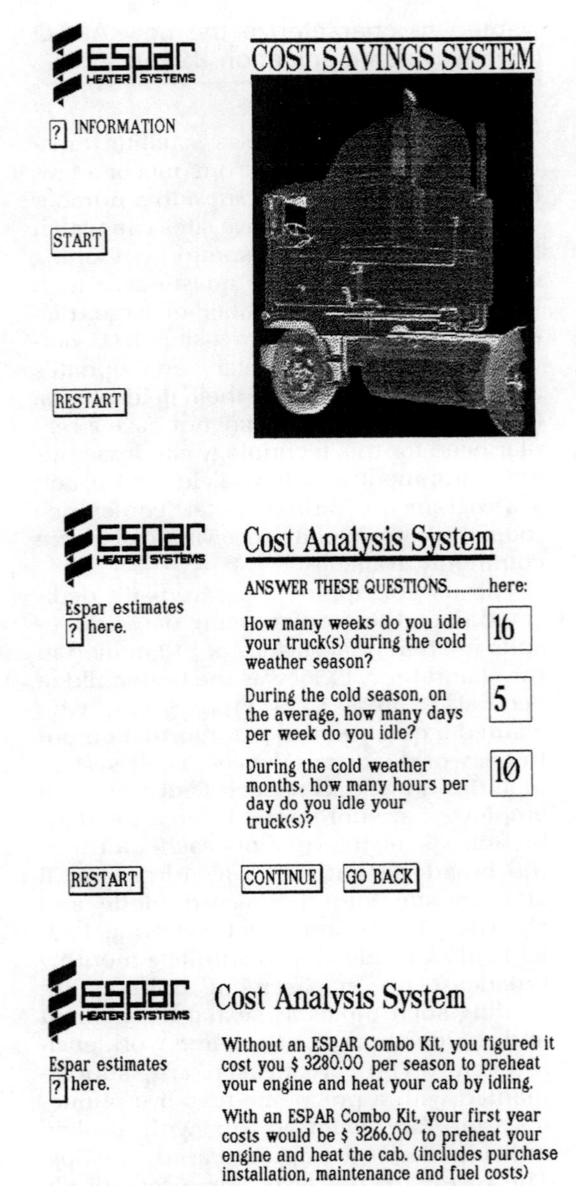

FIGURE 5.18
This computer-based sales support system helps sales representatives and customers quickly figure the benefits of using an Espar heater. (Courtesy of Espar Products and OmniCom Associates.)

minders, more sophisticated and powerful mediated technologies are filling these needs. *Performance support systems* (PSS) are computer-based interactive job aids, which help users perform their jobs by providing general help, brief tutorials, examples, on-line information, or aid in problem solving. For example, consider a bank customer service representative who may need to access rules and regulations for dozens of bank products and apply them to a range of common and not-so-common circumstances. It might be easy to remember how to establish a certificate of deposit account, but what happens when a CD is jointly held and one of the holders dies? The representative might try to find the pertinent regulation in the binders full of regulations in the back room; that is, if those policies manuals have been regularly and carefully updated. Or, she could consult her on-line PSS. Our firm, in fact, produced such a system, called TermBank, in the early 1980s for Marine Midland Bank.

Other performance support systems might include a data base of contacts, work samples, models, or a set of forms. For example, OmniCom developed a Supervisors' Support System for New York State Electric and Gas (NYSEG), starting with procedures for handling first-step union grievances. When faced with a potential grievance situation, a manager can call up this program on her PC, click on the menu topic that relates to the question, see examples of correctly completed paperwork, and actually fill out an electronic form using a built-in word processor. It is easy and inexpensive for NYSEG to update this system and for managers to refer to it.

Not all performance support systems are used for policies and procedures; some are used to support marketing and sales. For example, our company produced a sales analysis system for a client who markets diesel heaters for use in trucks. The system can be used in a sales call or at trade shows, and asks potential customers some questions about their truck routes, prices for diesel fuel, and idling habits. Based on this, the system calculates the cost of idling in terms of actual fuel usage and in terms of engine wear, provides the cost of an Espar heater, and determines the payback period. All of this can be printed out for a customer's subsequent review. The system shows graphics of typical installations and helps to calculate what-if situations, making it easy to create a business case for buying heaters. (See Figure 5.18.)

Another class of computer-based tools are called *expert systems.* These interactive programs solve problems or make recommendations based on users' responses to a set of questions, a knowledge base, and a system of if-then rules. Based on documenting and canning the declarative and procedural knowledge of experts, the goal of these systems is to allow novices to make decisions as well as the experts. Rather than teaching the user how to do something, the expert system solves the problem for them.

For example, American Express' Authorizer's Assistant pulls together the experience of their best authorizers and makes it available via computer at each workstation. It gathers information on cardholders from 14 different data bases, displays it on two screens, and makes routine decisions while making it easier for humans to make the more difficult ones. The result is a 45% to 60% annual return on investment and reduced poor decisions, which, in the past, led to bad debts (Harris, 1989). DuPont's 600 expert systems are reportedly saving more than $75 million a year; Digital Equipment reports that their expert systems, which configure computer systems for customers, save them $200 million annually (Alter, 1992).

Although expert systems are usually viewed as internal performance support tools, they can also become powerful marketing devices. DuPont provides an expert system called the Packaging Advisor to potential customers. It helps them evaluate alternative packaging materials, determine the quantities they need to meet their requirements, and estimates cost.

The Packaging Advisor runs on a personal computer and replaces traditional marketing communication techniques to inform both customers and field sales staff about new and existing products. It was developed through rapid prototyping, with an initial version shown to users after three months, and numerous refinements made over the next five months before deployment. It takes into account important features of the customer's products such as length of shelf life, temperatures it will encounter, sterilization methods used, oxygen tolerance during shelf life, humidity inside and outside the package, and pertinent Federal Drug Administration (FDA) regulations. Management believes that 30% of DuPont's resin sales in the following year were attributable to accounts where Packaging Advisor helped open the door. (Alter, 1992, p. 547)

Media Tools: A Summary

Although it's easy to become excited about the power of media to improve corporate communication, it's essential to remember that technology should be subordinated to people and aligned with business goals. Organizations should use information systems to "remove walls, not wire them, so it is easier for people in every job to make decisions and connections. Consider the information flows at a prosperous Dutch retailer with U.S. operations: any in-store clerk can see scanner data instantly; a video with best-practice cases arrives at every store weekly; data from supermarket scanners in Connecticut goes right to the president's home computer screen in the Netherlands" (Kanter, 1991, p. 9).

Even though many case studies of effective applications of media are extant, we cannot be too optimistic about the pace at which changes due to technology may come about.

The walls set up in organizations—between managers and workers, between functions, between work and home, between company and society—constrain change, and managers are slow to dismantle them. Only when the walls begin to crumble do companies feel the full potential of technology. In my own work helping companies restructure their operations, I have found that it can take a decade or more for managers to create large enough openings in the walls for technology to change the content of work radically. (Jaikumar, 1991, p. 100)

MICRO-LEVEL INTERVENTIONS

We've reviewed the strategic application of rules and tools to create a corporate communication infrastructure. Once these systems are in place, it is possible to develop individual interventions. Based on an analysis of the problem, a definition of expectations, and a description of the setting in which an intervention might be applied, a business plan is developed, including choice of media and strategies.

Criteria for Choosing Media

Criteria for choosing media include

- *feedback potential*—Can the medium support two-way communication (i.e., e-mail) or is it essentially a one-way medium (i.e., video or print)?
- *synchronous versus asynchronous*—If the medium does support two-way communication, can all parties contribute at the same time (i.e., in a telephone conference), or do they take turns (i.e., leaving messages on voice mail)?
- *bandwidth*—What is the capacity of the medium to support various modalities, such as text, audio, still pictures, or moving video?
- *confidentiality*—Are messages private, or is the medium public or accessible by system operators or managers?
- *encoding/decoding ease*—How difficult is it to produce for the medium, and how difficult is it to use the medium (e.g., it's rather difficult to produce a good video program, but it's easy to use one)?
- *cost*—How much does the basic support technology cost? How much does it cost to encode (produce) messages? How much does it cost to decode (receive) messages, especially in comparison to other alternatives?
- *social presence*—Does the medium communicate one's personality or warmth (e-mail is low in social presence while video conferencing is high)?
- *indexibility/random access*—Do users have access to selected parts of a message (i.e., searching out a particular face in a hypertext data base), or is the medium essentially linear (i.e., a videotape)?
- *individual or broadcast*—Is the technology essentially a mass or large-scale medium (i.e., satellite business TV) or is it sent individually (i.e., memos)?
- *speed*—How fast can messages be prepared and updated for this medium, and how quickly can they be available to the intended audience (e.g., bulletins on a computer e-mail system can be easily prepared and edited and immediately made available, while videotapes generally take longer to produce and distribute)?
- *gatekeepers*—Are they necessary or possible? Is there a requirement or a possibility for designated individuals to act as managers or screeners for messages on the system (i.e., electronic forums have system operators, called SysOps, who have the right to screen inappropriate messages) or is it essentially a free and open system (i.e., telephones)?
- *critical mass*—Is there a minimum number of people needed to use a system effectively (i.e., BTV or e-mail, which would not be practical without a large number of people with the equipment and skills to use the system)?

Although scores of research and dissertation projects have endeavored to demonstrate that one medium was more effective than another, this is a dangerous line to pursue. As educational researcher Richard Clark (1983) found when he examined this body of research, most of the differences could not be clearly shown to be due to the medium, nor could the results be generalized. Reasons for this include

1. Certain media are better for particular kinds of messages, although they are not more effective per se. For example, interactive video might be a more efficient and effective means for teaching machinists how to set up a new lathe, but it may not be any better than a simple manual or oral presentation at explaining a new sales strategy.
2. It is not the medium but, rather, what kinds of techniques a medium can present, or how a program is designed, that makes the difference. For example, a video program dramatizing and demonstrating an effective way to conduct a performance appraisal is a lot different than a video program that attempts to teach managers how to do this by way of a lecture or a discussion.
3. In many of the studies that show one medium to be superior to another, the difference was really due to the use of sophisticated applications of communication principles by professional de-

signers and producers, and a difference in the amount of time and effort that went into preparing the two programs that were compared. For example, many studies compare a traditional classroom presentation with a video or interactive video program and find the mediated programs to be as effective, or more effective, than the conventional approach. However, the classroom presentations were generally developed without the assistance of an instructional designer (or any other professional besides the lecturer), while the media program was generally the result of many months of work by professionals who were doing their best to show off the new medium. If the classroom presentation were given the same amount of attention and support, perhaps it would have been found to be superior.

4. The novelty effect of new media was also an uncontrolled variable. Users of a new medium pay more attention or are attracted more to something new. As the Hawthorne Studies demonstrated years ago, enhanced performance can be caused by almost any change in the environment, or just additional attention being paid to the subjects. (See Chapter 2 for a discussion of the Hawthorne effect.)

5. Sentimental preferences may lead designers to assume that certain channels, such as opinion leaders or field agents, are the only ones who can really effect persuasion and change. Therefore, other approaches are not tried. Some scholars, such as Hornik (1989), challenge this conventional wisdom and encourage communication professionals to try new methods.

Certainly, a given technology may have a particular display mechanism or bandwidth that makes it preferable for a particular kind of task. For instance, it is quite difficult to explain to someone the differences between the Chippendale and Queen Anne styles of chairs using an audiotape or print alone; a sketch or picture would do the job much more easily. Similarly, video is probably not the best medium to use to present the concepts of philosophical analysis. Convenience may play a major part in media selection. If your organization has videocassette players in each branch, it's much more feasible to field a video news program than if that equipment were not in place. In most cases, professional communicators use a mix of direct and mediated communication, even within individual interventions. For example, a self-study course we developed to teach supervisors how to conduct performance appraisals included a videotape with interviews, lecture segments, and dramatized scenarios and a workbook with rules and regulations, good and poor examples of appraisals, and work sheets that users filled out while watching the scenarios. An orientation system for a major bank included an overview video, an interactive information system on benefits options, and a computer-based employee data form.

Direct, human communication may be preferable in situations that are ill-defined or equivocal. Daft and Lengel (1986) make the distinction between uncertain situations (in which there is an unknown answer to a known question) and equivocal situations (in which even the questions are not known or adequately framed). They found that "managers reduce equivocality by defining or creating an answer rather than by learning the answer from the collection of additional data" (Daft & Lengel, 1986, p. 556). In ill-defined situations, managers tend to prefer to communicate through meetings, contacts, and professional associations. In more certain and analyzable situations, they prefer management information systems, reports, and written surveys. This points to the importance of (or at least perceived preference for) informal and direct personal communication in management communication, especially in situations of equivocality, which are common in contemporary organizations.

These preferences can be better understood when we recognize that media have "varying capacities for resolving ambiguity, bringing multiple interpretations together, and facilitating understanding" (Trevino, Daft, & Lengel, 1990, p. 75). They can be characterized as "rich or lean based on their capacity to facilitate shared meaning. Rich media have the highest capacity to facilitate shared meaning. Lean media have the lowest capacity" (Trevino et al., 1990, p. 75). This richness, which is closely related to bandwidth and social presence, is based on

- availability of instant feedback
- the ability to transmit multiple cues (i.e., body language, tone of voice)
- the use of natural language rather than numbers
- the personal focus, or extent to which the sender of the message is salient

Most scholars rate face-to-face communication as the richest channel, with video, e-mail, and memos falling below it. Managers tend to use rich channels for sorting out problems, and leaner media for unambiguous and routine tasks. Rich media are needed to help establish shared meanings and a common language. Once these symbol systems are in place, other kinds of media can be just as effective and more efficient. In fact, one study of managers found that a person's ability to select appropriate media for equivocal and unequivocal tasks and a person's rating in terms of overall managerial performance were highly correlated (Daft, Lengel, & Trevino, 1987).

Although it can be tempting to choose media on the basis of its efficiency, it is important to recognize the message that is sent by the choice of media itself. For example, although some instructions may be able to be communicated effectively by a video program, a manager makes a symbolic statement when he instead decides to visit each of 10 plants and meet with work groups individually. Likewise, the choice of an elaborate corporate theatre production sends the clear message that the company has money to spend and is doing so on behalf of the particular audience and event. An executive's use of e-mail may set the stage, implying that all managers should adopt this new technology and that traditional ways of communication and supervision may be changing.

Mixing and Matching: Case Studies

Most contemporary organizations apply a range of communication techniques and technologies (See Figure 5.19.). Allied-Signal uses a media mix for employee communications, including two video programs, two regular publications, information centers with posters and video monitors, an electronic bulletin board, and a phone system that allows employees to call in and leave a comment or question. Similarly, SmithKline produces weekly and monthly video news programs, a daily phone news service with stock quotes, quizzes, and new production information. They average 600 calls a day on that service (Carlberg, 1991).

Wal-Mart, a leading discount retail chain, uses a satellite communication system that provides real-time data, voice, and video links to over 1500 stores. Besides transmitting sales data, it is used as a private television network for employee training, updates, and managerial communication. For example, a buyer can demonstrate how particular new merchandise should be displayed and what items are selling well (Alter, 1992).

Case 1: Cancer Alert. In the mid-1980s, Lawrence Livermore National Laboratory discovered that its employees had an unusually high incidence of malignant skin cancer. While they investigated the possible causes for this situation, the lab wanted to ensure that employees were taking every possible means to protect themselves. The communication objective was to emphasize urgency without creating panic; therefore, the content of their messages emphasized the high rate of survival when melanoma is detected and treated early, and the need for employees to take prompt action to examine themselves. The campaign, "Spot Check '84," consisted of three phases. One phase was composed of meetings in which management and employees learned about the program. At these meetings, all managers received an information packet and were instructed to hold meetings with their work groups and encourage them to respond to the program. Second, print materials that employees could read in the privacy and supportive environment of their own homes were distributed. These materials contained information and instructions about how employees should examine themselves for suspicious moles. Also provided were

- articles in the laboratory publications
- a telephone hotline

- local media coverage and press conferences
- a poster series
- a lunch-hour forum featuring medical professionals
- a video program on how to conduct self-examinations

Follow-up surveys found that 53% of the lab employees sent response forms to the medical department and that an additional 20% had examined themselves but had not sent in a reply to the lab. More than 95% of the lab employees responding to a survey said they knew of the melanoma situation and the Spot Check program. The hotline had also been used extensively (Tackling the difficult subject, 1989).

Case 2: Teaching Reps to Handle Difficult Customers. NYSEG employs hundreds of customer representatives; these employees answer questions about bills, accept in-person payments, and also make payment agreements. A payment agreement is arranged when customers are not able to meet their financial obligations to the utility company. Under law, the utility may not just turn off a customer's electricity and gas if their account is in arrears, but must attempt to work with the customer to see if terms for a schedule of payments can be reached.

Dealing with payment agreements is one of a representative's most difficult jobs. They are often dealing with customers with a wide range of emotions; anger, frustration, and confusion. They may be elderly and having difficulty making ends meet; they may be middle-class businesspeople who have found themselves out of work or bankrupt; they may be college students who have never had to pay bills before and are having difficulty sorting out which roommate pays for what. NYSEG wanted to give these representatives training, not only in the technical aspects of concluding a payment agreement, but also in dealing with the interpersonal aspects of the job.

Although there are many generic customer relations programs, NYSEG wanted something that was targeted to their particular audience. In our analysis of the situation, we found that reps needed more than simple rules and pat answers. They needed to have the support of their colleagues and to understand that there were many good approaches to handling payment agreements. The intervention we created is a training program led by an experienced customer representative. Once the procedures are explained (i.e., how to calculate a minimum monthly payment, how to help customers plan a budget, etc.), a videotape, custom-produced for the utility, is shown.

In order to give reps a taste of what they may run into, the videotape consists of short dramatized scenarios of difficult customers. After each segment, the question, "What would YOU do?," appears on the screen, along with directions to pause the video. The class then talks about how they might (or actually do) deal with such situations. After the discussion, the video is resumed. The next segment consists of interviews with actual NYSEG representatives who offer their advice. It's often funny, always candid, and rings true to the participants. Newly hired reps feel an instant comradeship with their video mentors, and the training facilitators find it easy to lead discussions—even if they're not experienced trainers. A number of effective practices in design and production that had been developed by branch offices were also documented and shared.

Case 3: A Campaign to Promote School Bus Safety. The following analysis and proposal, *Reducing Bus-Stop-Law Violations Campaign*, is one example of a public awareness campaign that uses a variety of media. [Reprinted with permission. Courtesy of Robert Sherwin, Portland (OR) Public Schools.]

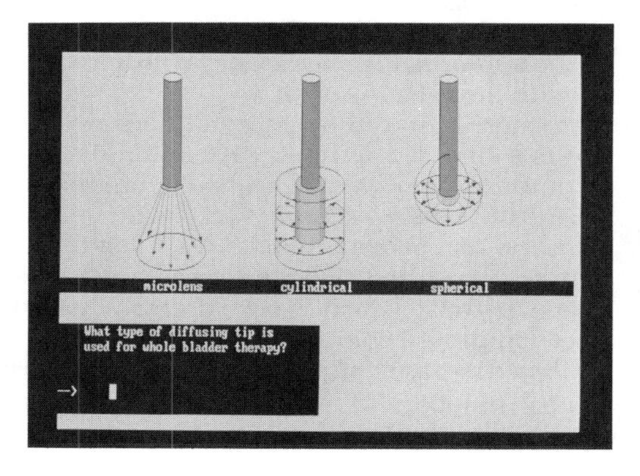

FIGURE 5.19 CBT programs, such as this one, are used to supplement manuals to train pharmaceutical sales representatives. (Courtesy of OmniCom Associates.)

REDUCING BUS-STOP-LAW VIOLATIONS CAMPAIGN

prepared by
Robert M. Sherwin
Public Information and Communication Dept.
Portland (OR) Public Schools

Problem

The number of Portland motorists violating the school-bus-stop safety law nearly doubled between the 1986 and 1988 school years.

The dramatic increase in drivers ignoring that law poses a serious threat to the physical safety of Portland's 14,000 students who ride 400 school buses daily.

The blatant disregard for the children's safety alarms the Portland School District's Student Transportation Advisory Committee and Student Transportation Department's administration.

Description

Portland School District and contract school-bus drivers filed 787 *written* observation reports on motorists ignoring flashing red lights of stopped school buses' through the third quarter of the 1988–89 school year. By comparison, at the same point in past years, there were 422 violations in 1986–87 and 420 in 1987–88.

There are at least two to three times that number of violations each year, which are not reported because drivers are unable to meet stringent reporting requirements, according to Student Transportation Department officials.

To report a violation to police, the school-bus driver must be able to identify the type of vehicle (van, sedan), license number and state, violator's direction of travel, weather conditions and more. While noting that information, the driver also has to maintain control of his vehicle and student passengers.

Reported violations, however, are followed up by the appropriate police agency with a warning letter to the vehicle's owner.

Violators observed by police officers are subject to a citation carrying a $110 minimum fine; 16 of those cases were successfully prosecuted during the 1988–89 school year.

Streets or neighborhoods with high incidences of reported violations receive increased police patrols, which have successfully produced improved compliance.

Certain sectors of the school district have greater numbers of violators and some sectors have had more dramatic increases in violations.

AREA	*1986/87*	*1987/88*
N.	39	35
N.E.	149	109
S.E.	95	144
N.W.	15	14
S.W.	124	118

AREA	*1988/1989*	*PERCENT GROWTH**
N.	109	179%
N.E.	324	117%
S.E.	216	127%
N.W.	12	−20%
S.W.	126	16%

*Between 1986/87 and 1988/89 school years for the *first three quarters* of each year only.

Goal

Implement a public-education campaign to decrease the number of reported and observed violators of the school-bus-stop law.

Objectives

1. Educate automobile drivers about the law through a multimedia public-education campaign.
2. Recruit the public-affairs assistance of at least one Portland radio station and its traffic reporters to assist in educating drivers about the importance of the law.
3. Secure a corporate sponsor for a paid advertising and promotional campaign.

Strategy

The campaign focuses on educating motorists about the law when they're driving—not when they are at home or work, watching television or jogging. Drivers are the ones who need to change and the message should be delivered while they are behind the wheel.

Three approaches are recommended for carrying the message to drivers:

1. Paid advertising campaign.
2. A special public-affairs appeal to radio traffic reporters.
3. Traditional news-media relations.

Four primary ways can deliver messages to drivers: radio, billboards, exterior transit advertising and bumper stickers/auto-window signs.

Radio public-service announcements are inexpensive but don't get "air play" at specific times. Consequently, a corporate-sponsored budget is recommended to purchase drive-time radio advertisements, Tri-Met bus advertising, billboard advertising and some form of "back-of-auto signs."

A public-affairs approach paying special attention to radio traffic reporters is recommended—commuters listen closely to their reports, they *are* on-air personalities and *can* inform motorists about the law.

The traditional media-relations approach has the shortest duration and probably would least affect motorist habits. The advantage to having news conferences, media events and numerous news releases is that they formally launch the campaign, are inexpensive and introduce drivers to the topic and tell them "this is important."

Message

The campaign's objective is simple: following and oncoming motorists will stop when school buses display red flashing lights. The campaign message should be equally simple—

Join Us
STOP
When Red Lights Flash.

Sponsor(s) listed here

Complexities of the law should be downplayed. Divided highways, flashing yellow lights, etc.—all part of the law—seem to confuse many motorists. When people are confused about something, they generally just do it their own way—right or wrong.

1. Paid Advertising Campaign. A long-term (from several months to a year) public-education campaign is recommended using paid advertising and promotional materials targeting drivers.

Most of the budget would pay for drive-time radio advertisements, with the balance to be used for billboards, Tri-Met outside bus cards and "window signs."

The campaign requires a sponsor with a direct link to "people on wheels." These companies should be approached:

- Tri-Met
- Broadway or Radio Cab companies
- Consolidated Freightways
- United Parcel Service
- U.S. Postal Service
- Metropolitan Disposal or Waste Management (garbage haulers)
- Hanna (car washes)
- Freightliner (trucks)
- Les Schwab (tires)
- Texaco Mini-Mart gas stations
- Ron Tonkin (auto sales)
- U-Haul, Ryder, Budget or Hertz (rental trucks)
- State Farm (or other large auto insurers)
- Lile, Bekins or other moving companies
- Auto after-market retailer (GI Joe's or Fred Meyer's)

Estimated costs for the recommended advertising campaign are:

1. Radio: Morning and afternoon drive time 60-second spots daily, once-a-week on each station for four weeks—8/28–9/22

KEX	$250 per week
KGW	$130
K103	$310
KINK	$297
KXL	$275

Total $1,262 per week 4 weeks = $5,048

2. Exterior Bus Advertising: 65 buses (50% of fleet), 2-color "tail-light" advertisements.

Production	$ 850
Full Rent	8,050
NonProfit Rent	0
Placement Labor	1,625
Total Full Rent	$8,900
Total NonProfit	2,475

3. Billboards: 10, 2-color line-art boards for one month

Production	$ 650
Space Rent	4,850
TOTAL	$5,500

4. Auto Window Sign, Static-Cling Stickers: 2-color, 5″ × 5″ @ $.645 each/1,000
5,000 copies = $3,225
TOTAL BUDGET:
at Full Transit-Ad Rental Rate $22,673
at NonProfit Transit Ad Rate 16,248

2. Radio Public Affairs—Traffic Reporters. Enlisting the support of radio traffic reporters would give higher visibility to the campaign and a boost to the paid-advertising segment.

A three-step process is recommended for enlisting those highly visible traffic spokespersons:

1. Invite all radio traffic reporters and their station's morning and afternoon DJs to drive a school bus.

   PPS would host a 9:30 a.m. "donut & coffee" session, providing guests with an overview of the school-bus-stop law and the problems that Portland Public Schools and other metro-area school districts have with the law's observance.

   To make the session more fun, the PPS Student Transportation Dept. would set up a course of pylons, normally used for training school bus drivers, then challenge the radio personalities to successfully navigate the course.

2. Every two weeks send all orientation-session attendees some new fact about the law, about school buses or other "exclusive" information to stimulate their on-air talk.
3. Attempt to involve on-air radio personnel in distributing promotional materials—bumper stickers/window signs, posters, etc.—in support of the campaign.
4. Seek creative ways bus-riding students could communicate to radio traffic reporters. The objective would be to tie the safety of those young bus riders with broadcasters' comments supporting the stop law.

3. Neighborhood Public Affairs. Portland School District's Student Transportation Department will organize a speakers' bureau to promote the campaign message.

The speakers' bureau will target civic, community, business, neighborhood and other groups residing or meeting in neighborhoods with the highest incidences of violations (North, Northeast and Southeast Portland).

The speakers' bureau would be composed of members of the school district's Student Transportation Advisory Committee and key transportation department administrators. Executives from sponsoring companies would be encouraged to join the speakers' bureau.

The Public Information Dept. will assist in speech preparation, any necessary speech training, developing audiovisuals and handouts and providing other technical assistance, as needed.

4. News Media Relations. Conduct a news conference when school opens to launch the advertising and public-service campaign.

The news conference will announce the "shocking" doubling of stop-law violations. A school bus at the news conference site will offer news media personnel a visual to tell their readers/listeners/viewers how the law works and the penalties for disobeying the law.

The news conference will announce the public-education campaign and thank the corporate sponsor(s) for making it possible.

Special news releases will be produced for neighborhood newspapers serving areas with rapidly increasing numbers of violations—North (*St. John's Review*), Northeast (*The Skanner, Observer* and *Hollywood Star*) and Southeast (*Sellwood Bee* and *Woodstock News*).

In addition, versions of those news releases would be mailed to editors of corporate-employee, association-membership and other newsletters with special readerships.

Evaluation

Compare quarterly the number of reported and observed violations with those of the past three years.

Use media-relations strategies to announce trends with specific emphasis, when possible, on neighborhood newspapers.

We've moved now from the concepts and strategies used to analyze communication problems, to designing interventions, and to developing an infrastructure of rules and tools for communication to specific examples of corporate communication interventions. However, success within interventions has two parts: 1) how well the intervention is designed and produced, and 2) how well the project is managed. Let's go on to consider the management of communication functions and projects.

REFERENCES/ SUGGESTED READINGS

Albrecht, K., & Zemke, R. (1990). *Service America!* New York: Warner Books.

Alter, S. (1992). *Information systems: A management perspective.* Reading, MA: Addison-Wesley.

Bassett, G.A. (1968). *The new face of communication.* New York: American Management Association.

Be ready. (1991, August). An advertisement for L.C. Williams and Associates. *Communication World,* 11.

Bowen, W. (1989). The puny payoff from office computers. In T. Forester (Ed.), *Computers in the human context* (pp. 267–271). Cambridge, MA: MIT Press.

Carlberg, S. (1991). *Corporate video survival.* White Plains, NY: Knowledge Industry Publications.

Clampitt, P. (1992, March). The pros of voice mail. *Communication World,* 13–14.

Clark, R.E. (1983). Reconsidering research on learning from media. *Review of Educational Research, 53*(4), 445–459.

Communicating on-screen . . . and on-line. (1989). *Case studies in organizational communication: 2* (pp. 39–44). Elmhurst, IL: Council of Communication Management.

Daft, R.L., & Lengel, R.H. (1986). Organizational information requirements, media richness and structural design. *Management Science, 32*(5), 554–571.

Daft, R.L., Lengel, R.H., & Trevino, L.K. (1987). Message equivocality, media selection, and manager performance: Implications for information systems. *MIS Quarterly, 11*, 355–366.

D'Aprix, R.M. (1981). Communicators in contemporary organizations. In C. Reuss & D. Silvis (Eds.), *Inside organizational communication* (pp. 17–32). New York: Longman.

Eddy, C. (1992, May). The future of multimedia: Bridging to virtual worlds. *Educational Technology*, 54–60.

Foltz, R.G. (1981). Communication in contemporary organizations. In C. Reuss & D. Silvis (Eds.), *Inside organizational communication* (pp. 5–16). New York: Longman.

Fritz, M. (1991, February). The world of virtual reality. *Training*, 45–50.

Gayeski, D. (1992, May). Making sense of multimedia. *Educational Technology*, 9–13.

Goldhaber, G. M. (1983). *Organization communication*, 3rd ed. Dubuque, IA: Wm. C. Brown.

Harris, C.L., et al. (1989). Office automation: Making it pay off. In T. Forester (Ed.), *Computers in the human context* (pp. 367–376). Cambridge, MA: MIT Press.

Haynes, G. (1981). Organizing and budgeting techniques. In C. Reuss & D. Silvis (Eds.), *Inside organizational communication* (pp. 62–76). New York: Longman.

Hornik, R.C. (1989). Channel effectiveness in development communication programs. In R. Rice & C. Atkin (Eds.), *Public communication campaigns*, 2nd ed. (pp. 309–330). Newbury Park, CA: Sage.

Jaikumar, J. (1991, September/October). The boundaries of business: The impact of technology. *Harvard Business Review*, 100–101.

Jensen, A., & Chilberg, J. (1991). *Small group communication: Theory and practice*. Belmont, CA: Wadsworth.

Johansen, P. (1990, May/June). How the communication profession evolved from the house organ. *Communication World*, 22–26.

Kanter, R.M. (1991, September/October). Service quality: You get what you pay for. *Harvard Business Review*, 8–9.

Keeping calm in the crisis. (1989). *Case studies in organizational communication: 2* (pp. 30–34). Elmhurst, IL: Council of Communication Management.

Knight, C.F. (1992, January/February). Emerson Electric: Consistent profits, consistently. *Harvard Business Review*, 57–70.

McGoon, C. (1992, March). Putting the employee newsletter on-line. *Communication World*, 16–18.

Meltzer, M. (1981). *Information: The ultimate management resource*. New York: AMACOM.

Parkinson, C.N., & Rowe, N. (1977). *Communicate: Parkinsons' formula for business survival*. London: Prentice Hall International.

Putting a communication program to the test. (1989). *Case studies in organizational communication: 2* (pp. 9–13). Elmhurst, IL: Council of Communication Management.

Rice, R. (1984). *The new media: Communication, research, and technology*. Beverly Hills, CA: Sage.

Ruch, R.S., & Goodman, R. (1983). *Image at the top: Crisis and renaissance in corporate leadership*. New York: The Free Press.

Schrage, M. (1990). *Shared minds: The new technologies of collaboration*. New York: Random House.

Sobkowski, A. (1992, May/June). Damage control when a crisis hits. *Executive Female*, 67–68.

Stevens, L. (1991, December). From a distance: New tools for remote publishing. *Publish*, 94–100.

Tackling the difficult subject of cancer. (1989). *Case studies in organizational communication: 2* (pp. 50–55). Elmhurst, IL: Council of Communication Management.

Tapping into employees' views. (1989). *Case studies in organizational communication: 2* (pp. 45–49). Elmhurst, IL: Council of Communication Management.

Tichy, N. (1983). *Managing strategic change*. New York: John Wiley & Sons.

Trevino, L., Daft, R., & Lengel, R. (1990). Understanding managers' media choices: A symbolic interactionist perspective. In J. Fulk & C. Steinfeld (Eds.), *Organizations and communication technology* (pp. 71–94). Newbury Park, CA: Sage.

Two-way communication on 'PrimeTime.' (1989). *Case studies in organizational communication: 2* (pp. 19–23). Elmhurst, IL: Council of Communication Management.

White, E. (1986). Interpersonal bias in television and interactive media. In G. Gumpert & R. Cathcart (Eds.), *Inter/media,* 3rd ed (pp. 115–124). New York: Oxford University Press.

6 Managing Communication Resources and Projects

We have reviewed the nature of communication within organizations and looked at the many opportunities for persons skilled in developing communication strategies and interventions to enhance organizational functioning. Now that we know what can be done, we need to focus on how it gets done—by whom, where, and with what resources.

INTEGRATING PROFESSIONAL COMMUNICATION

In looking at the big picture in corporate communication, we've seen the diverse ways in which people can make contributions that can be categorized as communications. Although in most enterprises, there are currently many different people working and departments existing in these areas, there is an increasing need to integrate communication functions (advertising, public relations, employee communications, training, MIS, information processing, and telecommunications). Some progressive companies are taking a multidisciplinary approach to corporate communication planning by placing functions in one department, or at least setting up formal mechanisms for their collaboration and unified control. This can achieve more impact for less cost, and eliminate various departments that work at multiple functions by emphasizing different messages.

Unfortunately, this integration is not yet common. Departments or outside agencies tend to fight over budgets rather than come up with overall strategies and tactics. Advertisers recommend advertising solutions; trainers recommend training solutions. Companies end up reinventing the wheel, paying for the development of the same graphic at three different agencies, or investigating interactive multimedia in five different departments. The skills and projects of one group may be greatly needed by another.

For example, one of our clients in the training department in a large international service organization, expressed his frustration at not being able to get a better handle on which employee behaviors directly affect customers' perceptions of the company. They had an insufficient staff and budget to perform such studies. However, we found that the advertising department called 100 customers each night, asking them about their experiences with the institution and about what practices made a difference to them. The advertising group was open to adding particular questions that might be helpful to the training group in their needs analyses and assessment activities. If it were not for my playing social director, people in these departments would not have met!

Consolidating the various communication resources in organizations comes as a radical suggestion to most people. However, many companies will have gone through a similar process if they integrated the previously separate departments of computing, office systems, and telecommunications (phone systems). These were separate islands as recently as the mid-1980s, but have been largely collapsed into an integrated information systems department (McKenney & McFarlan, 1982). Data processing was originally centrally controlled and needed complex and expensive equipment typically provided by just one vendor (an example is large IBM mainframes). Office systems consisted of typewriters and file drawers, rather primitive devices purchased from multiple suppliers. Phone systems were not controlled internally, but by a national monopoly. By the mid-1980s, all of this changed. Data processing could be done locally, but there was a need to standardize office systems with it. There emerged many alternatives to phone systems, and these had to be managed locally as computer systems. Until these systems were brought under coordinated control in information systems departments, chaos reigned.

Similarly, media production activities were generally spread out in most organizations in the late 1970s and early 1980s, but were consolidated into corporate media groups during the mid-1980s. Marlow (1981) commented that

> Our culture is becoming increasingly entangled in a network of communications. At the heart of the matter is communications technology. . . . It is time to plan for the future internal and external communications needs of the organization by centralizing media operations and making them accessible to the entire organization. . . . On a broader scale, the organizational need to better use communications systems and better control overall organizational communications activities may result in the centralizing of other communications functions. It is possible that by the next decade, more organizations will have merged their training, employee communications, advertising and marketing, press relations, government relations, communication relations, media center, and other 'communications systems' such as telecommunications, into one 'macro' communications activity. (pp. 49, 143–144)

Hughes Research Lab is the R&D arm of Hughes Aircraft. Their formal communication activities consist of internal communication and the development of presentations and reports, since all public relations activities and advertising are done by their parent organization. The lab has taken an integrated approach to information management. For example, the information services department consists of three functions: (1) information systems (computing and telecommunications); (2) the technical library; and (3) technology presentation (the group that produces reports, proposals, reproduction, and presentation support materials). According to Brian Manning, their director of information systems, the lab's approach to communications is broad and forward-looking, consolidating all professional activities that relate to the development and dissemination of information. These include phone communications, e-mail and voice mail, support of local area networks, on-line bibliographic tools, desktop publishing, conference room support, and graphic art.

If we are to develop such an umbrella concept of corporate communications, what do Renaissance communications managers actually do?

- work with executives to develop and articulate the organization's culture, image, strategies, and policies—particularly those related to information and instruction
- track, implement, and manage information technologies
- develop communication goals, systems, and skills
- supervise the design and production of internal and external communication interventions
- devise performance improvement systems for organizational units
- develop and implement communication training programs
- use assessment procedures to evaluate and fine-tune communication interventions and systems
- manage the financial aspects of communication practices and programs
- continually market the services of the communications staff

The advantages of consolidating communication functions are clear—less duplication of efforts, increased continuity and consistency, and tighter control of information acquisition and provision.

The Education Services Department of Lederle (a large pharmaceutical company) is responsible for all communication to its sales force. Its staff, consisting of training, editorial, and graphic specialists, handles all announcements, memos, and training, which are sent and utilized by their staff in the field, thereby ensuring consistency and reducing data overload. Rob Livesay, the manager of communication, says that the department offers a number of services to those who wish to communicate with the sales force—these balance out any inconvenience that their gatekeeper role might otherwise cause. For example, the education services department will take rough information, decide on how it should be communicated (voice mail, e-mail, bulletin, etc.), edit it, develop graphics and layout if necessary, and most important, coordinate reviews by the appropriate medical, legal, and regulatory agencies and departments (see Figure 6.1). The department can classify and prioritize messages and help the sales force maintain up-to-date information. They do this by providing each salesperson with an indexed binder into which each memo or bulletin gets placed. The department staff then keeps track of and announces which old items should be discarded. In addition to managing announcements and memos, the department also handles all training, employee development, and meeting planning for their division.

prio1

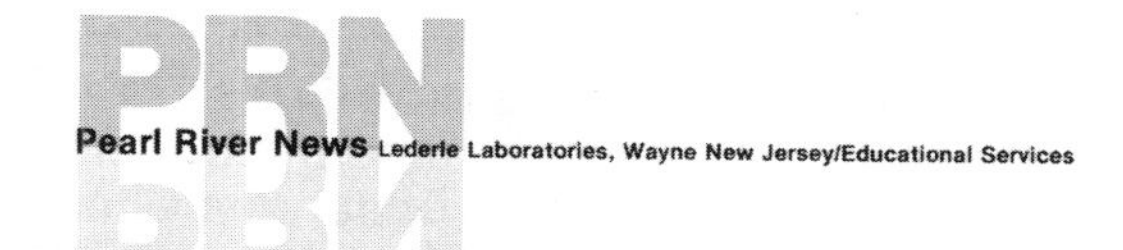

FIGURE 6.1
These employee bulletin formats are one way of coordinating and prioritizing messages sent in the field. (Courtesy of Lederle).

In order to integrate and elevate corporate communication, it's essential to gain support from top management. Communication is a responsibility that everyone in an organization shares; however, professional communicators are the leaders. Most researchers find that to move communication to a strategic level, the CEO must be committed to achieving excellence and control within formal communication functions, and the chief communication officer must be a member of the executive team.

In a recent survey of the United Kingdom's top 1000 corporations, researchers found that the majority of managers believed that they needed to have consistency in communication and that they were doing a pretty good job in achieving it. However, few had the management functions or procedures in place to really support or evaluate it. The researchers felt that part of the problem was that communication professionals were rarely represented in top management and, therefore, the communication function was left out of the strategic planning process. "The problem is compounded by fragmentation, with both budgets and responsibility for communication spread across a number of different departments or divisions within the organization so that strategic planning is virtually impossible" (Communication function, 1991, p. 11.).

In organizations with established islands of communication, reorganization will probably not come easily. As Meltzer (1981) states,

> Top management must act before the battle lines are drawn in the struggle over who is to be director of the information department. . . . What is needed is an information manager at a level equivalent to the finance director, the personnel director, the manufacturing director, and so on. . . . Company policy statements must clearly explain that the function of the new hierarchy is to manage the information resources of the organization to attain stated goals and objectives. The policy statements must be so succinct that no one inside or outside the company can misinterpret information management as information manipulation. (pp. 112–113)

NCR (now a division of AT&T) recently combined their documentation and education departments under one vice president of a newly constituted information products division. Marketing publications, documentation, and training had formerly worked together at NCR's training center, called "Sugar Camp," but throughout the 1980s had decentralized. Many of the writing functions went out to the individual plants, while training stayed at corporate headquarters. During the decade, says Maggie Flynn, an information products consultant at NCR, these groups lost track of each other. They developed their own layouts, writing styles, even logos, until several executives and the communication professionals themselves remarked that the materials didn't look as if they were from the same company, and often duplicated each other. The results were wasted efforts and confused customers who had to plow through several booklets of documentation and instruction to use an NCR product that contained parts developed at different sites.

In the early 1990s, a task force was formed to examine the situation, and they recommended that these functions be regrouped. One of their major efforts was to hire a consultant to develop their Quality Documentation Process, a set of standards for writing and laying out documentation. This standardization is now being extended into training materials, based on the recommendations of the internal task force. NCR hasn't consolidated all of its communication functions—there are still separate groups that handle information systems and stakeholder relations. Media production and internal and external publications are part of this latter department. However, communication professionals are continuing to look carefully at how to better coordinate their efforts and how to recommend structures that can support this goal. Stephen Bean, assistant vice president of the Information Products Division, is intent on creating a common language, a set of development procedures, and a measure for evaluation in order to increasingly professionalize his staff.

One department being responsible for information and instruction is, of course, not the only alternative (see Figure 6.2). In many nonprofits, the communications department and the development office are merged. For example, the Guthrie Healthcare System has a vice president in charge of public affairs, employee communication, and fund-raising. Professional communicators may also be placed within line operating units. A variant of this approach is to house them in one main communications department, but to use the account executive approach of being assigned to serve one or more specific groups. The debate over centralization versus decentralization is a continuing one. "What is needed is centralized control and decentralized authority, with the information manager assuming ultimate responsibility and accountability" (Meltzer, 1981, p. 114).

Models of Operation

Whatever the formal structure, the consolidation of various departments is a key to effective management. Communication and training are often kept so far away from the line organization that they have the reputation for not really understanding what's going on. They are literally afterthoughts. There are various strategies to overcome this isolation. Southern California Edison Company, for example, has a Corporate Communications Advisory Committee comprised of people throughout the company. This is one way for managers to keep tabs on communication projects and priorities (Berko, Wolvin, & Curtis, 1990).

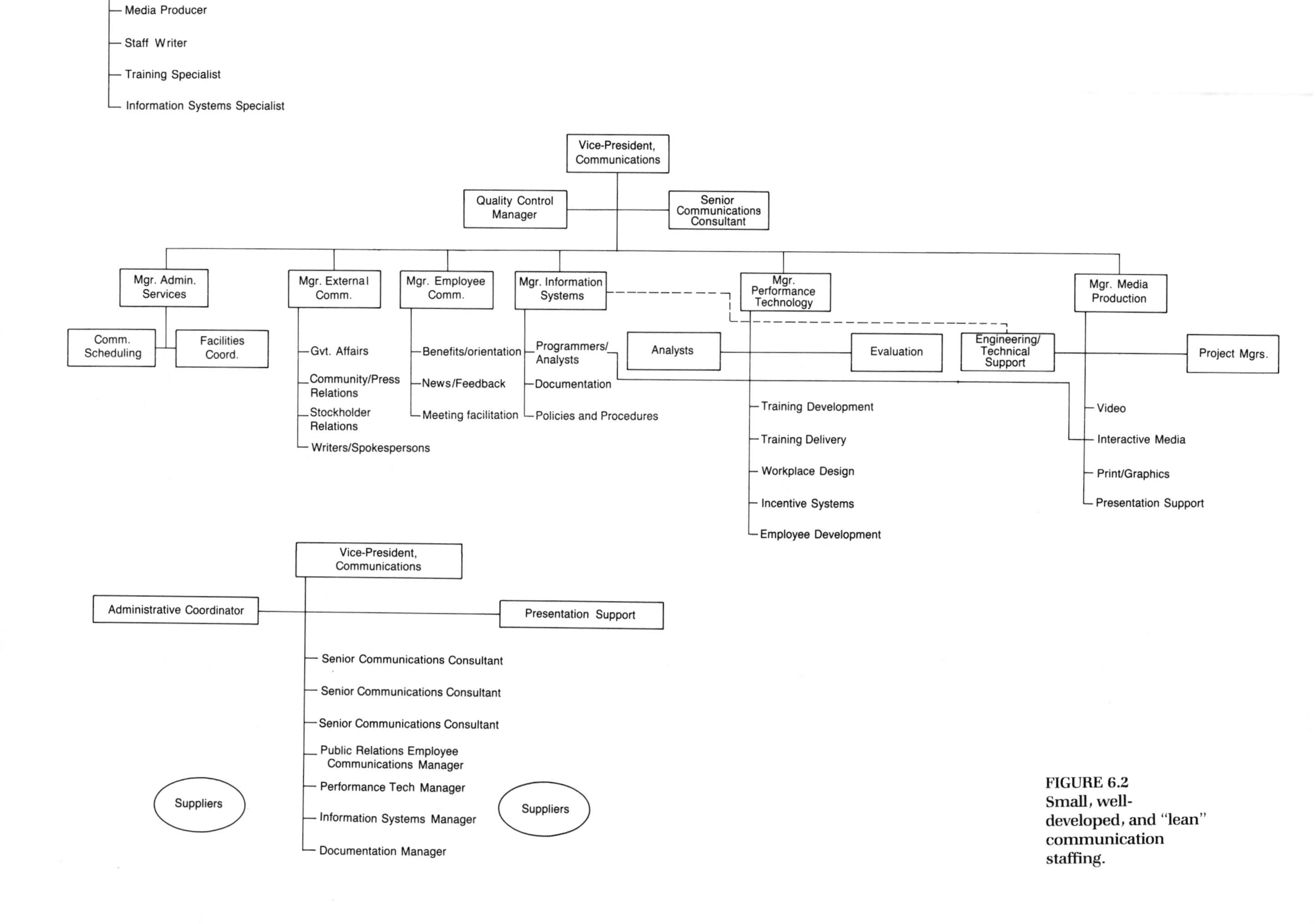

FIGURE 6.2
Small, well-developed, and "lean" communication staffing.

Other enterprises assign staff to particular divisions.

Southern California Edison and Toronto-based Ontario Hydro restructured their communication departments. They established a single contact for company clients using an account executive team concept. Ontario Hydro found that senior staff were frustrated by having to deal with many people within corporate relations. They wanted a high-level, multiskilled communication generalist to work with them to provide overall corporate relations advice, planning, and project management services. They now have seven corporate relations advisers with their own accounts to serve. These professionals work from beginning to end on projects and are involved in formulating comprehensive communication plans that are aligned with the company's basic business strategies (Lennon & Sheldon, 1992).

Reorganizing Communication Functions

When starting or reorganizing a communication department, there are several crucial questions to be resolved. Following are some of those, with guidelines for creating individual responses to them.

What Will It be Called? This may depend on what names already exist within the company. When merging departments, it's better to choose a new name rather than use an old one in order to mark a change and avoid the political problem of having it look like some areas were subsumed under another. Some possibilities are simply the communication department, information services, information systems, performance technology, information and instruction, strategic communications, or information management. My partner, David V. Williams suggests the title, Department of Symbolic Performance.

What Will be Its Functions or Business? An integrated communication department should analyze individual and organizational performance problems and determine which are amenable to communication interventions, develop rules and tools to solve them, manage the technologies and facilities for formal communication and information processing, and design and produce interventions, including marketing strategies and materials, advertising, public relations campaigns, computer systems, employee communication vehicles, crisis communication plans, feedback systems, formal meetings, and formal instructional programs. The department should see that formal and informal communication supports the organization's goals and culture and ensures that policies and programs are consistently aligned with them.

To Whom Will It Report? Ideally, the communication department should be headed by an executive vice president who reports directly to the CEO.

What Will be Its Financial Structure? It should be given a yearly budget for system-wide development and maintenance of technical systems and policies. Money for new capital equipment should be allocated on the basis of proposals. Resources for individual projects, such as large meetings, training programs, and data systems, should be provided by the business unit(s) who request and benefit from them, as shown by the business plan for each intervention.

Who Will Staff It? It is essential that the department be led by an executive with a broad background in human and mediated communication, organizational communication theory, and business. This person should not be perceived as a narrow specialist in any given intervention, such as data processing or training, but rather should be able to detect performance gaps and to recognize opportunities to improve communication systems. Most of the staff should have a formal educational background in one or more specific areas, such as instructional design, public relations, and marketing. Some of the staff should also be technical or creative specialists, such as video engineers or graphic artists. To strengthen links between the department and its potential sponsors, some personnel should be seconded by the line, on 2 to 3 year rotations through the department. These people need to be given a thorough orientation to the department's procedures, standards, and analytic tools, and can be given further professional development through workshops, conferences, or assistance from consultants.

DEFINING COMMUNICATION'S FUNCTION

Role and Purpose

What are clients really thinking when they call upon a professional communicator? Are you a contractor (technician, a layout artist, a wordsmith, or a computer whiz who simply does what's requested), a collaborator (someone who can help them think through and solve problems), or a cop (the gatekeeper or communication prevention specialist)?

An important part of managing a communication department is managing clients' expectations of it; of course, this is a typical chicken-and-egg situation. If clients come to you unwilling to consider alternatives and don't give you the resources to conduct proper needs analyses and assessments, you'll never get the opportunity to prove your capabilities. However, unless you can demonstrate your ability to bring real value to the table, you can only be considered a technician. As potential clients or sponsors approach the communication department, you need to assess their motives and expectations. Here are some possibilities:

- There's a problem, so let's throw some persuasion or training at it—the knee-jerk reaction.
- If a little communication is good, then more is better.
- If we communicate it or deliver training on it, we can't be said to be negligent.
- Communication and training is fashionable.
- Producing media or planning meetings seems like fun.
- This is a good way to get recognized or look good.
- If I use up more of my budget, I can request more next year.
- My colleagues or competitors are doing it, so I'll do it too, or even better—beat them at it!

There is an interesting fable about the "CEO's New Studio," which is a take-off of the "Emperor's New Clothes." The CEO overhears that the competition's CEO has just installed a new video studio, so he won't be outdone. He and his consultants ensure that the company gets the latest and greatest technology—except when it comes time for the cameras to roll, then he has nothing to say. Most of the people who watch his program won't admit that his message is empty or that the program is boring. The only one who reacts authentically is the janitor, who sees the program playing on a monitor, shakes his head, and tries, in vain, to turn it off (Brush & Brush, 1981).

Departments can operate on a reactive or proactive basis. In the former, professionals wait for others to identify a need or for events to happen, and then react. This is basically the stance of a journalist who merely responds by writing accounts of events once they've occurred or been formally announced. When this happens, often the story has already been told through informal and formal channels, and it may be too late to avoid dissonance, speculation, and a mix of perceptions and beliefs. The consequences of performance and communication gaps may have already become quite severe. In the more progressive proactive mode, professionals seek out problems and opportunities, and address them before the situations get out of hand and while they can still be solved in a coherent and efficient manner.

Rothwell and Kazanas (1989) recommend that human resources professionals develop strategies based not just on what is and what should be, but on the expected future external environment. That is, we should not just measure performance gaps based on what's needed now, but on what will be needed to perform efficiently in the future. In conducting such an analysis, they use the acronym **WOTS-UP.**

- **W**eaknesses
- **O**pportunities
- **T**hreats
- **S**trengths

Considering these factors, both at the organizational and departmental level, managers can decide whether the need is to build upon present strengths and take advantage of new opportunities, or to take a more defensive position of remedying weaknesses and diversifying so that possible threats will not have such serious consequences.

To establish a proactive mode of operation (see Figure 6.3), you need to align department goals with organizational goals. As corporate goals become more concrete and as values are articulated, the goals of the communication department need to be sharpened. Then, the vantage point from which you view your work needs to be determined. These include

- *logistical*—coordinating supplies and resources necessary to produce an intervention
- *tactical*—specific planning and activities undertaken to produce an intervention
- *strategic*—analysis of problems and planning of sets of interventions that will be used to meet organizational goals
- *policy*—creation of mission statements that define the basic purpose of the department and require action programs
- *environmental*—identification of external needs and ideals that can be sought by an organization through its policies and strategies (Gilbert, 1978)

These determine where you come in and the role you play:

- *analyst*—identifying problems
- *prescriber*—responding to problems by identifying solutions
- *implementer*—producing solutions by identifying and employing techniques and technologies

This brings up several major policy issues: Should it be up to clients/sponsors to monitor and identify their problems and assess the results of solutions, or is this the role of the communication specialists who will develop the interventions? Is the communication department charged with developing specific quantitative and qualitative statements of problems and the

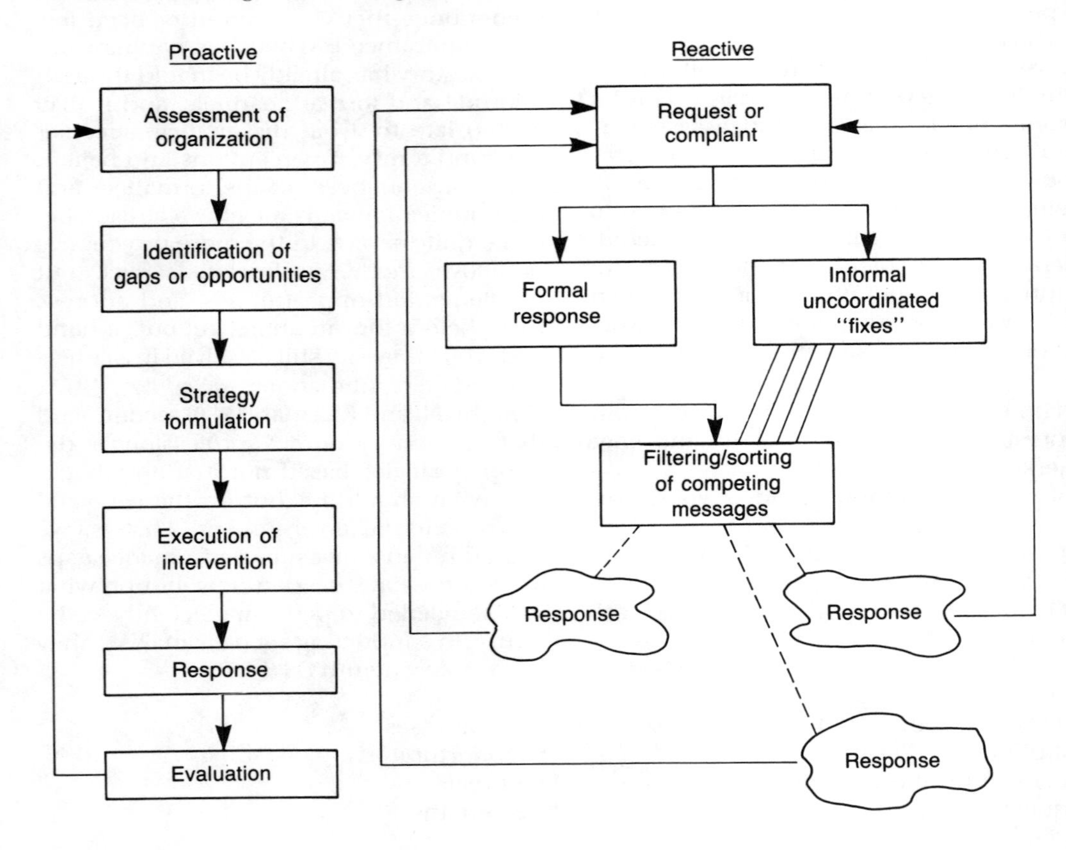

FIGURE 6.3 Proactive versus reactive models of practice.

effectiveness of their remedies, or is this a function of the line organization? Does the communication department actually implement solutions, or does it analyze problems and manage the technical execution of solutions by outside specialists? In many organizations, a small department of highly trained professionals serve as analysts, policymakers, consultants, and designers. They contract with vendors in software programming, media production, and writing to produce interventions. Other departments have the human and technical resources to do most production activities in-house. These broad missions and responsibilities must be developed and understood clearly by both the communication professionals and their potential clients or sponsors.

Based on an articulation of the roles and responsibilities of the communication department, a mission statement must be developed. This might read: *The information management department supports ABC's strategic goals and values by managing and developing its formal communication systems, policies, and technologies*, or *the communication department enhances the effectiveness and efficiency of communication with internal and external audiences of the XYZ Corporation through strategic analysis, message design, and application of communication techniques and technologies.*

In developing a department's mission statement, its functions must be made explicit and any possible conflicts with other departments must be ironed out. These roles can include

- employee communication
- employee skills training
- employee education and development
- employee orientation
- communication skills training for all employees
- customer communication
- customer training
- marketing communications
- advertising
- public relations
- liaison with stakeholders
- government and community liaison
- trade shows
- coordination of formal meetings
- media production and distribution
- coordination of communication-related special facilities
- crisis communication
- feeder for mass media stories
- change management
- total quality management facilitation
- producer of generic information or training programs related to the industry
- supplier of communication services to outside firms

In order to move out of the technician role and develop policies and strategies, professional communicators must develop or operate from power bases. These include

- *expertise*—one's education and experience
- *information*—the access to and control of information that others need
- *political*—the ability to see and influence those with power and to reward or to punish
- *staff support*—colleagues' or clients' trust and willingness to cooperate
- *tradition*—long service to or affiliation with an organization
- *credibility*—one's competence as assessed by outsiders, or one's reputation in the profession at large
- *charisma*—one's perceived physical and personality traits that lead to her being liked and trusted (Duncan & Powers, 1992).

Articulating Goals

Once a department has been established and its mission statement formulated, its goals must be articulated. There will probably be a half a dozen or so of these, and they will change somewhat from year to year depending on the organization's needs. Some examples include

The information management group will place top priority on the promotion of increased sales in the new aerospace materials division.

The corporate communication department will produce interventions that will decrease turnover in the nursing staff by 25%.

The college information office will not only publicize faculty accomplishments and ensure that they are noted in the regional newspapers at least once a week, but also provide information to national news media so that the college is mentioned favorably in at least five national publications or broadcasts each year.

Of course, individual department goals need to be aligned with corporate goals. If corporate goals haven't been articulated clearly, look around. What is the status of your company within its industry? Are more people being hired, or are new technologies or processes being introduced? If so, more training is probably going to be needed. However, if the company is in a slump, perhaps marketing materials will be most helpful. If the organization's style or culture is being transformed or if the company has been bought out or expanded through acquisitions, there will be a need to provide extensive information and role-modeling to employees.

Maintaining Value and Credibility

Scott Carlberg, video manager at Phillips Petroleum, says that his group has been a permanent fixture, even through two hostile takeover attempts, increased debt-to-capitalization of more than 75%, and a drop in oil prices from over $30 a barrel to under $10.

> The reason? Flexibility. Phillips' video is light on its feet. Moves like a cat. The key to success is understanding and anticipating change. The program format or company business line that is popular today may not be popular tomorrow. Communication needs change, markets change, managements change and so must corporate video in order to remain a welcome part of an organization. (Carlberg, 1991, p. 5)

To succeed in operating at a higher level in organizations, it's important to project a professional image. Clients must not perceive the communication department to be synonymous with the audiovisual storage closet or the typing pool. Often, potential internal sponsors don't even know what in-house resources are available. This can be accomplished through selective newsletters or bulletins on its activities, open houses, brown bag luncheons during which projects or facilities are shown, and announcements of staff accomplishments in employee communication vehicles. Often, establishing a good reputation in the field will help managers sit up and take notice. Communication professionals can gain this credibility through leadership roles in professional organizations, through publishing articles and books, through performing service to community organizations, or through leading professional workshops.

MANAGING STAFF

It goes without saying that the most important resources of a communication department are its human resources. In a mature unit, there will be three types of roles:

- *craftspeople*—artists, writers, photographers, graphic artists, audiovisual technicians
- *project or area managers*—media designers, editors, audiovisual managers, instructional designers, advertising/PR directors
- *executives*—analysts, policymakers, spokespeople, and the department head, who serves as the linking pin with the CEO

Unfortunately, most communication departments are currently not staffed with professionals who have enough knowledge or power to become strategic resources. D'Aprix (1977) describes how many people wind up in corporate communications in his tale of "Gretchen Greensleeves." Gretchen was an English major who, after futile attempts at a career in writing, becomes a typist for a senior vice president. He is impressed with her performance, since she does a good job at correcting his correspondence. A few years later, Ed, the man who edits the employee magazine retires. Gretchen gets the job, partly to satisfy equal opportunity objectives in promoting women, but mostly because she can spell and type well. Since she doesn't know anything about employee communication, she's told to do it "just like Ed did." Ed produced the typical folksy magazine with welcomes for new employees, announcements of retirements, columns on "know your benefits," and pictures of the company's softball team. Meanwhile, the tellers are organizing a union, managers share no information with subordinates, and turnover is high. The employee communication system does nothing to address these problems, but Gretchen thinks she's doing a great job. In fact, she wins the local editors' group's gold medal.

Clearly, communication professionals need a much more thorough education, broader experience, and enhanced credibility in order to formulate and execute communication policies and interventions. They need to be analysts, planners, and managers.

Once the right people are brought in to staff the communication department (more easily said than done, I fear!), decisions about how to organize the work need to be made. A study by the software firm, Aldus, identified three types of communication work groups:

- *whatever it takes group*—In this model, the staff doesn't have specific roles; rather, they are generalists who do whatever needs to be done at the moment. The department feels like a family because people ask others to do things on the basis of personal favors. It functions on mutual respect and strong interpersonal relationships.
- *central hub*—This model takes a spoke and wheel configuration, with generalists as the spokes and a traffic cop manager at the hub coordinating activities. Workloads are predictable, since the manager keeps work flowing at a manageable rate. The key here is control.
- *well-oiled machine*—These departments are highly structured. Staff have specific roles and specialties, and the focus of management is on planning and coordinating activities (Rosenzweig, 1991).

In mature departments, managers will need to devise specific performance objectives for individual people and jobs, standards for their analysis and design activities, and career and staffing plans. When people are brought in from the line on a rotation through communications, formal methods of orientation and training are needed.

MANAGING BUDGETS

How a communication group gets the money it needs to do its job is an important consideration. It forms the basis of not only what it can do, but how it is perceived by others in the organization. There are several common ways in which allocations are granted.

- *lump sum budget*—a given amount of money is allocated to run the entire department's functions
- *line item budget*—different lines or accounts of money are allocated for different on-going and anticipated expenses
- *formula budget*—the budget is determined by performance of other functions in the organization, such as a percentage of sales or profits
- *partial charge-back*—the department receives allocations to cover its basic operating budget, such as staff and overhead, but charges back out-of-pocket expenses to its clients
- *performance budget or full charge-back*—the department receives no funding of its own, but meets its expenses through charging other departments' line operating budgets for its services
- *program budgeting*—each department or program in an organization gets a budget for communication that is provided to the communication department as its services are used (Meltzer, 1981).

There are obviously advantages and disadvantages of each approach. When communication departments are given a fixed annual budget, it becomes a first come, first served system for clients. If a client requests a large project early in the year, perhaps most of the department's resources will be spent on it, leaving little flexibility for other interventions. Clients who use communication services most obviously benefit most. Other less-served departments may resent the fact that an overall budget is provided to it regardless of the constituencies it serves. Finally, this sort of funding makes it easy for the communication department to become lax about its service orientation and hesitant to go through the trouble to make a genuine business case for its services. Since they are seen as free to clients, there's little control or assessment of solutions. Given this, it's also more likely for the budget to be cut or the department to be completely abolished if the organization needs to cut spending. For these reasons, many communication groups are now supporting themselves on a full or partial charge-back system.

Many communication departments (especially audiovisual production) are being spun-off as independent businesses or they are being severely cut as corporations pull back to support only their core business. Some keep producers on their staff, but insist that they pay for their own operation, either from garnering business from departments inside the firm or from outside clients. For example, Reliance Electric, a Cleveland-based supplier of a wide line of motors and other production tools, spun off a corporate video department, which made the name change from Reliance Electric Visual Services (REVS) to AVID, and is now an independent firm with Reliance as a major client. In a similar situation, after 34 years at Owens-Corning Fiberglass, the director of corporate relations was let go. He and the director of video communication left to form their own corporate media firm, although Owens-Corning rented them their old facilities back at an attractive rate. They wound up doing much of the same work as they did before, except now they were responsible for selling their services as well as numerous other supporting tasks, which formerly would have been handled by the purchasing or facilities departments (Kanter, 1989).

Taking on outside clients is a controversial decision. Most in-house communicators have not developed the skills or resources to effectively market themselves to the public. It is difficult for them to compete with production houses and consulting groups who have full-time marketing staffs. Furthermore, although it may seem attractive to executives to be released from funding a communication department, they may find it quite distressing when they need to produce an urgent video or write an important speech and find their communication department busily committed to a project sponsored by an outside company.

The budgeting process has a number of levels. One is involved in planning—forecasting the probable costs of overhead and program costs; the other level is controlling—documenting and managing actual expenditures for each project. More information about these specific activities is included in the sections on managing projects and on evaluating communications. In order to determine what to charge for your services, first you have to determine your typical expenses. These include

1. personnel
   a. employee salaries/overtime
   b. employee benefits/payroll taxes
   c. overtime
   d. contractors (technicians, production houses, free-lancers, consultants, actors)
   e. royalty fees
   f. legal/accounting counsel
2. capital equipment
   a. furniture
   b. media production/distribution equipment
   c. office equipment (fax, copiers, phones)
3. supplies
   a. stationery/office supplies
   b. maintenance and repair items
   c. blank tapes and diskettes
   d. software/updates
   e. stock photos/graphics/music
   f. duplication of print/media materials
4. communication
   a. telephone/fax
   b. on-line communication/searches
   c. mail
   d. courier
   e. shipping
5. travel
   a. travel for required meetings
   b. travel for productions/consultation
6. professional development
   a. dues for professional organizations
   b. books/periodicals
   c. travel/registration for conferences
7. business development (for independent firms)
   a. advertising/public relations
   b. client/prospect calls/entertainment

Based on an analysis of ongoing expenses and typical project costs, communication managers develop rate cards that document prices per item or per hour. These rate cards may be quite firm and published, or they may be guidelines for internal use only. For example, a media production center might publish a rate card for typical technical tasks, such as slide duplication or straightforward video documentation of an event. However, generally per-hour charges of consultants and writers, as well as complex production activities, are not published, since they may vary greatly depending on the job. For example, a print production unit might charge $75 per hour for desktop publishing, but if a sponsor wanted a large-scale project that involved weeks of writing and page layout, the fee per hour might be quite a bit less.

Remember that, when developing rate cards, not everyone on a communications staff will be constantly generating billable hours, or time that can be charged to a client project. For example, some managers or sales and marketing associates may generate no billable hours. Even production personnel will spend time in initial client meetings, in general facilities upkeep, and in professional development. When a consulting firm charges $200/hour for a senior partner, it may initially sound unreasonably expensive. Most of us start multiplying $200 per hour times 40 hours a week times 50 weeks a year, and assume that this person is making $400,000 per year. However, such a person is only worth that kind of money to a client if he spends a great deal of time in research and professional development. Senior partners spend many hours generating new business, mentoring junior associates, and contributing to their professions through association leadership and publications. So, they may only generate 10 billable hours per week, or $100,000 per year. Since benefits alone are generally calculated at 30% of one's salary, this brings the salary down to $70,000. Also, remember that even to do straightforward consulting and reporting, this person will likely need a part-time secretary and an equipped office, as well as coverage for communication and travel expenses, professional dues, and conference registrations. It's easy to imagine that this support would add up to $20,000. Don't forget that some of those revenues need to go toward the firm's profit margin. That leaves less than $50,000 for take-home salary—a rather modest amount for a senior person.

MANAGING FACILITIES

The communication department is generally responsible for some specialized facilities, not only for its own work, but also for other formal communication activities of the entire organization. This facilities management includes purchasing and maintaining specialized equipment, designing and scheduling meeting rooms and auditoriums, staffing public areas (i.e., information centers and training resource rooms), and creating and updating displays, bulletin boards, and information kiosks. While some of these areas may be located in the communication department, many of them may be dispersed throughout a site or even across the world. The major categories of communication facilities to be managed may include

- *meeting areas*—specially equipped conference rooms, teleconferencing areas, auditoriums, and classrooms
- *specialized media production areas*—video studios, desktop publishing areas, graphics and photography labs, and screening and editing booths
- *communications systems and information processing units*—the central data processing and telephone systems, and satellite up-links and down-links
- *public access facilities*—libraries, self-service media production systems, learning labs, and display rooms
- *storage and maintenance areas*—audiovisual equipment check-out, media and print libraries, organizational archives, and hardware repair centers

Rapid changes in communication technologies, as well as individual organizations' situations, require flexible space. For example, yesterday's cavernous TV studios are often empty today now that most production is done with portable equipment in the field. Classrooms may need to be turned into meeting rooms or display rooms; large, centralized computer systems may be replaced by dispersed local area networks. As high-tech display systems become the norm for use in meetings and training, the communication department often becomes responsible for recommending and maintaining the hardware infrastructure.

C.L. Price is the technology manager in the Employee Development & Quality Performance Department of Dow Chemical. In 1991, he established a new learning center, consisting of classrooms and an auditorium designed for multi-media support, all controlled using Macintosh computers. Each room has flexible seating and an array of display devices, ranging from slides and overheads to computer-generated images, interactive videodisc, and videotape playback—all of which are displayed on a rear-projection screen. The instructor or group leader uses a software system (a series of menus selected by clicking a mouse) to control room lights and select and cue-up presentation media. Each place has an audience response system hooked into the computer, which can be used to display the results of a vote or the responses to a question. Groups also use the Macintoshes for recording and arranging notes during brainstorming, presenting electronic slide shows, and displaying spreadsheets and graphs. Originally designed for training purposes, now people want to use these facilities for holding meetings—all kinds of meetings: briefings, planning sessions, and executive speeches. They find they are more productive, and executives like the ability to generate and display their own computer graphics (See Figure 6.4.)

FIGURE 6.4
A high-tech training and meeting room. (Courtesy of Dow Chemical.)

It's challenging enough to keep up with the latest hardware, acquire it, and keep it running within your own department. However, in many companies, communication managers are responsible for distributed facilities, such as specialized audiovisual equipment or meeting rooms located in field offices. In fact, one of the important policy questions to be determined is who is responsible for what kinds of communication systems. As audiovisual equipment becomes easier to use and to maintain, the trend is for people at each site to retain control. This follows the pattern of data processing departments that used to purchase, maintain, and control all computer equipment in a company. Today, an information systems department might establish standards to ensure compatibility and may coordinate mass purchases to reduce costs; but, generally, individual departments buy their own computers and find ways to keep them running.

So, the pattern is for communication departments to set standards and to make initial large purchases, such as VCRs for each branch office or computer systems for each customer service representative's workstation. If more exotic equipment, such as an interactive video system, is to be placed in remote sites, it's important that a person be made responsible for its upkeep and be provided with training on how to troubleshoot and to maintain it. We have seen many interactive video projects fail because one by one, systems in the field develop small problems and no one there documents or fixes them, so they develop a reputation for being unreliable or just plain broken.

An important part of the budgeting process is forecasting what kinds of hardware will need to be purchased in the next 3 to 5 years. As technology changes, the half-life of most computer and video systems is only a few years. Equipment or rooms for playing back or receiving electronic communication may need to be developed. Finally, it is the role of the communication manager to propose and to implement new technologies as they are warranted by various analyses. Since these are capital purchases, they need to be made in conjunction with consultation from the chief financial officer and other accountants who can determine the best strategies for amortizing and writing off equipment.

MANAGING PROCEDURES AND STANDARDS

Communication managers need to develop and enforce policies, procedures, and standards relating to their own staff, as well as formal communication activities for all members of the organization. For example, it must be determined who speaks in an official capacity for the organization. Standards for the visual appearance of the company's logo and trademarks must be developed and policed (one advertising manager I know calls himself the logo cop). Procedures for routine and emergency activities also need to be put into place.

When Hurricane Gloria hit New England in 1985, Boston Edison was prepared. Their Major Emergency Plan of Operation (MEPO) went right into effect, and communication is one of its core systems. MEPO ensures that the company can receive fast and accurate information and provide around-the-clock information updates to the media and public officials. MEPO documents the company's response for the record. During the disaster, the communication staff created a variety of materials, including statistical comparisons of major storms and line art that were used widely by the mass media, consumer fact sheets and safety tips, two safety-oriented full-page newspaper ads, two special editions of internal publications, and a 12-minute videotape about the company's response to the storm. MEPO helped Boston Edison to "deploy its resource to advantage, obtain prompt assistance from outside sources, and communicate effectively with its audiences" (Keeping calm, 1989, pp. 31–33).

Part of developing internal standards is the creation of forms that are used to track jobs. These can include needs analyses forms, budgeting sheets, and systems for charting out project schedules, such as Gantt and PERT charts. An easy way to implement standards is through the use of software. Templates can be created for spreadsheets, so that budgets are all prepared in a consistent fashion. Style sheets can be developed for word processors and desktop publishing software so that correspondence and manuals have a consistent look and feel. This consistency not only gives a coherent look to final products, but it makes it easier for internal processes to be managed and for staff members to easily understand and pick up someone else's projects.

Standards can cover processes (what professionals actually do during each step of a communication project) and products (what the output should look like). They can take the form of algorithms or rules, which must be followed precisely, and of heuristics or guidelines. Finally, they can specify an entire process, such as the design and creation of instruction, or just parts of the process, such as standards for test-item writing. According to Silber (1992), standards generally result in increased client satisfaction and understanding of the process, higher quality and increased consistency of products, easier handing off of projects among staff, and improved productivity. Standards can also be associated with problems. It may take a lot of time and negotiation to build standards. They may result in cookie-cutter products or adherence to procedures or formats that are inappropriate, and they take time to explain and update. Obviously, every organization needs to decide which communication standards are appropriate for them.

Marlow (1981) advises that communication groups formalize and document their policies and procedures in an operations manual. This will provide a reference for new managers and staff, as well as provide the basis for a thorough examination of and agreement to methods of working. He recommends that such a manual include

- a description of the center's purpose and objectives
- a breakdown of each unit within the center
- a formal job description for each position and the chain of command
- measure of performance standards for each job
- the rate card
- a formal organizational chart
- organizational policies (i.e., sick leave, holidays)
- departmental policies
- work procedures for typical assignments
- all the forms the department uses
- lists of frequently used vendors
- copies of contracts
- procedures for using vendors
- statements regarding use of copyrighted materials

Marlow also suggests that media centers develop catalogs of their programs, divided by subject and media format.

CONSULTANTS AND CONTRACTORS

Although communication professionals have been stereotyped as producers and writers, increasingly their jobs are putting them in the role of consultants, especially as they develop in their careers. Consultants can be external contractors or in-house experts, and the functions they serve are quite varied.

The word consultant sometimes has a negative connotation; in that it usually means a person brought in from the outside to tell people what they should do, who collects a large fee, and then leaves it to the organization to figure out how to do what was recommended. However, consultants are more than "beltway bandits." Bell and Nadler (1985) define consultation as "the provision of information or help by a professional helper (consultant) to a help-needing person or system (client) in the context of a voluntary, temporary relationship which is mutually advantageous" (pp. 1–2). They distinguish between the client—who is the person responsible for managing a consulting project—and the *client system*—which is the organizational unit that will be affected by the consultation. The consultant is always external to the client system but not necessarily external to the organization; for example, a consultant may work within the communication department of a computer company whose printer manufacturing division may become the client system for one or more projects.

AT&T spends about $600 million for employee training each year, including training development and delivery, travel, and student salaries. They estimate further that the company may be spending up to six times this amount (over $3 billion) on post-training support, including coaching and documentation. To assist the company in transferring the latest in training technology to its business units, they have established the Corporate Training Support Group at company headquarters. The professionals in this internal consulting group include specialists in human performance and instruction who research new information technology concepts and systems and assist business units in applying them. They have undertaken pilot projects in electronic performance support systems, videodisc, and teleconferencing in order to assess their potential impact on the company's instruction and communication efforts.

Argyris (1985), a noted management theorist and Harvard University professor, provides us with some guidelines as to what consultants are and what they do. He writes that consultants must have technical expertise, intellectual competence, and interpersonal effectiveness. He maintains that they must remain separate from line management and set their own salary scale as do physicians. In effective relationships, consultants must never be dismissed for being open and authentic, but may be fired if they are judged to be incompetent by their peers (Argyris, 1985).

Unfortunately, anyone can call herself a consultant, especially in communications situations where it's difficult for a client to judge one's competence. Many people who are otherwise unemployed just set up shop and offer their services. Redding (1979) one of the recognized founders of the academic field of organizational communication, decries this situation and calls it "playing God for a fee." He observes that although most so-called consultants are very specialized, many claim to cover a broad range of skills and interventions. Redding says few have specific training for consulting, which makes them incompetents at best and frauds at worst.

Consultants' Roles

If you intend to assume the role of a consultant, or to use the services of one, it's helpful to keep in mind several models of consultation and ensure that both parties in the project have a clear idea of each other's expectations.

Provider or Technical Expert. In this role the consultant is a specialist who is hired to provide a specific end-product, such as writing a press release, installing a new local area network, or writing the specifications for a new computer graphics workstation. Most of the front-end analysis has probably already been done, and there is a clear mandate as to how the deliverable should appear. Organizations hire providers or technical experts when their staff needs specific expertise or additional help on a short-term basis. For example, my firm has written and produced videotapes for organizations who don't have in-house resources. We have also developed computer-based training programs for clients whose staffs were too busy to take on another project.

Investigator. Often managers want hard data or observations about some situation, but either they don't have the staff resources to do it or they feel that an internal person would not have the objectivity to successfully conduct such an investigation. In these cases, outside consultants are hired to perform tasks such as conducting an organization-wide communication audit, evaluating the efficiency and effectiveness of a particular business unit, or documenting existing hardware and software resources. These situations often involve extensive data collection and analysis, which is a task that can be done more easily by a consultant or consulting group that specializes in this. For example, one of my first contracts was for a library system that wanted to know what kinds of resources, such as video playback systems and satellite dishes, were already in place in member institutions to support distance education. In some cases, the consultant often has an existing standardized product, such as a survey form or methodology, that may be customized for the client.

Advocate. Consultants are also brought in to persuade people of the merits of some new policy or technology. They may do this through presentations at meetings or reports. For example, Xerox invited me to be a guest on a mock satellite telecourse program during which I spoke about how this technology could be used effectively in training. Trainers from their field sites participated in the telecourse from conference rooms in the same facility, and could also walk into the studio to see what it was like to appear on TV. I also prepared a booklet on how to effectively use satellite TV in training, which recounted success stories of other clients.

Prescriber. In this model, a consultant is referred to diagnose a problem that has been identified by the organization and to recommend, but probably not implement, solutions. The organization has likely developed a general idea of their wants and needs, and may have already conducted an investigation to gather data. They then call in someone with more technical or conceptual expertise to help confirm the source of the problem and to recommend potential solutions. These situations do not involve the use of just one method of investigation, such as conducting a survey, and are typically quite specific to the organization. For example, a communication department may feel that they would like to develop a stronger image in the national press for their research and development efforts, but may not be sure how to go about this. The consultant will use a number of methods to investigate the situation and call on his background in similar situations. The deliverables in this model are usually written recommendations.

Teacher or Stimulator. When a staff wants to develop particular skills or stay abreast of changes in their fields, consultants are often brought in to offer workshops or courses, either on a one-time or a continuing basis. For example, I've been hired quite often to offer workshops in interactive media for communication and training specialists who want to use new media or present briefings on current issues in communications management for executives. Alternatively, some consultants offer public seminars to accomplish the same ends. Companies often use experts in business writing or work force diversity to present workshops that supplement those offered by internal trainers.

Change Agent. When organizations have identified a direction in which they'd like to change, they often bring in a person with experience in advocating and establishing similar new policies, procedures, or tools. This person may make presentations about the benefits of the change, can help individual managers put new systems in place, and can work alongside staff as they learn to deal with the new situations. For example, if it was decided that the training unit of a company wanted to move from doing primarily classroom-based courses to distributed instruction by way of multimedia, a change agent who specializes in this new technology might be brought in on a long-term basis to support and encourage the trainers as they learned new skills and changed their operating procedures, and to provide help or facilitation as needed.

Colleague/Collaborator/Reflector. In this situation, a consultant with broad knowledge of the field develops a long-term relationship with executives in an organization. She serves as a sounding board and professional confidante, bringing in an outside perspective and a breadth of experiences to a more general situation. This model spans individual problems or interventions. For instance, an organizational psychologist may be placed on retainer with a firm to help to screen candidates for executive positions, look over plans, or give advice in difficult personnel situations. My firm serves as the communication agency for one particular client; while some of our associates perform technical tasks, my partner and I spend many hours each month discussing general matters, from using computers for communication through staffing problems, with the CEO.

Consulting gets a bad reputation not only from a handful of incompetents who use the label consultant, but from clients who don't know when and how to use a consultant's services. Some common problems in consulting include

- the person who requests services doesn't have the authority to carry out the consultant's recommendations, so nothing gets done
- the client has a preconceived notion of what the results or end-product should be (even if she admittedly needs a consultant to help with the situation!) and won't accept anything else
- the consultant applies old solutions to new problems
- the client is not specific about the desired intervention
- the client is not totally open with the consultant about organizational problems and constraints
- the client and consultant are too rigid about following plans and end up producing something that they both know isn't what the organization really needs
- the consultant uses experts to sell the job, and then assigns the work to inexperienced staffers
- the consultant is biased in terms of ap-

proaches or is secretly trying to sell her or others' services or products
- the consultant takes on a job for which he (or the consulting group) is not qualified
- the client tries to change or add to the job midstream
- the client tries to use consultants as the scapegoats or hatchet people
- some people in the client system feel that they are more qualified than the consultant, and resent her presence
- too many consultants with too many different approaches are brought in
- the client (or several people key people in the client system) doesn't really trust the consultant
- the consultant tries to build dependence, not independence
- the consultant does not keep confidential matters secret

Whether you serve as an internal or external consultant, or you hire consultants, it is important to try to avoid such problems. This can be done on the client's end through careful interviews, detailed descriptions of the environment and the desired results, the development of a consensus on the need for and selection of a consultant, and maintenance of a trustful and hands-off stance during the project. On the consultant's end, problems can be avoided by accurately assessing and describing your competence and approaches, being selective about taking on projects, asking probing questions about the organizational and political situation before agreeing to an assignment, and then ethically and thoroughly applying your expertise to help the client system function on its own in the future.

One of the best compliments our firm has ever received was the way one of our clients introduced us to one of his colleagues. He said, "they show us where the shoals are so it looks like we walk on water." This, in my mind, defines the ideal client-consultant relationship.

The Communication Agency

As organizations are being forced to integrate and upgrade their communication activities, while, at the same time, reducing headcount and overhead to focus on their core business, more of them are turning to outside communication service providers. It is quite common for organizations to use outside advertising agencies. They typically recognize the need for specialized high-level skills in media buying and in creative development. Many also rely on public relations agencies. Finally, most enterprises occasionally contract with vendors of production services; however, these individual outside islands are even harder to control in terms of costs and consistency than are separate internal departments. Managers spend an excessive amount of time finding and orienting new vendors. Moreover, it is difficult, if not impossible, to get them to work together to share ideas or resources, since many feel that they are competitors. To address these issues, some firms are emerging as communication agencies who can handle all of a client's information and instruction needs.

Espar Products is the North American distributor of diesel heaters for trucks, buses, off-road vehicles, and boats. This small organization had tried rather unsuccessfully to develop its own training programs for use by distributors, and had contracted with a series of advertising agencies, none of which really understood their product or market. After having developed some well-received training programs for them, my firm was asked to become their advertising agency as well. We can leverage information and materials—for example, photos and graphics produced for spec sheets are also incorporated into the training programs and company-specific proposals we create. Their sales staff has to call only one office, and because of the diversity and extensiveness of our work for them, we can dedicate one person to their account. This allows us to develop a deeper understanding of the company and its environment, and reduces by a factor of 3 or 4 the number of different organizations with which their president must interface.

"The communication agency of the future will enter a quite new partnership with its clients, a seamless interface with the marketing company where all possible expressions of brand communication needs are automatically included in the team skills used in the earliest market analysis and creative briefing" (Sharpe, 1988, p. 33). This integrated approach, naturally, will call for a much different mix within agencies. Since organizations will only use outside contractors when it's clear that it's justified in terms of superior skills or reduced costs, agencies will need to hire high-level, broadly trained communication professionals and operate very efficiently. However, the potential market for this one-stop approach to communication services is quite large, since many organizations don't want to, or know how to, manage formal communication programs in-house.

The In-House/Out-of-House Mix

During the past decades, the pendulum has swung between establishing comprehensive communications resources in-house and relying on outside professionals to perform those tasks. Throughout the 1980s, large communication and training staffs were built, including rather elaborate media production facilities in many organizations. The impact of mergers, downsizing, and a flagging economy in the late 1980s and early 1990s reversed that trend. Obviously, most organizations follow some middle course, using internal resources for recurring projects and external contractors for unusually large, sophisticated, or unique projects. Both sets of professionals have advantages and disadvantages. (See Figure 6.5.)

As organizations become increasingly hesitant to make long-term commitments in terms of staff and facilities and as communications technologies continue to develop at a rapid pace, we are seeing more creative mixes of internal and external resources. For example, ISVOR-FIAT is the wholly owned subsidiary of the Italian car manufacturer, Fiat, which is responsible for all of Fiat's management and technical training. Since Italian labor rules make it very difficult to fire staff and there are few sources of experienced instructors in Italy, they hire hundreds of contractors to develop and deliver instruction. Most of these contractors are newly retired high school teachers in their mid-40s who have university degrees, extensive experience, and the desire for part-time work. These contractors, in turn, use production houses to actually create media.

I've found it useful to use contractors to help internal staff develop new skills. In this way, the experience of outside producers or consultants can be transferred to the organization quickly. This improves the quality and timeliness of the work (especially when first dealing with new technologies, such as satellite broadcasts or interactive multimedia), while reducing long-term dependency on outside consultants. Although the fees of outsiders may look disproportionately large in comparison to internal staff members' salaries, remember that contractors can often do the work in a fraction of the time, and they don't consume the overhead expenses of supplies, communication costs, and office space.

FIGURE 6.5 Comparative strengths of internal and external resources.

In-house	Out-of-house
Fixed expense	Only paid as needed
Knowledge of company	Broad experience
Ease of access	More customer-driven
Known quantity	Access to diverse skills
Can depreciate equipment	Keeps up with new technology
Always there	Can shrink or grow as needed

When Troy State University first began work in interactive video, they used a grant to pay faculty to produce projects during their summer vacation and to put on a series of preparational, educational workshops. Our firm provided them with reading material, a two-day intensive hands-on workshop, and follow-up consultation by phone and by mail. We launched them into their projects by recommending a hardware/software system that we had used successfully, got them started in their programs, and answered a few questions about program bugs over the phone. After a few weeks, several interactive video programs were ready for use in the classroom and their project has continued for almost a decade. These faculty are now teaching other trainers and educators to develop interactive multimedia through workshops that they now offer themselves.

Following are some guidelines for creating effective collaborations:

- Use consultants to help select appropriate new technology systems, but be sure they are not affiliated with any particular manufacturer. Select people who have actually set up and used such systems themselves.
- Contract with professional coaches to assist internal staff in producing new types of interventions. This will reduce the learning curve and will ensure that first programs are effective and it will yield two outcomes—staff training and one or more usable interventions.
- Invest in your internal staff. Often, they know more than clients or managers give them credit for. Sometimes by simply paying a small retainer for a consultant to be on call to answer questions or review project drafts, their work and skills can be enhanced.
- Don't become entirely dependent on outside sources. They may go out of business, be unable to respond in a timely manner to your requests, or become unaffordable. Make sure that your own staff has control over projects that are produced by external sources, so that simple modifications can be made. For example, obtain computer files of graphics or desktop published documents, and be sure you know how to update and edit them (Gayeski, 1985).

PROJECT MANAGEMENT

A large part of the work of communication managers is administering projects; in fact, most professionals find that an average of 75% of the time spent on interventions is focused on activities other than writing and production. These activities include

- meeting with potential sponsors
- scoping out the project
- writing proposals
- getting approvals
- setting up budgets and schedules
- meeting with and supervising team members and vendors
- setting up reviews
- developing project updates
- generating invoices
- handing off the intervention

Proposals

An initial step for most projects is writing a proposal. This kind of document may follow up a client request or may be in response to a formal request for proposal (RFP) from an organization or government agency. Proposals may be as simple as a memo outlining the project and stating the budget, or may consist of hundreds of pages documenting approaches, past experience, project personnel, budgets, and deliverables.

Often, proposals can become the basis for the client's issue of a purchase order, so they should be carefully worded and comprehensive. Not only do you want to sell the job, but you want to make sure that you are making the limitations and conditions clear so that when you get it, you won't be sorry. Proposals offer the communication manager the opportunity to think through a complete approach to a project, including the probable products and processes. A typical proposal includes

- a description of the client problem or opportunity as you understand it, giving a background of how you learned about it
- a brief description of your approach to solving the problem, describing the deliverables in as much detail as possible
- an overview of your staff's capabilities and experiences
- a schedule, including a timeline with key bench mark dates
- a statement of who will be handling each task
- a detailed budget, including when billing will occur
- a statement of the conditions under which the project will be done, such as what the client is required to provide
- a statement about who will evaluate the work and with what criteria
- the procedures that will be followed if either party defaults on the agreement (See Figure 6.6.)

FIGURE 6.6
Sample contract for a video production. (Courtesy of OmniCom Associates.)

Agreement for the Development of a Videotape on XYZ's Flow Meter

OmniCom Associates (OmniCom), a New York joint venture and XYZ Systems, Inc. (XYZ), a New York corporation, agree to develop an informational videotape on XYZ's new flow meter.

XYZ shall develop a draft fifteen-minute script, including words to be spoken and visuals to be shot, providing an overview of XYZ, Inc., applications of the new flow meter, and a demonstration of how to set up and use this meter. The script shall include segments to be shot at XYZ headquarters as well as up to three segments of pre-existing videotapes of clinical procedures, and up to five computer-generated graphics but shall not specify any video special effects (such as dissolves, wipes, or animation) except for the superimposition of titles.

OmniCom, in consultation with XYZ, shall modify the draft script to maximize its effectiveness for the video medium. XYZ will then approve the final video script in writing; XYZ retains complete responsibility for the content accuracy of the script.

Once the script is approved by XYZ, OmniCom shall schedule two work-days (a total of 14 hours) which are mutually acceptable during which to shoot the video. All video shooting shall take place at XYZ headquarters in Ithaca. XYZ shall provide and designate in advance the facilities in which to shoot the program, satisfactory to OmniCom for video production in terms of lighting, acoustics, background, and space. XYZshall provide one of its employees, mutually satisfactory as competent and fluent in demonstrating the use of the flow meter and in narrating the program, following the approved script. XYZ shall also provide all necessary equipment for the demonstration. OmniCom shall provide a fully experienced and qualified video crew to shoot the video, using professional 3/4" U-Matic equipment. The liason shall review and approve the accuracy and suitability of the video segments during the videotaping sessions.

Once the video has been shot and audio has been recorded, OmniCom shall create the graphics and edit the videotape according to the approved script. OmniCom shall select, in consultation with XYZ, music for the introduction and conclusion of the videotape, and shall license it for non-broadcast display. All editing shall be done using professional 3/4" U-Matic equipment. The videotape is "work for hire" and is therefore copyrighted by XYZ. OmniCom shall provide one 3/4" U-Matic edited master and one VHS copy suitable for immediate playback and advise XYZ regarding further duplication. Once the editing has been completed, XYZ shall also take posession of all the original unedited videotapes, etc.

Scheduling

A number of common software programs can support project management. Some systems produce GANTT and PERT charts. These charts allow you to visualize the various components of a project, see which can be done concurrently and which must wait for a previous step to be completed, and assign staff members to various tasks. As the project goes on, you can compare the projected schedule with the actual schedule and make adjustments before the project gets too far off track (see Figure 6.7, which presents a sample computer-based training project). By merging project charts, you can easily see if you're overcommitting some person or facility.

Project Budgeting

One of the most important jobs of a communication manager is to develop budgets for projects. A number of software pro-

It is further agreed that:

- XYZ shall provide OmniCom with a draft script no later than July 15, 1993.
- OmniCom shall provide the video script for XYZ's final approval no later than July 30, 1993.
- XYZ shall approve the video script in writing no later than August 15, 1993.
- OmniCom shall provide two working days of video shooting within six weeks of final approval of the script.
- OmniCom shall provide the completed program within six weeks of the conclusion of shooting.

XYZ shall pay OmniCom $xxxx. in the following installments, net 30 days:

Upon signing of this agreement	$xxx.
Upon final approval of script	$xxx.
Upon completion of shooting	$xxx.
Upon completion of editing	$xxxx.

Time is of the essence of this agreement.

Changes in this Agreement may only be made in writing and may require changes in the schedule and/or payment due OmniCom.

This document contains the the entire agreement between the parties and shall be interpreted according to the laws of New York State.

Each signer personally warrants that he/she is authorized to enter his/her respective organization into such agreements.

____________________ For OmniCom Associates

____________________ For XYZ Systems,Inc.

________________ date

________________, date

grams facilitate the creation of budgets, especially for large audiovisual productions. Spreadsheets make it easy to keep running totals of various categories. Typical costs of communication projects include

- *direct costs*—materials, equipment, travel, producers'/consultants' salaries and benefits
- *indirect costs*—clerical support, project management
- *development costs*—writing, production, and editing of materials; piloting program; research
- *overhead costs*—equipment, maintenance, space, phone and mail
- *dissemination costs*—materials duplication, the proportion of the audience's salary taken up by the intervention, travel, meeting space

Some critical questions you should address when you are developing a cost-justification for a project include the following.

FIGURE 6.7 A computer-generated schedule for computer-based training project. The top portion shows activities, a priority code, the initials of the person responsible, and scheduled and actual dates of the beginning and the end of each phase. The bottom portion is a visual depiction of the process on a timeline. (Adapted from a plan by Mary Lou Kish).

ACT NO. --1--	ACTIVITY DESCRIPTION ---2---	PTY CDE -3-	RSP -4-	SCHEDULE START ---5---	SCHEDULE END ---6---	ACTUAL START ---7---	ACTUAL END ---8---
1	Initial client meeting	1	all	01/07/93	01/07/93	01/07/93	01/07/93
2	Write proposal	1	mlk	01/23/93	02/13/93	02/07/93	02/13/93
3	Client proposal approv.	1	jsb	02/14/93	02/21/93	02/25/93	02/25/93
4	Write script outline	1	mlk	02/14/93	02/28/93	02/21/93	02/28/93
5	Client outline approv.	1	jsb	02/28/93	02/28/93	03/01/93	03/01/93
6	Tracking chart	1	mlk	02/27/93	03/06/93	03/03/93	03/03/93
7	Write flowchart/script	1	mlk	03/01/93	03/20/93	03/11/93	03/16/93
8	Budget	1	mlk	03/20/93	03/27/93	03/26/93	03/26/93
9	Client script approval	1	jsb	03/20/93	03/20/93	03/25/93	03/25/93
10	Author / debug program	1	mlk	03/20/93	04/30/93	03/27/93	04/30/93
11	Present program	1	all	05/01/93	05/01/93	05/03/93	05/03/93
12	Final report / evaluation	1	mlk	05/08/93	05/08/93	05/08/93	05/08/93

* * * END OF FILE * * *

```
MAXIMUM TASKS: 2000   | CORTLAND |            APRIL '93
CURRENT TASKS:   12   |          |   S   M   T   W   T   F   S
                                                     1   2   3
F1-SORT         F6-04/25/92     Ctrl-D       4   5   6   7   8   9  10
F2-SEARCH       F7-SAVE FILE    DUP FIELD   11  12  13  14  15  16  17
F3-ADD          F8-EXIT                     18  19  20  21  22  23  24
F4-DEL/UNDEL    F9-UTILITIES    Alt-D       25  26  27  28  29  30
F5-CALENDAR     F10-HELP        DUP & TAB
```

```
CORTLAND                          CHRONOLOGICAL ORDER                          04/25/92
-----------------------------------------------------------------------------------------
:                          :      JANUARY'93       :      FEBRUARY'93        :      MARCH'93        :
:       DESCRIPTION        :0  0  1  1  2  2  3:0  0  1  1  2  2 2  :0  0  1  1  2  2  3:
:                          :1  5  0  5  0  5  1:1  5  0  5  0  5 8  :1  5  0  5  0  5  1:
:--------------------------++++++++++++++++++++++++++++++++++++++++++---++++++++++++++++++:
:Initial client meeting    :   X                                                         :
:Write proposal            :            I--------------X-----X                           :
:Client proposal approv.   :                           I------I  X                       :
:Write script outline      :                           I------X------X                   :
:Tracking chart            :                                     I-------X--I            :
:Client outline approv.    :                                      I   X                  :
:Write flowchart/script    :                                          I---------X----X---I :
:Client script approval    :                                                       I   X :
:Budget                    :                                                       I-----XI :
:Author / debug program    :                                                       I------X----:
:Present program           :                                                         :
:Final report / evaluation:                                                          :
:--------------------------++++++++++++++++++++++++++++++++++++++++++---++++++++++++++++++:
:                          :0  0  1  1  2  2  3:0  0  1  1  2  2 2  :0  0  1  1  2  2  3:
:       DESCRIPTION        :1  5  0  5  0  5  1:1  5  0  5  0  5 8  :1  5  0  5  0  5  1:
:                          :      JANUARY'93       :      FEBRUARY'93        :      MARCH'93        :
-----------------------------------------------------------------------------------------

        I---I  -  Scheduled date range
        x===x  -  Actual date range
          X    -  Actual start and end
                  on same day
```

- Who is the client and what is the business need for the intervention?
- What is the cause of the problem?
- What operational results do you want to track?
- What knowledge and skills are causally linked to the operational results you will be tracking?
- What is the total cost of developing and implementing the program?
- What information will you use to determine whether the desired results are happening?
- How long must you wait to determine whether the results are occurring? (Robinson & Robinson, 1989).

In order to create a business plan for an intervention, you must calculate its *return on investment* (ROI). The *return* is the expected gain or savings, in dollars, leading from the intervention. The *investment* is the total cost of developing and carrying out the intervention.

$$\text{ROI} = \frac{\text{Return}}{\text{Investment}}$$

For example, if a program on safety costs $14,000 and reduces accidents from an average of 24 per year to 16 per year and thereby saved the company $48,000 per year, the return on investment is 48,000/14,000 or over 300%. If a program on time management results in one more sales call every 2 days per representative, that equals 100 more sales calls per year per person. If you have 25 reps, and you know from past experience that they sell on the average of $500 per call, you may increase sales by 25 × 100 × $500, or $1,250,000.

When calculating investment, it's crucial to include the true cost of the intervention. This includes the time taken up by the communication professionals and sponsors (in terms of their salary per hour), any out-of-pocket expenses, plus the proportional salary of the audience of the message(s) for their time in reading and discussing it. For example, it may take 15 person-days from the communication staff to develop a brochure on safety, and $5000 to print 10,000 copies of it. However, you need to include the 5 days of the client's and SME's time, and factor in the salary of each person who takes time to read it. Here's what the total cost might be:

Comm. mgr.:	
5 days @ $600/day =	$3000
Writer:	
6 days @ $400/day =	2400
Artist:	
4 days @ $350/day =	1400
Client/SME:	
5 days @ $700/day =	3500
Printing:	5000
Mailing:	300
Reading:	
10,000 people × .25 hrs. × $40/hr. =	$100,000
TOTAL:	**$115,600**

Often, it's necessary to compare several communication methods in terms of costs. Systems that initially seem quite expensive in terms of up-front design and production costs turn out to be quite efficient in terms of delivery. For example, if typically 250 new hires go through an orientation program at corporate headquarters each year, it may turn out to be cost-effective to produce and use an interactive video orientation program available at each site instead (see Figure 6.8). Even producing an interactive video program costing $150,000 and leasing interactive workstations, each at $2000 per year, the total cost of mediated training comes out to be $163,700 versus a total cost of $311,937 for traditional classroom training. Of course, most of this comes from saving trainee's time (and the proportion of their salaries), as well as travel expenses. Even excluding their salaries, the classroom training turns out to be more expensive.

Reporting Systems

Once a project is approved, it becomes a part of the larger scheme of department activities that must be managed. Most managers track not only individual projects, but the aggregate work of the communication department. This can be done through monthly budgets, schedules, and reports, which document expenditures, work accomplished, timeliness, and requests for other projects. (See Figure 6.9.)

In addition to keeping budgets and schedules, it's essential to document project activities and discussions. Often, things get mentioned in a casual conversation and are then forgotten. This leads to poor communication and performance. These problems are exacerbated if someone on the project—be it the client or one of the communication team—leaves the project. This happens on a regular basis even in the most stable organizations. Also, the larger the team, the greater the chance that some bit of important information won't get shared with everyone. It's essential to keep complete files and to document the outcomes of meetings and phone conversations. Some project managers recommend keeping notebooks, with separate tabs for the proposal and contract, minutes of meetings, the schedule and budget, correspondence, and logs of telephone calls.

FIGURE 6.8
A computer-based analysis of training methods for 250 people shows that classroom sessions would cost $311,937, while interactive video would cost $163,700 when one considers time and travel costs. (Analysis was created with Interactive Video Training Cost Model Software, copyright © 1988, Lang Learning Systems.)

```
Interactive Video Training Cost Model
Copyright (c) 1988 Lang Learning Systems
                                                              Break Even

Number of people to be trained in group              250          *
Number of actual working days per year               200
Average annual cost (salary plus benefits)         30000          *
Length of training session (in days)                   1
Related travel time (in days, there and back           2
Average cost of travel                               500
Average cost of accommodation                        200
Annual cost of trainer (salary + overheads)        45000
Classroom course development cost                   6000          *
Student package cost (replication)                    55
Classroom trainer / student ratio                     15
Working days per year for trainer                    160
Interactive Video course development cost         150000          *
Student / workstation ratio                           50
Annual workstation lease charge                     2000          *
Session length reduction (percent)                    66
Number of times course is run                          1

F1-Help    F2-Run Model    F3-Break Even                    F10-Quit
```

```
Training Cost Model - Based on 250 people trained for 1 days

CLASSROOM TRAINING                       INTERACTIVE VIDEO TRAINING

Salary costs:          112500            Salary costs:            12750

Development costs:       6000            Development costs:      150000

Direct costs:          193437            Workstation usage:         950

TOTAL COST:            311937            TOTAL COST:             163700

Cost / employee + salary:    1247        Cost / employee + salary:     654

Cost / employee - salary:     797        Cost / employee - salary:     603

                                         Lease cost per year:    10000
Number of times
course is run : 1                        Equipment usage: 9%

F1-Help  F2-Restart Model  F3-Repeat Course  F4-Barcharts      F10-Quit
```

ESTABLISHING THE VALUE OF COMMUNICATION

Few professional communicators, or their clients, know if their rules and tools really work. Individual projects may be well-received by the audience or by peers, but there is usually no data about how this links up with organizational performance. In order to prove communication's worth to the organization, we must document the critical success factors of communication for the enterprise (what things must go right to succeed). Then we need to identify indicators or measures of performance that can be tracked, and develop systems for collecting and using this information. Finally, we need to link up communication and training interventions with changes in performance and overall success factors.

In the previous section, we looked at

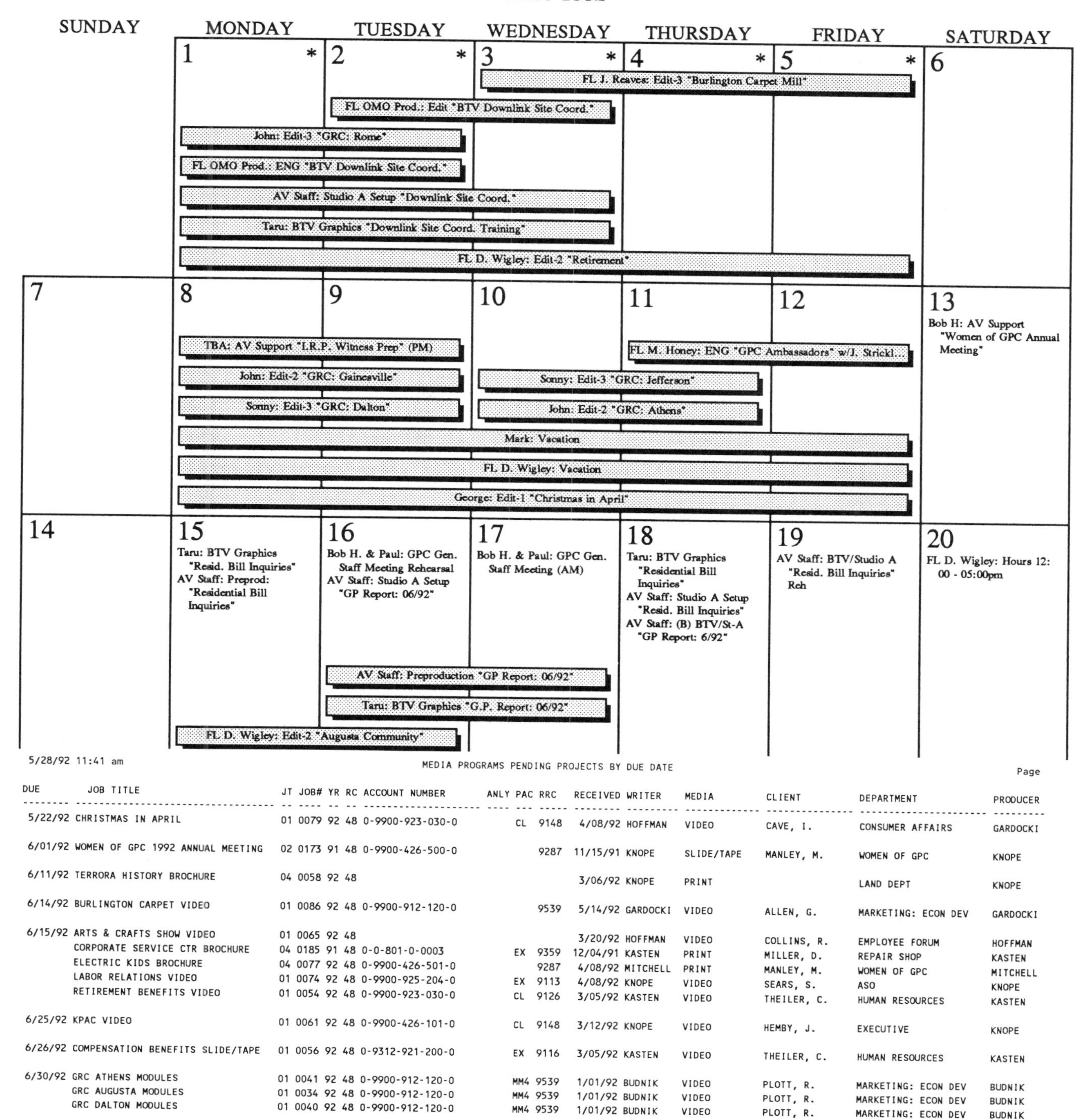

5/28/92 11:41 am

MEDIA PROGRAMS PENDING PROJECTS BY DUE DATE

Page

DUE	JOB TITLE	JT	JOB#	YR	RC	ACCOUNT NUMBER	ANLY	PAC	RRC	RECEIVED	WRITER	MEDIA	CLIENT	DEPARTMENT	PRODUCER
5/22/92	CHRISTMAS IN APRIL	01	0079	92	48	0-9900-923-030-0		CL	9148	4/08/92	HOFFMAN	VIDEO	CAVE, I.	CONSUMER AFFAIRS	GARDOCKI
6/01/92	WOMEN OF GPC 1992 ANNUAL MEETING	02	0173	91	48	0-9900-426-500-0			9287	11/15/91	KNOPE	SLIDE/TAPE	MANLEY, M.	WOMEN OF GPC	KNOPE
6/11/92	TERRORA HISTORY BROCHURE	04	0058	92	48					3/06/92	KNOPE	PRINT		LAND DEPT	KNOPE
6/14/92	BURLINGTON CARPET VIDEO	01	0086	92	48	0-9900-912-120-0			9539	5/14/92	GARDOCKI	VIDEO	ALLEN, G.	MARKETING: ECON DEV	GARDOCKI
6/15/92	ARTS & CRAFTS SHOW VIDEO	01	0065	92	48					3/20/92	HOFFMAN	VIDEO	COLLINS, R.	EMPLOYEE FORUM	HOFFMAN
	CORPORATE SERVICE CTR BROCHURE	04	0185	91	48	0-0-801-0-0003		EX	9359	12/04/91	KASTEN	PRINT	MILLER, D.	REPAIR SHOP	KASTEN
	ELECTRIC KIDS BROCHURE	04	0077	92	48	0-9900-426-501-0			9287	4/08/92	MITCHELL	PRINT	MANLEY, M.	WOMEN OF GPC	MITCHELL
	LABOR RELATIONS VIDEO	01	0074	92	48	0-9900-925-204-0		EX	9113	4/08/92	KNOPE	VIDEO	SEARS, S.	ASO	KNOPE
	RETIREMENT BENEFITS VIDEO	01	0054	92	48	0-9900-923-030-0		CL	9126	3/05/92	KASTEN	VIDEO	THEILER, C.	HUMAN RESOURCES	KASTEN
6/25/92	KPAC VIDEO	01	0061	92	48	0-9900-426-101-0		CL	9148	3/12/92	KNOPE	VIDEO	HEMBY, J.	EXECUTIVE	KNOPE
6/26/92	COMPENSATION BENEFITS SLIDE/TAPE	01	0056	92	48	0-9312-921-200-0		EX	9116	3/05/92	KASTEN	VIDEO	THEILER, C.	HUMAN RESOURCES	KASTEN
6/30/92	GRC ATHENS MODULES	01	0041	92	48	0-9900-912-120-0		MM4	9539	1/01/92	BUDNIK	VIDEO	PLOTT, R.	MARKETING: ECON DEV	BUDNIK
	GRC AUGUSTA MODULES	01	0034	92	48	0-9900-912-120-0		MM4	9539	1/01/92	BUDNIK	VIDEO	PLOTT, R.	MARKETING: ECON DEV	BUDNIK
	GRC DALTON MODULES	01	0040	92	48	0-9900-912-120-0		MM4	9539	1/01/92	BUDNIK	VIDEO	PLOTT, R.	MARKETING: ECON DEV	BUDNIK

FIGURE 6.9 Management reports show project activity during each month. (Courtesy of Georgia Power Company.)

how to justify the cost of interventions by comparing the cost of current problems or lost opportunities with the estimated cost of the intervention. Once a project has been fielded, managers need to follow-up and track the actual results. This should be done in two ways. The first evaluations are generally accomplished by measuring performance right after the intervention has occurred, such as assessing the effectiveness of training by a post-test or by observing simulated on-the-job performance within the classroom setting. Next, measures of actual on-the-job performance, both short-term and long-term are taken. By doing these, we can link communication programs with business results.

"There are a number of ways to keep score in organizations. Almost all organizations have at least one way—to keep financial score. Some keep *only* financial score. Beyond the HPT (human performance technology) rhetoric, it is fairly clear that the HPT profession has hardly ever kept financial score" (Swanson, 1992, p. 603). Swanson and Gradous (1988) have developed the FFB (forecasting financial benefits) model. In it, they measure performance goals, the dollar value of each performance unit, and the time and money needed to reach the performance goal. Out of this, they calculate total performance value gain. One of the factors they point out is the amount of time it takes workers to get up to speed or to meet production goals. While most people eventually become proficient by trial-and-error learning on the job, and while this approach avoids the costs of developing and attending training sessions, money is meanwhile lost through mistakes and inadequate performance. In many situations, performance technology or communication interventions can help employees meet goals within 2 to 3 days instead of several months.

If it's not possible to measure the cost of performance gaps in terms of individual behaviors, another approach is to look at the desired skill level of individuals, rate their average current skill level, determine what percentage of their job is related to using those skills, and then compare what percentage of the salary of the ideal performer the company is actually not getting a good return on. Let's say approximately 30% of a manager's job is to lead meetings, and he rates a 2 on a scale of 1 to 5 in terms of his ability to do so. The total company cost for this manager (his salary, benefits, and overhead) comes to about $100,000. In addition, $30,000 of his salary is being spent on leading meetings. If he did this exceptionally well, we would expect that the company would be getting the full $30,000 worth of his performance. But he only does this at about a 40% level of performance; therefore, the company is losing 60% of $30,000 or about $5000 each year. If there are 20 such managers in a company, it's losing $100,000 in performance gaps. This makes it quite easy to justify the cost of some sort of solution to bring them up to full efficiency and productivity.

Evaluation of projects takes several areas into account:

- the quality and timeliness of specific events or deliverables
- statements about how the audience and clients liked the project
- the overall results in terms of long-term goals
- the relationship of the communication provider with the client
- an analysis of the cost and benefits of the project

Unfortunately, few communicators and human performance technologists ever really evaluate programs. This leads to a vicious circle—no evaluation is done, so there is no data to prove the worth of projects (see Figure 6.10). Communicators don't have much information to share with management about their effectiveness, so they are gradually removed from the mainstream. As this happens, their strategies and programs become less effective, and they become more defensive. As they become more defensive, they fear evaluation and don't do it. And so the cycle goes.

In 1989, Sears Merchandise Group found that the cost of hiring and training their 119,000 new sales associates during just 1 year (due to their high rate of turnover) was $900 each, or more than $100 million. This was 17% of the group's total income for that year!

Costs of this magnitude often lead managers to make cuts in training. The explanation is the absence of relevant information: while wages and training costs are universally measured and known, the return on these investments in employee development is not because the incremental value of better service has long been considered unknowable. Now, however, that assumption is breaking down. Managers are looking for measures that will help them evaluate the relationship between training and employee retention, for example, or the value of the consistency of service that comes from lower turnover (Schlesinger & Heskett, 1991, p. 76).

Other organizations have started to measure the links between communication, turnover, and customer retention. Marriott Corporation reduced employee turnover by 10%—this yielded savings that were greater than the total operating profits of two of its divisions. Merck & Company found that the total costs of turnover are one and a half times an employee's annual salary. And Ryder Truck Rental found that increased training led to decreased turnover (Schlesinger & Heskett, 1991).

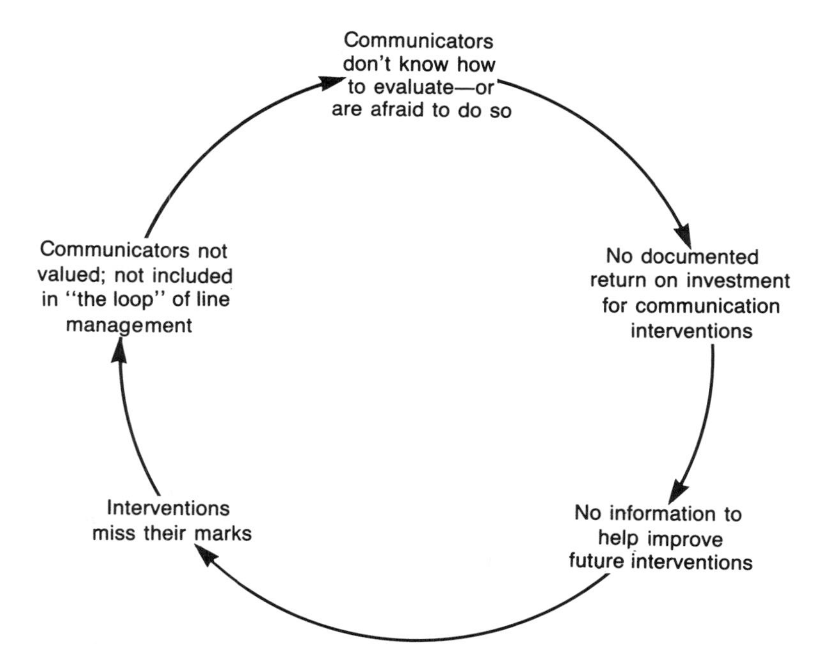

FIGURE 6.10
The "vicious circle" of no evaluation.

Communication needs to be visible on the spreadsheet. What is the cost of meetings when we factor in the salaries of everyone present? What is the value of a highly trained employee? How can information be counted as an asset? For example, what is the worth of two decades of customer information and research studies in the trucking industry? How can three boxes of videotapes shot at plant locations all over the world, worth several thousand dollars in terms of the resources that went into shooting them, appear in a statement of the company's net worth?

Managing formal communication in organizations is a complex job—it calls for leadership, skills in influencing and negotiation, inspired but precise planning, and astute financial analysis. Too few professional communicators have the tools or the expertise to move beyond writing, talking, and producing. For communication to be elevated to, as communication scholar and consultant David Berlo calls it, major product status within organizations, we need to manage it just as carefully and scrutinize it just as closely as we would any other company product. We need to determine the market for it, produce it as efficiently as possible, and decide whether it will provide a good return on investment. Until we can do that, we are, indeed, a frill.

REFERENCES/ SUGGESTED READINGS

Alter, S. (1992). *Information systems: A management perspective.* Reading, MA: Addison-Wesley.

Argyris, C. (1985). Explorations in consulting-client relationships. In C. Bell & L. Nadler (Eds.), *Clients and consultants: Meeting and exceeding expectations,* 2nd ed. (pp. 22–48). Houston, TX: Gulf Publishing Co.

Bell, C., & Nadler, L. (1985). *Clients and consultants: Meeting and exceeding expectations,* 2nd ed. Houston, TX: Gulf Publishing Co.

Berko, R., Wolvin, A., & Curtis, R. (1990). *This business of communicating,* 4th ed. Dubuque, IA: Wm. C. Brown.

Brush, D., & Brush, J. (1981). *Private television communications: Into the eighties.* Cold Spring, NY: HI Press.

Carlberg, S. (1991). *Corporate video sur-*

vival. White Plains, NY: Knowledge Industry Publications.

Communication function needs better management. (1991, November). *IABC Communication World,* 11.

D'Aprix, R.M. (1977). *The believable corporation.* New York: AMACOM.

Duncan, J.B., & Powers, E.S. (1992). The politics of intervening in organizations. In H. Stolovitch & E. Keeps (Eds.), *Handbook of human performance technology* (pp. 77–93). San Francisco: Jossey-Bass.

Gayeski, D. (1985, September). The role of the consultant: In-house vs. out of house—The case for interaction. *Chicago P&I.* Chicago Chapter, National Society for Performance & Instruction.

Gilbert, T.F. (1978). *Human competence.* New York: McGraw-Hill.

Kanter, R.M. (1989). *When giants learn to dance.* New York: Simon and Schuster.

Keeping calm in the crisis. (1989). *Case studies in organizational communication: 2* (pp. 30–34). Elmhurst, IL: Council of Communication Management.

Lennon, K., & Sheldon, K. (1992, January). From Canada to California: New directions in corporate communication. *IABC Communication World,* 43–49.

Marlow, E. (1981). *Managing the corporate media center.* White Plains, NY: Knowledge Industry Publications.

Meltzer, M. (1981). *Information: The ultimate management resource.* New York: AMACOM.

McKenney, J.L., & McFarlan, F.W. (1982, September/October). The information archipelago—Maps and bridges. *Harvard Business Review,* 109–119.

Redding, W.C. (1979). Graduate education and the communication consultant: Playing God for a fee. *Communication Education, 28,* 346–352.

Robinson, D.G., & Robinson, J. (1989). *Training for impact.* San Francisco: Jossey-Bass.

Rosenzweig, S. (1991, December). Which type are you? *Publish,* 10–11.

Rothwell, W., & Kazanas, H. (1989). *Strategic human resource development.* Englewood Cliffs, NJ: Prentice Hall.

Schlesinger, L., & Heskett, J. (1991). Breaking the cycle of failure in services. *Sloan Management Review, 32,* 17–29.

Sharpe, J. (1988). Corporate size and the individual consumer: Marketing's challenge for the 1990s. *Multinational Business, 4,* 32–34.

Silber, K. (1992, March). Corporate ID/PT standards. *Performance & Instruction,* 6–10.

Swanson, R. (1992). Demonstrating financial benefits to clients. In H. Stolovitch & E. Keeps (Eds.), *Handbook of human performance technology* (pp. 602–618). San Francisco: Jossey-Bass.

Swanson, R., & Gradous, D.B. (1988). *Forecasting financial benefits of human resource development.* San Francisco: Jossey-Bass.

7 Where Do We Go From Here?

We're nearing the end of a journey that's taken us from the roots of communication in clans, courts, and corporate cultures to archetypes of contemporary organizational communication practice, with stops along the way to admire some examples of new forms of corporate information architecture. There are signposts out there, pointing to our individual and collective futures, but like most highway markers, they offer more options than directions.

WHAT DO ORGANIZATIONS AND INDIVIDUALS WANT TO ACHIEVE?

As we've seen, formal communication systems are powerful means for acquiring and disseminating information that can lead to dramatic organizational change. Communication rules and tools can create a culture that's oriented toward predictability, order, and formality, or one that's characterized by innovation, unpredictable amounts and types of information, and spontaneity. Modern theories and technologies can encourage us to work in physically isolated environments aided by communication devices, or to embrace a more personal style of management by walking around. There are forces that impel us to control the quality and quantity of messages, and equally powerful ones that advocate a loosening of bureaucratic channels and chains of approval. Executives and communication professionals in each enterprise will need to decide how to build a system that works for them. But some similarities among information-age organizations do exist.

Every organization wants to enhance its performance. More and more, communication is being touted as the key.

> Quality performance depends on two premises. The first is that people will perform at desired levels if they are capable, have clearly defined job roles, know what is expected of them, have the tools to do the job, have the knowledge and skills to perform, receive feedback on how well they perform, and perceive and receive rewards for good performance. Together, these factors make up a performance system. The second premise is that managers and supervisors can improve the performance of their employees by clearly defining and communicating an organizational purpose, and defining and aligning the organization's values, aims, and goals, the key job roles, the objectives, and the financial, management, and other systems behind the purpose or mission of the organization. (Powers, 1992, p. 247)

Again and again, we see the essential roles of formal communication systems when

- articulating and communicating the organization's goals, values, culture, and image
- defining and clearly explaining the roles and activities necessary to reach the organization's goals
- interviewing and selecting the right people
- providing them with the necessary information and skills to do their jobs well
- giving feedback and coaching on individual and group performance
- creating a culture that nurtures open, honest, fair, and multiway flows of communication and collaboration

- collecting and analyzing key performance indicators, strategic plans, and policies
- managing information flow rate to optimize an individual's ability to use it
- establishing standards and policies so that internal and external communication is audited and aligned with organizational values and goals

Managers tell me that what worked for them a decade ago just doesn't work anymore. Leaders are frustrated at not being able to drive these clumsy, large beasts we call corporations, yet seem to feel that they need to change direction and move quickly. Assuming that these common statements have any ground in reality (or make good, old-fashioned business sense), how can we as Renaissance communicators improve the situation? Tichy and Devanna (1986) use the metaphor of a play to outline the challenges facing today's executives. In Act I, leaders must make their organizations more aware of external competition through formal presentations and communication with peers from other organizations. In Act II, leaders must create new visions for the organization and somehow communicate that the new way is better than the old one. In Act III, executives must find ways to support the new culture by designing new systems and dealing with the redistribution of power.

No matter what the business or culture, organizations need to develop, as Zuboff (1988) calls them, *informated environments* in which workers at all levels will have access to the data necessary for decision support and to tools for collaboration and communication. Moreover, we need management structures that consolidate and give leverage to communication rules and tools. The question now is who will create and manage these systems. As we've seen, corporate communication is currently an archipelago of (sometimes warring) islands of practice, each using different languages and artifacts. They're seen as rather amusing, colorful, and sometimes useful colonies by the mother country of the line organization. To gain full citizenship, in most cases, will take quite a bit of development.

WHO WILL PROVIDE THE DIRECTION?

It is apparent that professional communicators need to create a new image, values, and identity—not just for their clients—but for themselves. Too often stereotyped as hack journalists, equipment operators, or smooth-talking orators, we need to establish closer links with line management, develop skills in management and in policy-making, establish standards, and generally get our acts together.

Who will be in charge of this professional development—universities who prepare professional communicators, professional associations, accreditation committees, or individual practitioners? It seems as though each one of these constituencies is waiting for the others to make a move. Professors contemplating curricular changes ask, "But where are the jobs for these new Renaissance communicators?" CEOs ask me, "Where can I look to hire communication strategists—not just press-release writers or camera operators?" Practitioners are waiting for their organizations to create a post of vice president for communication systems, and for the job to somehow be handed to them. It's a typical chicken-and-egg problem. Obviously, someone has to break the cycle . . . and some are.

We cannot wait to be told what to do, either by our professors or by our CEOs. In fact, this reactive stance of waiting for orders is exactly what characterizes the stereotypical professional communicator. As a group, we tend to wait for an event to happen, and then report on it. Or we expect clients to identify performance problems, brainstorm solutions, and ask us to produce interventions. We read ads, go to trade shows, and listen to vendors to determine what to buy, and then recommend technologies based on their engineering specifications rather than as solutions to problems. Likewise, we wait for executive-level jobs to be presented to us before we prepare ourselves for them.

Convincing a whole era of executives to rethink their implementation of formal communication systems and their relationships with professional communicators will not happen overnight, but obviously this must be the goal. What do they think about us now, and what are their hopes?

WHAT DO CEOs THINK ABOUT COMMUNICATION?

Many communication professionals find it difficult to get the ear of management and move out of their roles of producers or spokepeople. Brian Manning, director of information systems for Hughes Research, capsulized the frustrations of many of his peers when he told me, "The major problem is getting into dialogue with senior executives so we can discuss strategic opportunities." It seems as though when we're invited to participate in policy-level discussions, we are not perceived as making a contribution, and often, we're not invited to the party at all.

Carlberg (1991) says that 60% of the corporate video professionals he interviewed had not had corporate goals communicated to them by their supervisors. Most of these professionals thought they knew where to find them, but hadn't bothered. One producer said, "We tried to get hold of the company's goals, but they're kept like some deep, dark secret. I'm sure that my supervisor has never even seen them" (Carlberg, 1991, p. 9).

One president of a large human services organization said that his stereotype of communication professionals was that they weren't innovators. He didn't look to them to bring new ideas to the table, but rather to use tried and true vehicles and strategies. In his organization, he was accustomed to scientists and managers bringing him analyses of and proposals for new technologies, complete with financial plans and expected goals. He was aware that there were many new communication technologies that probably could improve the organization's internal and external communication, but no one ever came to him explaining or proposing them.

The situation today is not so different from three decades ago when Lundberg (1965), then chairman of the Bank of America, offered some penetrating observations at a conference of the Public Relations Society of America:

> Members of management are public relations oriented in the truest sense of the term. They have to be, for their jobs—effectively running a business—require that they consider every so-called public sector that exists in relationship to the whole company and its goals. Management must necessarily gauge each act of the corporation within this total perspective. If the public relations profession will make the necessary effort and commitment to achieve this same perspective—and it requires a lot of work to learn intimately the total thrust of a concern—then a place at the management conference table will be willingly, even eagerly, accorded. Conversely, if public relations remains trapped by its own specific perspective then it will be relegated to the status of a technical department (Lundberg, 1965, p. 3).

He is not alone in his assessment—a study conducted by *Business Week* in 1979 reported that 60% of CEOs didn't trust their own communications officers (The corporate image, 1979).

Many CEOs are already communication-savvy and will not wait for someone on their staff to propose new systems. Fred Smith, CEO of Federal Express, was the person who initiated their satellite TV and interactive video systems. Marcie Abramson, vice president of operations at U.S. Healthcare, created her own vision of a high-tech service center environment and actually had to convince her MIS department that her plans could become reality. The communication and training departments had little or nothing to do with building her dream.

Executives don't just want to hear good news, they want to hear analyses of the company's shortcomings and what can be done about them. CEOs also expect to be told when they're about to make a big mistake. One of the biggest stories of the 1980s was a communication faux pas made by General Motors when, at an elaborate breakfast, they celebrated the signing of a new union contract, which provided for extensive union concessions due to the recession. At 4 P.M. that same day they announced a new bonus package for their top executives. The outcry by union members and the public was tremendous. Although

GM employed professional communicators in their internal communication department at the corporate level and had, in total, about 300 people paid to concentrate on employee communication, the chairman says that no one spoke up (Ruch & Goodman, 1982).

Richard Rosenberg, CEO of the Bank of America in San Francisco, described what CEOs will look for in public relations professionals. CEOs will look for them to:

- be consummate professionals—master all aspects of a craft
- know the business of their firm—what, how, and why
- be concerned with profit and loss—manage the department as a profit center
- take the initiative, don't wait to be asked to do something
- give their best advice whether it's popular or not
- keep confidences
- be the best informed, most articulate, most persuasive people in the organization (CEO describes, 1991).

Harold Burson, chairman of Burson-Marsteller, one of the largest public relations agencies in the United States, describes the standards against which CEOs will be measuring communicators:

- intelligence
- leadership potential (presence and action)
- ability to get along with people—supervisors, peers, and others below them
- tuned in to what's going on within and outside the organization (one who reads a lot, listens, and observes)
- great credibility and integrity—trusted by others
- good communicator, verbally and in writing (CEOs set standards, 1992).

According to the 1992 IABC Research Foundation study "Excellence in Public Relations and Communication Management," CEOs said their communication departments provided a 184% return on investment. Their top communicators said it was 188%, but believed that their CEOs would rate the ROI at 127%. Interviews determined that CEOs want top communicators to be senior advisors, not just department managers or bridges to the public. But most of the time, they don't live up to these expectations. The authors of this study say that the root of the problem is top communicator's lack of evaluation, research, and environmental scanning techniques, and little knowledge of conflict resolution theories (Latest 'excellence' study results, 1991).

The 1991 Wyatt Communication and Training Survey of more than 2200 CEOs and executives in human resources and communications came up with similar results. More than one-half of the executives saw the role of communication in an organization as strategic, while only about one-third of communication practitioners themselves viewed their roles as strategic. Two-thirds still thought of themselves as craftspersons. "The question is, are we as professionals actually seizing upon our new-found respect of senior management and taking greater responsibility and exerting increasing influence in planning our organization's future? Or, are we waiting until our CEOs come to us and say, 'Where were you when we needed you most?' " The Wyatt Company, a human resources consulting firm, found that most communication professionals fail in planning and in measuring results. Fewer than one-third of the survey participants said that they have a formal communication plan, and less than 10% measure communication's return on investment! (The changing role, 1992, p. 38).

During my research for this book, I talked with Roger D'Aprix, a principal in the communication consulting firm William Mercer, about communicators' reputations in most organizations. He said bluntly that nobody has any respect for the communication people. In his role as a senior consultant, he says he typically deals with vice presidents of human resources or of quality. When I asked him how he rose to his former position of director of communications at Xerox and gained influence in that role, he said that he made a name for himself outside the company—then they listened to him. He advises communication professionals to reduce communication strategies to some simple solutions and make those solutions look like a product that management can understand and buy into.

CEOs need help in becoming their organizations' most influential and skilled communicators. When faced with the need to turn a company around, an executive often finds that the keys to that renewal lie in communication.

Preston Trucking underwent major management changes based on performance technology concepts in the early 1980s. The reason? They were in deep trouble in terms of competition and labor problems. Their CEO, Will Potter, commented, "It's amazing what you accomplish when you're running scared." Basically, Potter established clear and direct channels of communication. Preston now has required orientation and training programs, most of them conducted in small groups and featuring Potter who asks each associate (employee) to sign a commitment to excellence. When somebody breaches that promise, Potter is likely to call him up directly—even the lowest ranking driver—to ask why the person let him down. Performance on a wide range of behaviors, from safety to fuel efficiency to late shipments, are measured and displayed prominently in all the terminals, along with their mission statement and the commitment to excellence. Associates know what's expected of them, and get clear feedback on their successes and failures. After several years of these practices, both the financial problems and the labor unrest had vanished. The company won a productivity award from the U.S. Senate and was named one of the 10 best companies to work for in America (Pearlstein, 1989).

It is such practices as Preston Trucking's clear statements of expectations and evaluations, U.S. Healthcare's informated environments, and Federal Express' communication systems that link offices that characterize twenty-first century corporate communications worldwide. As we look toward the future, not only are we shaped by our organizational leaders, but we are also influenced by larger environmental trends.

MANAGING INFORMATION AND KNOWLEDGE SYSTEMS

Although I don't want to get into fortune-telling, there are some clear images of organizational information systems emerging (even without a crystal ball or other multimedia communication device!). These general societal and economic trends, centering around the concept of information, will mean a number of things.

- *More information will be available to us.* This will offer us many options but increases the chances of data overload.
- *Information is expandable.* That is, information is not used up when it's shared; however, the capacity of humans to process it is not infinitely expandable.
- *Information will be substitutable for capital, labor, and objects—even money.* As Toffler says in *PowerShift* (1990), first-wave money is metal or a commodity. Second-wave money is printed paper. Third-wave money is electronic pulses—it is information—the basis of knowledge.
- *Since information is the new money, our wealth can easily slip through our fingers.* Information is transportable and, in fact, diffuses and often leaks. We will need new safeguards for our resources.
- *Information will come faster.* We will need to scan, sort, and process it better.
- *A higher proportion of information will come from further away.* We will be challenged to learn new forms of communication and increase our skill in communicating with people quite different from ourselves.
- *More information will be sent point-to-point than point-to-mass.* This will give audiences more control over what they receive and give communicators more channels to use. It will also increase the number and variety of messages that we need to construct.
- *Acquiring and disseminating increasingly more information will not be considered desirable.* We are already facing information overload. We need to manage this system so that our organizations are not choked by it.
- *Information will be increasingly mediated*

by technology. Management will support new technologies because of the efficiencies they can provide. Employees will demand new technologies because of the control and opportunities for input they make available.

- *More tasks will be done by and with computer-based information systems.* This means that intelligent workstations will be a primary means for efficient interactive communication and training.

This last point should not be underestimated. It means more than learning how to use a keyboard and software packages—it has major implications for the discipline of communication. Most computer systems (unlike most people) require precision in communicating with them. This imposition of strict rules can be startling and important (Bassett, 1968). "It can be terrifying to realize that virtually every communication we pass along to others contains errors or ambiguities that may cause total misunderstanding; that, in fact, much of the information we have been transmitting and receiving is misunderstood and our assumptions about the accuracy of our communication are no longer acceptable" (Bassett, 1968, p. 87). Instead of being annoyed at this, communicators should welcome this. For the first time, comments Bassett, we'll be forced to face the fact that we don't know what we are saying. "Such a realization may go further toward solving basic human problems in the world than any other single factor" (Bassett, 1968, p. 87).

In fact, many communication scholars, such as David Berlo, use computer analogies when illuminating human communication challenges. He comments that we don't feed computers data in formats that their software systems don't expect—it must come at a rate and in a form that can be processed. We use defined and standardized words and syntax to request actions. Many social critics are afraid that organizations are treating employees like computers—if only we were that humane!

Knowledge is an organization's most important asset and learning is the critical technology of today's economy (Carnevale, 1992).

> Most competitive improvements can't be bought. They have to be learned. . . . If learning is at the heart of economic progress, then the processes of emulation and commercialization are its arterial system. They extend the reach and range of new knowledge. . . . Networks at all levels are primary structures for generating as well as disseminating new knowledge. Networks can be formal or informal relationships among individuals or organizational entities. Learning networks can be cooperative or collaborative. In cooperative networks, learning occurs in isolation and then is passed on to other individuals or institutions. In cooperative networks, learning and dissemination tend to be additive and dissemination tends to occur sequentially. Collaborative networks, in contrast, effectively create a common work space where learning and dissemination increase exponentially and dissemination is simultaneous. (Carnevale, 1992, pp. s1–s7)

Organizations must learn and build knowledge in many different ways. Argyris and Schon (1978) developed a concept called *double loop learning* or learning-to-learn. "When the error is detected and corrected permits the organization to carry on its present policies or achieve present objectives, the error-detection-and-correction process is single loop learning. . . . Double loop learning occurs when error is detected and corrected in ways that involve the modification of an organization's underlying norms, policies, and objectives" (Argyris & Schon, 1978, pp. 2–3). In organizational cultures dominated by single-loop learning, people play games to avoid admitting to mistakes and to curry political favor (Argyris & Schon, 1978). For example, a manager may always give good news before bad news, place the blame for allocation of resources on top management, or play down a failure. Alternatively, double-loop learning organizations extend the range of errors and failures they can detect, admit, and learn to solve. Members become more skilled at collaboration, reflection, and joint inquiry, and get better at learning and adding to corporate knowledge.

Organizations learn and build knowledge in a variety of ways, including hiring, educating, and training knowledgeable people, documenting knowledge in print and data bases, improving technology and procedures, and creating systems and standards. In fact, companies must learn efficiently in order to stay ahead of the competition. The

Boston Consulting Group model *BCG learning curve* is widely used as an indicator of how well a company learns. It plots the company's cost of producing a product against the market price. If a company learns well, it should be able to reduce its costs faster than the market price is reduced. Another formula for corporate learning is L ≥ C. In order for an organization to survive, its rate of learning must equal or exceed its rate of change (Revans, 1980).

To build knowledge, it must be surveyed, elicited, organized, documented, leveraged, and automated. In order to do this, communication professionals must

- survey the knowledge resources of the organization
- determine the knowledge and expertise required to perform functional tasks
- package and distribute that knowledge through training courses, manuals, or knowledge-based systems
- restructure the organization to use knowledge most efficiently
- create and monitor long-term knowledge-based activities, such as research and development efforts, strategic alliances, and hiring
- safeguard proprietary knowledge
- provide a knowledge architecture for the organization

Knowledge must be valued, rewarded, documented, and catalogued for easy retrieval, and organizations need to put a bottom-line figure on it (Wiig, 1990). As we become information-age organizations, our management of information and knowledge will be crucial.

Much is being written about corporate knowledge and organizational learning. It is interesting to note that these activities or concepts center around communication; in fact, the Renaissance communicator should be the person in charge of corporate knowledge—is this perhaps another idea for a position or departmental title? Dixon (1992) categorizes organizational learning into five areas:

- information acquisition
- information distribution and interpretation
- development of meaning
- organization memory
- retrieval of information

This might be a categorization of professional or formal communication as well. As you can see from Figure 7.1, Dixon includes in her model all of the functions we've been discussing in this book. In fact, it is an excellent summary of the management domains of the future professional communicator.

VALUES AND ETHICS

Even as we face economic challenges in the 1990s, organizations are turning, more than ever, to an examination of their values and ethics. As an example of this, General Electric executives Jack Welch (chairman and

FIGURE 7.1 A model of organizational learning. (Reproduced with permission from Dixon, N., 1992. Copyright © 1992, Jossey-Bass, Inc., Publishers.)

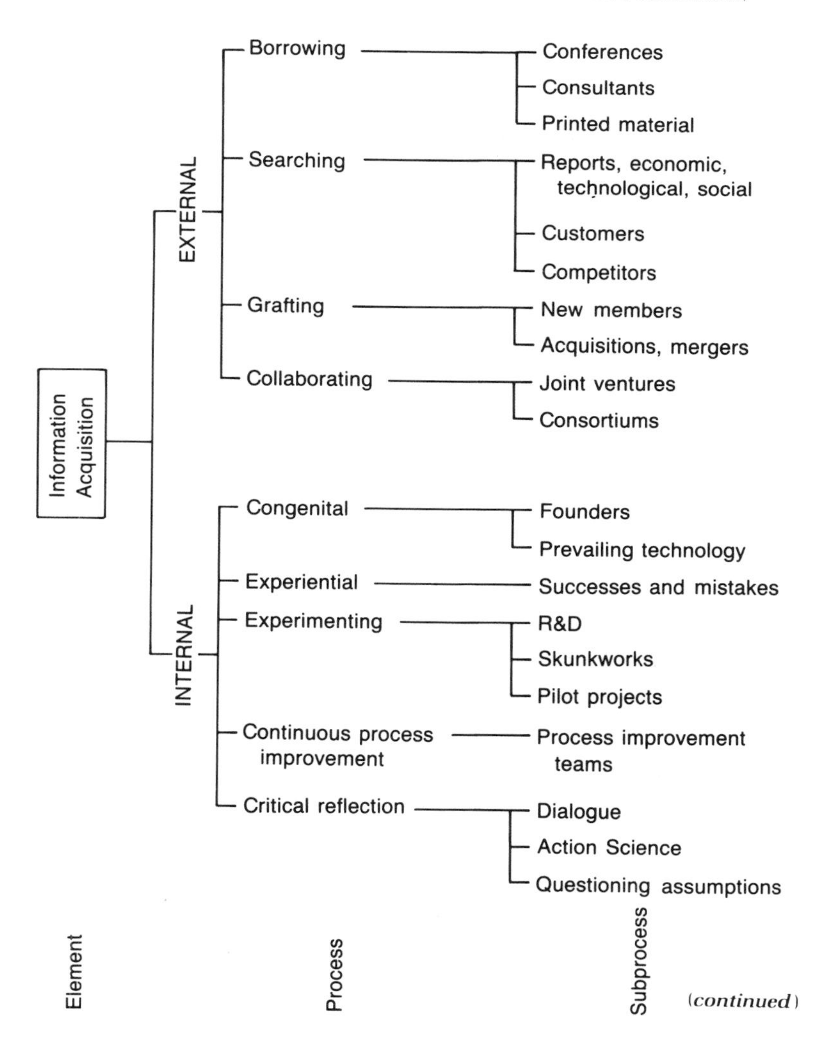

(*continued*)

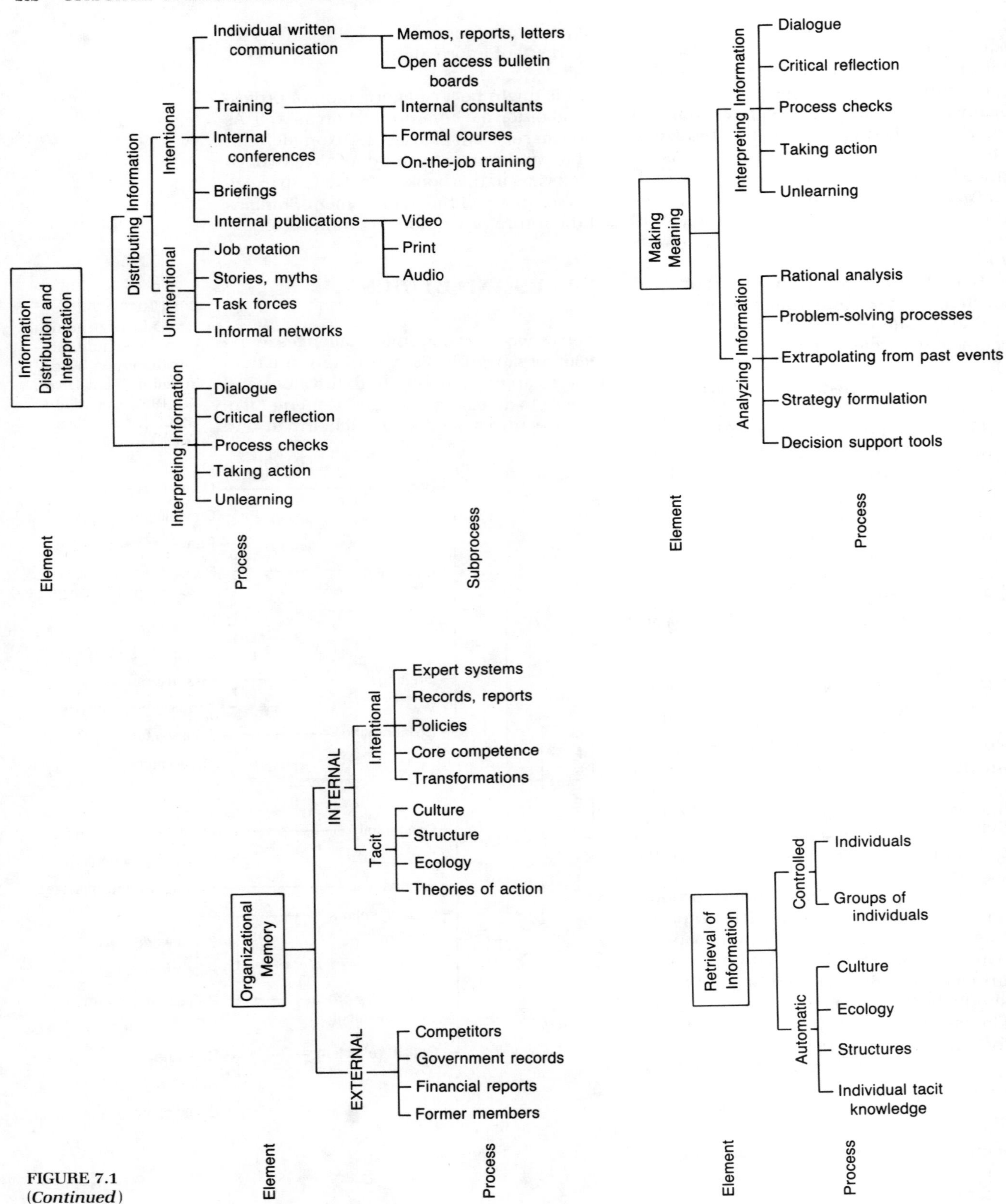

FIGURE 7.1
(*Continued*)

CEO) and Ed Hood (vice chairman) stated in their 1992 Annual Report, "For the past few years, those of you who invest in our Company, and share our interest in and affection for it, have patiently read letters from us that dealt only minimally with traditional business data and focused instead on soft concepts and values" (General Electric, 1992, p. 5).

How can professional communicators help organizations determine, articulate, and reify values? I'd like to share parts of a letter that was sent to me by Walter F. Giersbach (personal communication, May 1992), director of internal communication at the Dun & Bradstreet Corporation.

Businesses increasingly are coming under fire for actions that are ethically questionable, if not outright illegal. The Dun & Bradstreet Corporation faced this issue of business practices and values squarely because its 'products' are business ratings, marketing intelligence and other information that must be unbiased, impartial and totally honest.

A great deal of labor went into preparing a 'Statement of Values' that all employees (D&B calls them 'associates') could reaffirm [see Figure 7.2]. In mid-1991, more than 100 senior managers of the corporation and its operating units met to discuss D&B's business practices and the actions that they would take to make them a reality. The resulting 'Statement of Values' was issued in December along with 10 'Values Actions' that are the reality of business practices as they apply to customers, fellow associates and shareowners. Among its points, the Values Actions call for establishing a corporate office of business practises, twice-annual performance reviews and annual skip-level interviews with all associates, mutual goal-setting, attitudinal surveys and communication of D&B values to all associates, new hires and customers. Most of the Values Actions are now in place, and indeed were already practices by D&B companies.

The Values message was introduced to the more than 58,000 D&B people in 43 countries through a lead story in the December issues of D&B Update, the publication for all associates. The 'Statement of Values' in English and a total of 11 other languages and dialects was distributed to D&B offices worldwide.

A videotape—'Values at Dun & Bradstreet: A conversation with Charlie Moritz'—was the centerpiece of the communication effort. It focused on a candid and unrehearsed meeting in November 1991 at which D&B's chairman and chief executive officer, Charles W. Moritz, and five D&B people discussed what values mean in the business world. The tape, in English and translations—voiceovered U.N.-style—has now been presented to all D&B associates worldwide to stimulate discussion.

A CEOs or COO's worst nightmare is that a mission statement will end up being tacked to a bulletin board in the coffee room and ignored until it yellows with age. We were determined that this would not happen. To support the Values effort, poster-size versions of the 'Statement' were framed for reception areas, operating-unit employee publications carried articles on what values mean in their operation, and 4 × 6″ acrylic 'L-shaped' framed versions were produced for all associates participating in the discussion sessions.

The payoff has come from a reaffirmation of values and business practices that transcends nationalities, cultures and demographics. One person's response is typical: 'I now know what the expectations of the organization are.'

S T A T E M E N T O F V A L U E S

As the men and women who are
The Dun & Bradstreet Corporation, we are a team – One Company
united through shared values relating to our ethics, customers,
ourselves and our shareowners.

E T H I C S

We will practice the highest standards of personal ethics
and integrity so that in all our relationships
we can have pride in ourselves and our company.

C U S T O M E R S

We will strive relentlessly to exceed our customers' expectations
so that they will want to continue to do business with us.

O U R S E L V E S

We will respect and treat each other as individuals who want the
opportunity to contribute and succeed. Each of us
will be accountable for quality and continuous improvement
in all we do. We will work to be the best.

S H A R E O W N E R S

We will accept our responsibility to be effective stewards
of our shareowners' resources so that through our performance
shareowners are properly rewarded for their investment
in Dun & Bradstreet.

By our living and working in accordance with these values,
everyone who is affected by our behavior will say of us,
"Customer Focus is how they do business."

FIGURE 7.2 Professional communicators are influential in assisting their organizations to articulate their goals and values. (Courtesy of The Dun & Bradstreet Corporation.)

Contrast this with the typical confusion of employees about where the company is going and what is expected of them.

> In looking at the results of a study conducted by the Management Institute at Columbia University, more than 60 percent of the middle managers and professionals working in successful companies complain about lack of goal clarity. Few were able to tell us what the organization's strategic objectives were. These intelligent and well-educated people had few clues that would have enabled them to manage themselves—they had to be told what to do! They tended to spend an inordinate amount of time analyzing the behavior of those who advanced to see if they could divine the rules that led to success. (Devanna, 1992, p. 224)

Clearly, we need to help executives to formulate and to explain organizations' values and goals. Professional communicators will be kept busy if we do nothing more than this, especially in today's and tomorrow's rapidly changing environments; however, just turning on the information spigot is not the answer.

The role of the professional communicator is to design messages and provide information and communication conduits; most practitioners feel that their goal is to provide more information to more people. However, there are several important factors that need to be balanced.

Is Access to Information a Right or a Privilege?

We tend to confuse our democratic political ideologies, which hold that everyone is entitled to as much information and education as possible without endangering personal privacy or national security. However, organizations do not operate under these same rules. How much information should employees or the public expect from a corporation's management? Are organizations under the obligation to provide information and instruction that will allow employees to develop to their fullest potential and to participate in democratic decision making?

The Conference Board (1972) first noted some important trends of what we now call the information age, and pointed out the differentiation between the haves and have nots in terms of information access. Certainly, most individuals in developed nations cannot be described as information poor. In a given year, the average American will read or complete 3000 notices and forms, read 100 newspapers and 36 magazines, watch 2463 hours of television, listen to 730 hours of radio, read 3 books, buy 20 records, talk on the telephone for almost 61 hours, and encounter 21,170 ads (Lederman, 1986). And I've not heard anyone complain about a lack of memos, circulars, videotapes, or meetings at work.

However, we cannot assume that access leads to fairness. While it's easier to have access to information than to other resources, it does not substitute for budget or authority. Merely providing information will not assure equality. Many writers have observed that once informed, people's choices are maximized. Will the informed be harder to control (Cleveland, 1990)? And if so, are managers obliged to provide information? In fact, are they obliged, by virtue of their custodianship of the company's resources, to not provide such information that could create profit-reducing situations?

Another misconception is that we should constantly strive for complete understanding. Many communication experts would advise us against assuming that it should always be our goal. Strategic ambiguity may be good; in fact, it may promote cohesion by allowing people to concentrate on areas of commonality. For instance, many organizations use vague terms like "excellence" in their messages, and we may criticize the creation of such ambiguity. However, people will interpret according to their own individual styles, which is probably better than coming up with a more thorough, operational, and concrete message, which would not apply to or which might even offend some (Daniels & Spiker, 1991).

A related issue is access to technology. The old political slogan of "a chicken in every pot" has been modernized by many liberals to become "a computer at every desk." Winner (1989) lists several faulty assumptions made by computer romanticists:

- people are bereft of information
- information is knowledge
- knowledge is power
- increasing access to information en-

hances democracy and equalizes social power

Even in societies where there is no shortage of information, such as the United States, there are still many people who are functionally illiterate. They have not learned to translate and to use this information. So merely rewiring society will not make us more democratic or more effective as countries or as individuals. Moreover, since information is a competitive advantage, we must ensure that it is not leaked to competing industries or countries. In many cases, it is unwise to provide individuals with more information than they need to know. They are likely to go to work for the competition and unless stringent nondisclosure agreements were made in advance, along with them will go valuable corporate information. Privacy may also be compromised.

As communication professionals, we will be challenged to define what information is required, what is desired, and what is confidential and proprietary. We need to determine which positions have access to what information, and under what conditions. While balancing safeguards and controls, we need to encourage employee understanding, development, and participation.

What Are the Legal Issues?

Obviously, there are any number of confidential records that organizations maintain—its list of customers, proprietary formulas or procedures, personnel data, and strategic plans. There are national laws, which to some effect influence what we can do with such information, and increasing numbers of court cases, which are defining acceptable practice. For example, a Swedish law on codetermination at work requires that management keeps unions informed on the economy of the business and guidelines for personnel. The union has the right to examine the books and other documents held by the employer. In the United States, certain questions are illegal to ask of job applicants. A number of successful suits have discouraged personnel officers from providing anything except the most perfunctory statements about former employees when they are asked for recommendations.

A number of issues center around the use of new technologies to monitor employees' behavior. Such practices as listening in to phone calls or counting computer keystrokes are common means of supervision that are being called into question. Although these measures have typically been applied to those in junior-level positions in organizations, such as secretaries, telemarketing representatives, or data entry clerks, the extensive use of computer technologies might lead such monitoring to be applied to higher level work as well. For example, it is quite possible to determine the number of e-mail and phone messages one sends and receives and with whom one is communicating. Although supervisors have always watched employees' work in order to provide evaluation and coaching, the major problems with these measures is the possibility for such data collection to occur without an employees' knowledge.

We need to develop policies and procedures that make appropriate use of record-keeping when it is done with an employee's full knowledge. In many instances, people want their performance tracked because it provides objective data about their contributions. In some organizations, the number of claims processed or phone calls answered lead to direct incentives in terms of pay increases or commissions. Response files from computer-based testing can document an employee's mastery of skills and concepts, making her a more viable candidate for promotion.

However, we must avoid situations in which employees are being monitored without their consent and awareness, make sure that the data collected is accurate and that its interpretation is done correctly, and ensure that the data does not get into the wrong hands. We also need to overcome the natural resistance to and negative impacts of employees who feel that they are being watched, and make sure that the kind of data we collect doesn't lead to other performance problems. For example, if the number of orders taken is a measure upon which pay increases are given, can we be sure that this doesn't lead to curt behaviors and poor customer service? Moreover, we must retain the ability of people to function as humans, not machines, and avoid the creation of oppressive work environments.

How Can We Avoid Information Overload?

A balance must be struck between opportunity and overload (Schramm, 1988). We need to get better at determining the optimal information load and adjust the design and dissemination of messages accordingly. We have seen that more information is not necessarily better; it doesn't inevitably lead to happier or smarter organizational members. In fact, it can lead to frustration, burn-out, confusion, and poor performance.

Information load is a function of the volume, rate, and complexity of messages to be processed (Farace, Monge, & Russell, 1977; Shockley-Zalabak, 1988). Goldhaber et al. (1979) have said that information load is also dependent on an individual's cognitive complexity (ability). So, we have to look not only at the design, timing, and number of messages, but also to the ability of individuals to process them. This is a multidimensional problem that is obviously not easily solved.

The problem is only getting worse. In new organizations that are more innovative and partnership-oriented, there is an enormous communication overhead. There is more time spent in meetings and selling ideas than in commanding. In many organizations, people spend more than half their time in meetings, and produce and circulate an embarrassment of information riches (Kanter, 1989). Goldhaber et al. (1979) predicted that management would increasingly face its inability to "cope with and tolerate the ravaging effects of information overload" (p. 79). They point out that at least half of an organization's information requirements don't directly deal with the product or service of the organization; rather, government regulations and general record-keeping flood most organizations' file cabinets. The authors maintain that this glut of information may lead to disastrous communication results, such as important information being ignored and the making of arbitrary decisions, failure to take any action, misplacement of significant information, distrust of message sources, and more variations in the interpretation of messages as they pass through many channels and intermediary sources.

There are no easy answers to the question of how to deal with information overload. One approach is to decide between the nice-to-know and need-to-know. We must recognize that excess data can actually hinder the decision-making process. All information is costly, so the cost of collecting it must be scrutinized. "The managers who 'demerchandise' information and perform their tasks with the best and least—not necessarily the most—information will ultimately gain and maintain power" (Meltzer, 1981, p. 18).

Farace et al. (1977) provide some ways (from easy to difficult) to approach this task:

- ignore information
- work according to rules, despite the load
- allow waiting lines
- set message-priority categories and select high-priority ones
- destroy lowest priority messages
- allow clients to self-serve their own needs
- reduce performance standards
- establish new performance standards and new ways to meet them
- develop rules for chunking
- turn work over to subcontractors
- create branch offices
- create slack resources to handle peak loads
- create lateral relations within the organization
- create self-contained task forces
- implement MIS
- reorganize to structure a fundamental change in the organization

---

The Bank of Montreal's Corporate Policies and Procedures Department realized that the volumes of documentation and policies they were producing were not only costing thousands of dollars, but were so cumbersome that they were rarely updated properly or even used. As the department became involved with change management, they developed a method of documenting the capacity of the branches to absorb and to implement new information. One of the results is the change management calendar which plots out the cumulative changes which a branch is expected to handle, from simple manual updates (given a score of "0" or "1") through new product releases or system implementations (given scores of 15 or above). By charting out all the rollouts of training programs, new regulations and products, as well as holidays and unusually busy times (which decrease the capacity to deal with change), it is easy to see exactly how many changes are being put forth for potential execution. (See Figure 7.3.)

The calendar includes a critical line, which marks the capacity of the branches to deal with new information. The yell-line is the point beyond which people will literally yell from information overload. Systems such as this one make the information load on organizations concrete, and allow various departments to coordinate their introduction of new systems and documents so that the end-users can effectively deal with them.

Xerox uses an executive information system and new planning and control process to standardize and reduce the information provided in their annual planning meetings.

> Previously, the company's 20 strategic business units each generated plans in different formats, with different levels of detail and different terminology. People came to the annual planning meeting overwhelmed by information. Much time was wasted in meetings just agreeing on facts. With the new method, each business unit had to submit a five-page plan electronically in a specific format five days before the meeting. The same approach was pushed downward in the organization. In addition, the EIS [executive information system] fostered more consistent reporting of results and trends across international marketing units. (Alter, 1992, p. 446)

FIGURE 7.3
The Change Management Calendar, a tool to calculate the information load of new policies and programs on field offices. (Courtesy of Bank of Montreal.)

CHANGE MANAGEMENT CALENDAR - DESCRIPTION

CONSUMPTION POINTS

- CONSUMING THE CAPACITY OF A BRANCH TO ABSORB CHANGE.
- THE EFFORT REQUIRED TO ABSORB, OR **IMPLEMENT,** THE CHANGE. THE FOLLOWING ARE GUIDELINES ONLY-DISCRETION IS REQUIRED. INDIVIDUAL CHANGE IMPLEMENTATION CONSUMPTION POINTS WILL BE IMPACTED BY THE CHANGE'S UNIQUE CHARACTERISTICS. WHEN IN DOUBT, COMPARE THE SPECIFIC CHANGE TO OTHER CHANGES FOR RELATIVITY.

SCORES OF	MEAN:	(EXAMPLES)
"0"	• READ & FILE ONLY	E.G. CBA UPDATES INT'L MANUAL UPDATES
"5"	• READ, DETERMINE WHO & ADVISE OTHERS (BUDGET, MEETING) & FILE • VERY MINOR CHANGE, EVEN JUST AWARENESS OF POSSIBLE CHANGE. VERY MINOR PREPARATION REQUIRED	E.G. MOST CHANGES, MINOR PROCEDURAL OR SYSTEM CHANGE. NEW INTEREST TIER OR SERVICE CHARGE
"10"	• READ, DETERMINE WHO & ADVISE OTHERS (BUDGET, MEETING) & FILE • REQUIRES UNIT TO PREPARE FOR CHANGE AND CHANGE ONGOING BEHAVIOUR • REQUIRES LITTLE "MANAGEMENT" DURING IMPLEMENTATION	E.G. R-49 NON-RESIDENT TAX
"15"	• AS ABOVE, PLUS - REQUIRES DETAILED PREPARATION FOR CHANGE. • REQUIRES SIGNIFICANT MANAGEMENT DURING IMPLEMENTATION	E.G. PERSONAL OVERDRAFT ADMINISTRATION CCAPS RELEASE 4
"20" AND ABOVE	• UNIQUE & COMPLEX IMPLEMENTATION IMPACTS	E.G. PROJECT "A" P.P.R. MECH RELEASE 48 - AUTOMATION OF TRANSFER/ ROLLOVER PROCESSING

(continued)

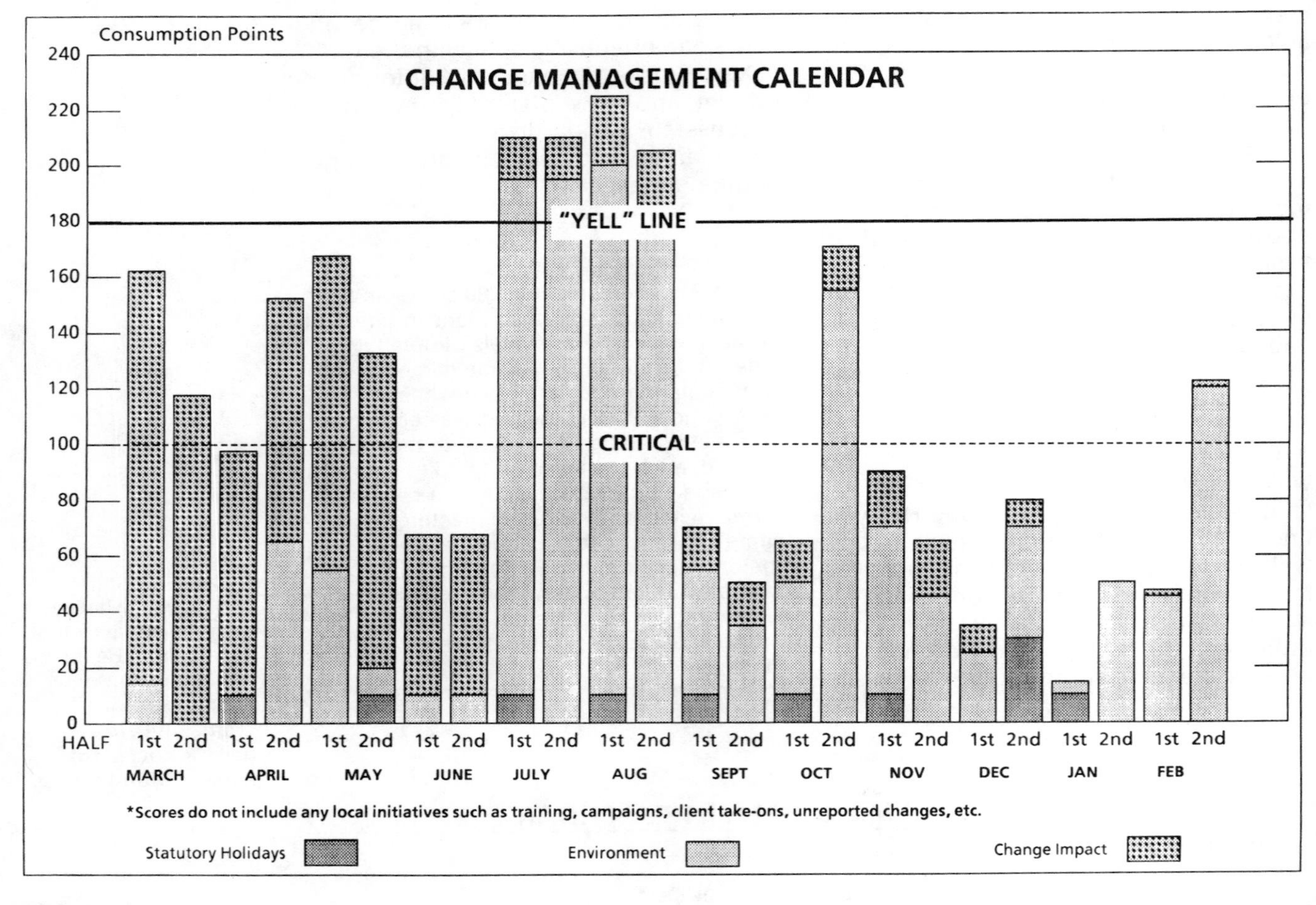

FIGURE 7.3
(*Continued*)

Standards and policies help reduce information overload. As we've seen in Chapter 5, Lederle's U.S. Education Department manages the dissemination of all information to field sales staff to eliminate redundant and contradictory information and to better control its rate. The Bank of Montreal and NCR have created standards for the materials they develop to achieve consistency in values, tone, and layout.

Other technological means can help preprocess data so that it is more easily categorized and digested. For example, Ocean Spray Cranberries uses an innovative tool called *CoverStory,* an expert system that extracts important data from their mainframe computer data base and generates automated news bulletins (Alter, 1992). E-mail systems, such as *The Coordinator,* provide formats so that it's easy to recognize the nature of a memo and respond appropriately without the need to develop and to type lengthy messages. Finally, interactive employee information systems, such as those being piloted at American Express, allow users to choose which messages they need, and when they are ready to receive them.

One final relief from our burden of information is education. David Berlo maintains that we need to reduce information by increasing knowledge. As stated in Chapter 2, information cannot be defined in isolation—information is the measure of uncertainty, or entropy, in a situation. The greater the uncertainty, the more the information. If a message is completely predictable, it contains no information. The more knowledge a person has, the less he will be surprised by data; therefore, the less information there will be.

As we will see, the costs of formal communication are staggering. If we can put a dollar figure on information development, distribution, and consumption, and get control of information overload, we may well be in a position to help our organizations save millions of wasted dollars and hours. Furthermore, we will take a significant step in reducing the high level of stress—sometimes even called information anxiety—found in so many occupations.

Are We a Tool of Management or of the People?

An almost universal dilemma of organizational communication practitioners is balancing out the needs and desires of various constituencies, especially those groups considered management and employees. Is our aim to strengthen the status quo or to upset it? Do we take a functionalist perspective or a critical one?

The old saying goes "he who pays the piper calls the tune." In our contexts, we cannot overlook the issue of who is paying the bill. Generally, top management is the client for organizational studies and interventions. We need to ask ourselves if we want our work to enhance their power, or to begin to share that power.

"In too many cases the communications professionals themselves are also confused about their roles and their responsibilities. They don't know whether they are journalists, the voice of management, representatives of the employee public or just hired guns" (D'Aprix, 1977, p. 167). Some think their job is to stand up to management and represent the needs and interest of the audience; others see themselves as the developers and the protectors of management's image. Unfortunately, most have the "I just work here" attitude and don't even question the ultimate result of their interventions. A more helpful perspective is to see yourself as a facilitator of change.

The role of communication practitioners, especially outside consultants, involves many potential ambiguities and tensions. The major question is who benefits from the intervention.

> No matter how cooperative the relationships between groups within a client organization, some groups and individuals will benefit from a diagnostic study more than others, and some may actually suffer in consequence of it. . . . This issue is illustrated by questions like these: Are consultants responsible for weighing the conflicting interests and values within an entire organization or is this solely the job of management? Should consultants be willing to have their findings feed into internal political struggles? Are consultants supposed to promote the narrow interests of the client who originally sponsored the study, or should they try to help 'the whole organization;' some broad stratum within it, such as top management; or all the members of the group whose problems were initially presented?" (Harrison, 1987, pp. 124–125)

Changing our organizations from their bureaucratic models to a more contemporary form will mean changing the communication and power structure; and providing advice will sometimes make clients look inadequate. There are times when our interventions or recommendations will eliminate or demote employees. There are no easy solutions to avoiding or dealing with such conflicts (this may be one reason why consultants get paid so well!). Here are some approaches:

- make sure your mission and role is understood
- ensure confidentiality of sources of sensitive information
- do not misrepresent the purposes of an assignment
- remember the organization is the client, not the individual who commissioned a project (it's the organization with whom you have the contract and that provides the money!)
- seek broad sponsorship (e.g., management and union)
- look for solutions that are win-win

Increasingly, management itself sees the need for sharing power and information, so clients in executive positions may themselves call for interventions that are more empowering of employees than employee groups might even wish. Although most people have heard stories of hatchet men called in from the outside to clean house and may at first distrust studies or consultation, these situations are few and far between. In fact, I have found that not having a foot in management or the rank-and-file allows me to be an advocate of both groups, and perhaps be an intermediary in situations where groups don't yet trust each other. By carrying messages back and forth and setting up situations in which differing groups need to work together in unusual roles, a neutral outsider can help to foster trust and commonality. We have found this especially evident when we produce video programs; therefore, we involve both managers and their subordinates (often, many organizational levels apart). If we threw out the tape at the end of the process, it still would have been worthwhile as a team-building exercise.

Is It Ethical to Experiment?

We practice our craft in an era of unprecedented development of new techniques and technologies. We need to somehow try out new techniques and interventions and see what works, but under what conditions are these acceptable? Can we charge for such experimentation? Is it ethical to try something that we don't know will work?

We can take a leaf from the book of medical research in which it is necessary to undertake clinical studies of the effectiveness of various drugs or surgical techniques. First, experimentation is done in ways in which it minimizes effects on real people or organizations. For instance, initial medical research may be done by computer modeling or animal studies. Second, after a given practice shows signs of being both safe and effective, it is tried out under real conditions, but only in cases in which patients agree to be subjects and understand the possible consequences. In many cases, drugs are only tried out on patients who are known to be terminally ill. Without the treatment, it is certain that they would die soon anyway. In these cases, we may lament that a medicine or procedure cannot be made more available, or used on patients whose conditions had not progressed so far. But it takes time for the level of proof to be reached to enable it to be approved.

In communication interventions, usually no one suffers fatal consequences; however, money and jobs may be lost. Still, we need to try out new approaches. The ethical way to do this is for both the practitioner and the client to take on an experimental stance—use pilot studies with a limited audience, duration, and budget. Be open about the unknown efficacy of the intervention, while making every effort to ensure success and to predict the outcome based on your knowledge of existing research and practice. Often, organizations will work with researchers or students in studies in which the client pays little or nothing for the intervention, while the practitioners hold out no guarantees of any particular outcome.

A corollary of this type of experimentation is the process of building skills and experience. How do consultants gain their experience? How do developers get the chance to produce a program using some new technology if they have no demo reel or track record? There are many cases in which consultants, producers, and contractors do misrepresent their capabilities and backgrounds, and this does nothing to build a solid image for the field.

Obviously, a common way to increase one's knowledge and competence is through formal instruction or continuing education workshops. Most people build their qualifications by gradually extending what they do—a producer may engage in developing some materials that will end up being modified for international use. From this experience and interaction with others in the organization who have done similar international work, she may become interested in the topic and read more about it. Before long, she may develop unusual expertise in intercultural communication. As clients or supervisors note your success in small projects, they will request your participation in those with a broader scope.

Some organizations establish communication research and development budgets that can fund experimental projects. Eliminating the need for a client to fund such an undertaking, as well as the possibility of missing the mark in aiding a really critical situation, pilot projects can help practitioners build skills while providing prototypes that may later be used to demonstrate the application of a technique or technology.

Is It OK to Manipulate?

Being a skilled communicator is a lot like being a black belt in karate. We have powerful tools for influencing others and, as we've seen from countless examples of journalism, we can paint a picture of reality that may be far from the objective truth. How we use words, compose photographs, choose examples, and illustrate data means the difference between telling and selling, or as David Berlo puts it, illumination versus indoctrination.

Obviously, public relations has as its intent the fostering of a good image of the client organization. In training, a company seeks to foster behaviors that it feels are worthy and will increase its productivity. In proposals or presentations, we want people to adopt our point-of-view. Persuasion is a part of almost all communication. Should we assume that our audiences know enough to beware of our messages and to understand that we are only telling what is in our interest? Or do we have an obligation to tell both sides of the story?

History has shown us that while we are not obligated to condemn ourselves, it's generally unwise to present too slanted a view. Inevitably, people will learn the other side to our stories and will learn to distrust our messages. It is our ethical duty to be truthful and complete in our reporting, and mindful of creating images that are representative and not staged.

How Can We Protect Intellectual and Artistic Properties?

Communication professionals are usually brokers for others' ideas and messages. We are rarely the experts on subjects about which we write or provide training. Neither are we the creators of innovative techniques. The old rules and laws of copyright and plagiarism are inadequate to cover the myriad ways to manipulate and steal the work of others. For example, it's easy to do an on-line computer search and pull together the full text of a number of encyclopedia or journal articles. The material can be easily stored and edited in a word processing format on a computer diskette and turned into a proposal, research report, or article. While there's nothing wrong with such research (it has always been done by way of manual methods) it is often tempting for authors to simply forget about crediting the proper sources. Furthermore, techniques to scan and edit images or record and manipulate sounds make it easy to copy and slightly modify an existing work and make it one's own.

In an era when information is precious and many people make their livings from creating it, the means for stealing it are also proliferating. How many of us think twice before running off 50 copies of an interesting article for our staff? Would you make a copy of your favorite clip-art collection on diskette so that your colleague could use it in his school newsletter? Would you feel funny in calling up a consultant or professor and asking for information that you later include in a report for which you receive a handsome fee?

Certainly, few of us would go into a grocery store and take an extra steak for our friend without paying for it, simply because it seems like the store has a lot of steaks and we know our friend is struggling to make ends meet. Nor would we ask for a free sample TV set and then sell it to a client. However, information doesn't come free to those who happen to create or discover it. Prices for literature and software are inflated due to this common shoplifting. In fact, many good writers, artists, and inventors are discouraged from going into the field because they are constantly the victims of white-collar crime.

We don't yet have good standards for exactly how much manipulation is needed to make a given work your own and, therefore, freed of copyright restrictions. We do generally have organizational policies against software pirating and academic rules about plagiarism. As professionals in the information business, we need to not only live up to these rules and standards, but we need to define new ones and teach others about the importance of adhering to them.

EXPANDING COMMUNICATORS' INFLUENCE

There are many people who would call themselves professional communicators, but is there really such a profession? Follett (1990) said, "The word 'profession' connotes for most people a foundation of science and a motive of service. That is, a profession is said to rest on the basis of a proved body of knowledge, and such knowledge is supposed to be used in the service of others rather than merely for one's own purposes" (p. 77). Most people consider professionals to have met strict educational requirements and professional certification, such as registered nurses (RNs), certified public accountants (CPAs), or lawyers. Most professions have one major professional association and accrediting body and rules as to who has the right to use a given title.

In comparison, most people who engage in public relations, employee communication, or training have had no specific formal background; in fact, they may not have majored in anything even similar to communications. We belong to a laundry list of organizations, some of whom do have accreditation exams, but none of which are recognized by the general public as being the group that sets the necessary standards for professional practice. Contrast this to the medical profession where one must have graduated from an accredited medical school and passed state exams, as well as continue to practice according to an agreed-upon set of rules, methods, and ethics. If one's license is revoked, one may no longer practice.

Not only do we not agree on standards, most practitioners couldn't even begin to articulate what those standards could be. Many people get into the field based on some perceived talent or stated interest. I remember vividly being taken on a tour of a million-dollar corporate television studio in one of the world's largest organizations. The video staff obviously was unable to make the equipment operate, even after having stayed up half the night before to rehearse for the meeting. They enthusiastically talked about the wonderful possibilities of the programming they would someday create, having purchased about 10 books and having been sent to a few one-day workshops held by equipment vendors. Their qualifications for being on the video staff were, in their words, how interested they said they were in learning to work with video. It's rather obvious to point out that we don't select our surgeons or engineers based on their interest; that is, having otherwise no qualifications!

Ruch and Goodman (1982) comment that many professional communicators are

- one dimensional—thinking only in print terms;
- not research-oriented;
- short on advanced educational training or thin on professional development;
- not interested in practicing a more holistic approach to communication in business;
- given to overstatement rather than the practice of a more conservative approach;
- too quick to assume that good communication is a panacea that can be used as a substitute for sound management;
- simplistic at times in ignoring some of the overall dimensions of the total corporation in favor of narrow tunnel vision;
- prone to believe their own hyperbole to the point where at times they lose sight of the horizon. (p. 81)

These authors and practitioners find that CEOs are calling on other disciplines in desperation to fill roles in decision making and communication strategy formulation. "It is high time to regroup, retrain, and restructure" (Ruch & Goodman, 1982, p. 81).

Most corporate communications are embarrassingly bad and, unfortunately, most managers and clients don't know they have a choice. This is because of "management's ignorance of what an effective staff can do if it is properly directed and supported" (D'Aprix, 1977, p. 97), as well as their lack of knowledge as to where to look for well-educated practitioners. D'Aprix asks how anyone could take the old house organ seriously, and notes that this is most people's impression of what professional communicators do. He says the answer to professionalizing the field lies in moving "Gretchen Greensleeves" back to her secretarial job and hiring a few professionals and executing a "total internal communication program matched to the business plan of the organization" (D'Aprix, 1977, p. 98).

Landes (1992) describes the traditional roles of communicators as

- reporters
- promoters
- corporate apologists

He says that to help organizations achieve the kind of excellence for which they're searching in many total quality programs, communicators need to rise above these tasks. If they continue to be seen as slightly more specialized typists, their futures are dim.

The IABC 1991 conference featured a panel that discussed sudden unemployment of communicators during the recession of that year. Panel members remarked

> Communication departments are on an as-needed basis. Companies now hire consultants when they need more staff.
>
> The higher you get, the more vulnerable you are. Be prepared. Even assume you'll be laid off.
>
> There's a trend toward managers being replaced by lower-level people with less experience. I should have noticed.
>
> Another survey said there's a 12 percent unemployment rate among communication professionals. That's well above the U.S. average. It's not just the recession. What is it?
>
> We've been working for companies where communication is not their business. Communicators will never be on the CEO track.
>
> It's not their business. They don't understand it.
>
> We don't understand them. It's the top positions, the people around this table, that are being eliminated. CEOs today have a better understanding of the value of communication.
>
> There's just as much communication, but technology does more. CEOs are enlightened and as a result, communication is diversified and spread out. (Kane, 1991, pp. 26–29)

The communication professional needs to move from being a frill to being a custodian of corporate knowledge, and needs to do a better job in cost-justifying jobs and interventions. How can communication and information become visible as an asset? How can we assess the costs of communication and information acquisition? People spend most of their time in meetings and reading memos. The cost of communication is more than the cost of the slides shown at the meeting or the duplication of memos—it's people's time. Communicators need to move from a focus on the production of information to a focus on the reduction of information.

Are We Pushers or Prescribers?

Why aren't professional communicators in the boardroom? Toffler (1990) says that few Chief Information Officers (CIOs) report directly to the president or CEO, but they allocate information and control who has access to it and owns it. As information is redirected, the organization gets restructured.

Too often, we push technologies or techniques for their own sake, rather than prescribe solutions to organizational problems. Ron Martin, vice president for employee communication at American Express, says that communicators need to make significant contributions to the organization. This means creating a culture committed to quality service and making sure that customers perceive it. Communication has to add value to the company's product. Martin told me he thinks that many professional communicators are too quick to blame senior managers for their own shortcomings. Greater influence in the company is largely a part of one's self-perception—one's failure to see how she can contribute. Today, he adds, communicators have to take the initiative, go to management with proposals, interpretations, and ideas, and to prove their worth. You can't just wait for the CEO to summon you to his office. As I was talking with Martin in December 1991, he was in the midst of managing a new research and development project, which brings video and computer information to employee multimedia-capable desktop computers.

Ed O'Brien, vice president of human resources at Corning Asahi Video, has similar thoughts. When he was given his former job to head up education and training at Corning Incorporated, he was only told that top management wanted high-impact training. Some people, says O'Brien, would have waited for clearer directions—marching orders. He didn't wait. He developed what he thought was right. It worked—their president says that corporate education and training is a world-class operation. Instead of promoting what the department used to do, he took a hard look at what the corporation needed. He determined that the training groups' former focus on skills training could be done just as well by the local community college and by outside contractors. He prepared his department to facilitate change by new communication methods with an almost completely new staff.

Expanding Our Influence

In order to elevate our profession and our own careers, professional communicators need to find out what the organization's most pressing problems are, study top management people, and work on tough problems. In order to do this, our managers and clients should

- provide access to all company planning documents and reports
- give communicators wide exposure to senior management
- make communications professionals feel like they're part of the staff
- not wince when a tough issue is being communicated and not deal exclusively with safe issues
- insist that the communication program operate from a clear and mutually understood philosophical base—don't just let them react (D'Aprix, 1977).

Ron Brown, supervisor of the audiovisual department at Amway, saw a new way to use his skills to improve the organization's performance. Inspired by a talk given at an ITVA conference, Ron became interested in the concepts of quality and service, especially as they relate to serving internal customers. He observed that Amway could benefit from applying some of the new ideas and techniques he heard about, and used part of his own budget to send himself to a highly rated course on customer service provided by Disney. Although employee communication was and is not part of Ron's official charge, he called up a group of managers like himself and asked if they'd like to volunteer their time to work with a Service Group. They identified a number of needs, such as better employee information, and used their own time and existing budgets to put on events, such as brown-bag seminars on how to lead better meetings and how to conduct communication audits. After a year or so, Ron's group got official funding and hired a consulting group to conduct a thorough needs analysis and recommend interventions to improve internal communication. Ron didn't wait for a new job description or more staff members to be handed to him. He proactively addressed some important company issues where he was. His leadership has not gone unnoticed.

Putting a Dollar Value on Our Performance

Although almost anyone would tell you that organizations spend much money on formal communication activities, few could tell you their own expenditures, not to mention any sort of bench mark or average figure. The numbers are perhaps too astronomical to even consider, lest our sponsors think twice about funding us! The noted performance technologist, Thomas Gilbert (1988), estimated the cost of corporate training in the United States alone to exceed the U.S. national deficit at the time of his writing, or to be more than $305 billion dollars. The design and development costs, which are only about $5 billion, are generally the only costs that appear in management budgets. What doesn't appear is the cost for the delivery of training and the salaries of trainees who are present throughout the training. He says that in order to make training better, you generally need to shorten it. "As far as we can tell, the so-called information revolution has greatly increased the quantity, but not the quality, of data and by so doing, it has reduced the quality of performance" (Gilbert, 1988, p. 20).

In a study of the costs of producing annual reports, Ruch and Goodman (1982) found that nobody in the 25 companies they surveyed (and, likely, in most companies) was keeping track of the executive time spent in developing these important information pieces. The "guesstimate" of senior officer time spent in preparing the average annual report was 1000 hours at $275 per hour. This doesn't include the approximately $100,000 spent in the actual design, execution, and printing of those reports. No accounting of executive preparation time for the reports appeared in the budgets of any of those projects, and none of the respondent companies had any research program to assess the effectiveness of the publications.

Farace, Monge, and Russell (1977) have lamented management's inattention to communication costs. They remind us that meetings are often held according to a schedule, not on the basis of need.

> We are definitely not arguing here for an end to meetings, nor advocating a "magic formula" for optimizing communication-resource allocation in an organization. We are arguing at a much more basic level—we want organizations to begin to recognize the importance of considering communication as a cost-related aspect of overall organizational behavior. We want organizations to rationalize and investigate the nature of communication activities in the organization. We advocate the concept of an information manager or a communication manager as a central part of organizational activity, whether this is an individual, a group, or a responsibility that is allocated among several members. (Farace, Monge, & Russell, 1977, pp. 44–45).

While we may not want to call attention to the large communication expenditures that are currently being incurred, our inattention to the bottom line has left us unable to cost-justify ourselves. One of the few attempts to do this, the 1992 IABC study "Excellence in Public Relations and Communication Management," found that CEOs who demand excellent communication reported a $300 U.S. return on every $100 invested. "Excellent organizations are characterized by participative cultures, organic structures, two-way communication systems, and high job satisfaction" (Culture is key, 1991, p. 15).

Psychologist and organizational consultant David V. Williams says that formal communication has been a wishing well into which management throws extra loose change in the vague hope that some unspecified salutary consequence will occur at some undetermined time. It has not been considered a new tool from which a return on management's investment is demanded. He also comments that communicators are often treated as children who are given gifts of funding because they ask for it—not as adults who are being given financing for a business proposition. We don't expect something in return for gifts, except perhaps a warm feeling in our hearts and some appreciation; however, we do expect bottom-line results from a business investment. We increasingly have to move into the latter category. As Larry McMahon, vice president of human resources at Federal Express, said during one of our phone conversations, "ROI [return on investment] has everything to do with everything."

Yet, there are challenges to putting a dollar value on communications. It is largely an enabling activity, especially when we are attempting to install new communications techniques and technologies. It is difficult to separate the process and tool from the content and outcome. Telecommunications ". . . provides an information highway system. Personal computers, office technology applications, electronics customer delivery, videoconferencing, document interchange, and the like are the traffic. It will be hard therefore to measure the impact of telecommunications in, and of, itself" (Fulk and Steinfield, 1990, p. 297).

Moreover, many professional communicators don't want to put their jobs on the line by taking the risk of assessing ROI. One vice president told me of a situation in which a documentation specialist in his organization was losing budget and space, while another department was growing by leaps and bounds. The other department happened to be the customer help line. My colleague was surprised to see that the documentation specialist didn't realize that he had the tools to curtail unbridled expansion and expense in customer service, and to create more satisfied customers as well, by creating better documentation. When this was pointed out to him and he was encouraged to make a pitch to management for more resources to solve the problem, he was unwilling to take the risk. The kind of response exhibited by the documentation specialist, said the vice president, will no longer be tolerated in his group. Although professional communicators probably chose their profession because they prefer to be the person behind the scenes who writes someone else's words, the field needs stronger leaders and more vocal advocates.

It will be unlikely that we will achieve the status of a profession or make the field of communication into a real organization, unless we establish and follow standards;

show the difference in performance between educated, experienced professionals and interested dilettantes; and demonstrate our contributions to the profitability of our organizational sponsors. We will continue to be the people with the fancy gadgets, the obsequious reporters, or the warmed-over schoolteachers in corporate classrooms.

THE INFRASTRUCTURE TO A NEW WORKPLACE AND ECONOMY

Toffler (1990) says the characteristics of our new economy and society are

- interactivity
- mobility
- convertibility
- connectivity
- ubiquity
- globalization

Training, internal and external communication, knowledge asset management, information systems, and telecommunication technologies are central to establishing these trends and to making organizations more profitable and enjoyable places to work. We need to find ways of increasing knowledge and collaboration among our increasingly far-flung and diverse organizational counterparts.

The corporation of tomorrow will resemble a switchboard—a small communication control center managing rapid transmission of information and a network of relationships with outside partners (Kanter, 1989). More and more, the role of professional communicators will not be to create messages, but rather to produce and to manage effective systems for every member of an organization to use. Similar to traditional data-processing departments, which were once the sole users of computing equipment, we need to move to being managers of information networks, analysts, and internal consultants.

Lederle International is the international pharmaceutical division of American Cyanamid. With 27 affiliates around the world and sales in over 140 countries, Lederle earned close to 1 billion dollars in 1992. In the mid-1980s, the division began investigating the ways in which it could use computer-based training to provide instruction to the international sales force. Early efforts using U.S.-produced materials and sophisticated technologies were less than successful. Merely translating programs with content created for the U.S. was not applicable to other markets in which there were not only divergent selling styles, but also different strategies and competition due to the distinct set of medical practices and legal drugs and procedures for each country.

In 1990, they tried a different approach. They contracted with our firm to develop a custom authoring system that would allow each affiliate to modify U.S.-created CBT courses and to easily produce their own. Instead of using high-tech interactive video equipment that is not commonly available or serviceable in other countries, we stayed with programs that could be run on ordinary PCs and even on laptop computers. The result was *LederLearn*™, their own software that makes it possible for nonprogrammers to create their own interactive tutorials and sales simulations (see Figure 7.4). The U.S. training office commissions the design and development of courseware to support major new products. Affiliates are given copies of *LederLearn*™, as well as the open source code to the programs. Their manager of international field sales training, Ed Nathan, gives week-long courses for managers on how to modify and create CBT, and within a year of its launch, *LederLearn*™ has been used to create and customize several major courses.

Now, affiliates don't feel as though the U.S. headquarters are dumping generic and inappropriate materials on them. Rather, they are providing the tools and guidance for local content specialists to use this new technology themselves. Their reaction has been enthusiastic, and interactive systems are now seen as an integral way to provide instruction and market support rather than as a toy (Nathan & Gayeski, 1992).

Other dispersed organizations, such as Amway, use print and video programs to provide information and training to independent entrepreneurs who sell their products. Senior management credits video for major business gains, and their distributors pay around $1 million out of their own pockets to buy Amway tapes and print materials. (See Figure 7.5)

Most people spend more than half of their waking hours at work, and more and more of that time will be spent in formal communication and information-processing activities. "Processes of communication are the central lifeblood of the organization. One view of organizations holds that they are information-processing mechanisms. That is, in order for organizations to combine the resources of technology, capital, and people into products or services they must use information as the dynamic glue" (Tichy, 1983, p. 91).

As communication technologies become more widely adopted, many social critics wonder whether technologies will emerge as employees' tools or as their masters. They wonder whether people will be liberated or enslaved by them.

> It is a certainty that change is occurring and that the workplace is being affected by new technology and the information explosion. The organization is changing in terms of where work is done (home or office), how decisions are made, and who in the organization participates in those decisions, and who the workers are and where they come from. These changes will create institutions which are different from organizations as they have existed since the Industrial Revolution. (Lederman, 1986, p. 321)

Communication researcher Sherry Turkle (1986) reports that in her conversations with computer hobbyists, their relationship with computers carries "longings for a better and simpler life in a more transparent society . . . small computers become the focus of hopes of building cottage industries that will allow the hobbyist to work out of his home, have more personal autonomy, not have to punch a time card, and be able to spend more time with his family and out-of-doors" (Turkle, 1986, p. 453). This dream has become a reality for many people, although working at home, like any situation, is not without its challenges and drawbacks. Turkle, who has spent many years developing and studying on-line communities, sees significant benefits in the adoption of these systems. She envisions a ". . . future where relationships with technology will be more direct, where people will understand how things work, and where dependence on big government, big corporations, and big machines will end" (Turkle, 1986, p. 453).

We are seeing a rapid blurring of traditional distinctions between technologies and those who use them:

- communication professionals versus ordinary users
- technicians versus artists
- mass versus narrowcast media

2

Computer-Based Training: An Important Tool for the 90s

by Ed Nathan, Manager, International Sales Training

Developing Computer-Based Training Using LEDERLEARN*, a seminar to help affiliate training staffs develop effective computer-based training (CBT), was recently conducted in Belgium.

From May 13 to 17, 1991, at the Intercontinental Hotel in Brussels, trainers from South Africa, the UK, Belgium, and Germany convened for an intensive workshop to gain practical CBT skills. Using an upgraded version of the LEDERLEARN authoring system—a custom-designed, user-friendly software package—the participants learned how to create quality CBT programs suited to their own market requirements and conditions.

The seminar guided the attendees through the planning and design processes necessary for effective CBT development and then provided extensive hands-on work with the computer. The highlight of this portion of the program was the trial run of the beta test copy of the updated version of the LEDERLEARN software, which proved very successful.

With the practical skills acquired at the seminar, participants will be able to more effectively evaluate a variety of CBT programs, build their own programs, or customize existing ones.

The Participants Speak Out

The attendees provided us with some valuable feedback concerning this new training program. Besides the quantitative results listed below, they gave some thoughtful written comments that will enable us to improve the course for its next presentation. When asked what they liked most about the program, the most common responses were:

- Sharing CBT ideas with other participants
- Learning and practicing the various elements that combine to make effective CBT.

When asked what they liked least, the most common responses concerned the first day of the program, which dealt with learning theory. The comments noted that:

- Day one was too technical and theoretical—even if it was necessary
- Sometimes the professional jargon was a problem.

* Trademark American Cyanamid Company

FIGURE 7.4 Lederle International uses their own authoring system to create computer-based training, which can easily be translated within each market. (© Copyright 1991 Lederle International Division, American Cyanamid Company. Reprinted with permission.)

FIGURE 7.5
Amway uses print and video materials to help their distributors learn the business. (Courtesy of Amway Corporation.)

- passive reception versus active control and transmission
- data processing systems versus audiovisual display systems

These systems have already had a significant impact on corporate structures. Electronic systems weaken the effects of traditional reporting structures and change communication patterns. They allow for an increased span of control, more informed workers, and, therefore, flatter organizations. We have yet to see whether these new structures will remain more effective than the traditional bureaucracy in terms of profitability and employee satisfaction. Furthermore, we are only beginning to see any real changes in the ways people communicate within organizations.

There is already some evidence that the traditional preference for face-to-face communication with one's supervisor is changing. Katherine Marshall, former director of internal communications at Unisys, reports that in a phone survey of their employees, their daily electronic newsletter, *Unisys News Network,* has supplanted immediate supervisors as the preferred source of information for employees. The influence of the grapevine has also been minimized by fast and direct corporate communication, such as their printed newsletter and corporate television. She comments that the current role of supervisors should be one of news interpreter rather than that of being the initial source of news (Marshall, 1992).

SOME FINAL THOUGHTS

Professional communicators need to see, and help others to see, information and communication as an asset and as a product—not just as a frill or an unavoidable process. We need to coordinate our efforts and broaden our perspectives and set of skills so that communication can become the umbrella under which training, MIS, marketing, employee communication, and public relations are conceived and managed. While these radical transformations will not happen overnight, in some enterprises it is already occurring.

Larry McMahan, vice president of human resources at Federal Express, comments that their company has ceased making any distinction between communication and training. The messages that they create and disseminate to their worldwide work force through meetings, print, FXTV, and interactive multimedia have some of each. For example, when employees sign on to take an interactive course, they will first see a few important news items sent by way of e-mail into the system. U.S. Healthcare has made data that were typically accessible only through executives' terminals on MIS systems prominent to all employees and visitors, making a public display of their progress toward goals. This is a powerful motivational message to employees and an interesting public relations tool as well. Dow Chemical's technologies originally developed for training have now been adopted as tools for brainstorming and strategic planning. Corning Incorporated has turned their training department into a change management resource. Progressive organizations are synthesizing technologies, techniques, and information.

Corporate communication will not be limited to large enterprises; in fact, the effects of leading-edge practices and technologies will help smaller organizations to flourish. The future will neither belong to big and intelligent technology, nor to large and powerful bureaucracies, but to human (individual and collective) organizations that will be able to set up human goals shared by the largest as possible fraction of people throughout the world, and to achieve them by using knowledge and technology, in the public and general interest (Petrella, 1990). Rather than being a tool of the bureaucracy, the communication cops, or the gatekeepers, Renaissance communicators will instead be radical change agents.

Robert Clingon's business card reads "Czar of Bureaucracy." In real life, he's the corporate vice-president of quality at CAE Link, the company that makes the famous "Link Simulators" for aircraft. He's leading his organization in a major cultural change toward total quality management and continuous process improvement. In doing so, his job is to eliminate needless bureaucracy wherever it's found. He says that everything affects communications. In his job, he needs to look at internal communication systems and work with customers to ask for their help and their patience as their organization goes through this difficult but necessary evolution.

Lest we become too caught up in our capabilities and visions of controlling communication systems in our organizations, we need to remember that our biggest job is to share our knowledge, skills, and technologies with others. The responsibility for communication must be diffused within the organization, not just within the formal communications department. SMEs and line workers need to learn to develop their own instruction and knowledge bases so that they can share their skills with others. Managers and team leaders need to recognize that their main job is communication. Organizations must search out and reward those who demonstrate excellent communication skills. As environments become more equivocal, the need for sensitive and frequent interpersonal communication is vital, and professional communicators can only provide the support for people to carry out these practices themselves. We need to help management set and enforce standards for appropriate and effective communication, both internally and externally.

Sharon Paul, the executive vice pres dent of public affairs for Labatt Brewei ies of Canada, observes that many orga nizations give lip service to communication but fail to recognize it in terms of bottom-line management, such as in remuneration, recruitment, promotion, and appraisal processes. She comments that in companies that truly understand the importance of communication, "communication skills proficiency is incorporated into position descriptions throughout the organizations, and employees are routinely recognized and rewarded for their communication expertise as one of the elements of their jobs" (Observation, 1992).

FIGURE 7.6
The Bank of Montreal's Public Affairs Department responded to their organization's new strategic plan by a series of television ads, posters, and employee communication pieces. (Courtesy of Bank of Montreal.)

Television Highlights

Bank of Montreal's television launch, Phase I, will hit the airwaves Saturday, May 4th, 1991, and run until the end of June.

That's just the beginning. Bank of Montreal will maintain a powerful television presence in the fall and beyond.

Advertising for the Bank will reach an astonishing 92 per cent of Canadians from coast to coast, in a combination of 30- and 60-second commercials.

Our messages will be seen in a variety of exciting programs, from soap operas and major movies to big sporting events. Watch for them in the Stanley Cup playoffs, the final episodes of "Dallas", and the hit comedy "Designing Women". You'll also see them on popular news shows like the CBC Journal and The CTV National News. And Bank of Montreal will be sponsoring "Mystery and Suspense Month" on CTV—three fast-paced mystery movies based on Dick Francis' best-selling novels.

We're Paying Attention to You

Also included in this package are employee feedback cards. Tell us what you think!

If you have any questions, please contact Corporate Advertising & Promotion:
Kate Cooke (416) 867-4965
David Lavoie (416) 867-4966

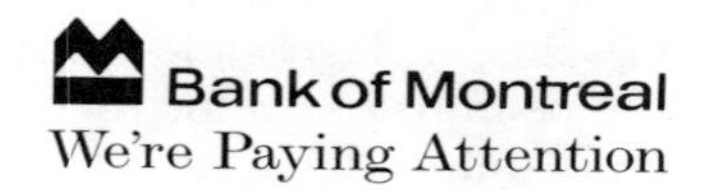

Our Mission

As organizations develop and transform themselves, professional communicators will be at the core of successful change management. Tichy (1983), theorist and practitioner, emphasizes the need for coordinated technical, political, and cultural approaches to change, using the following *change levers*:

- new and more sensitive environmental scanning and information processing capabilities
- clear statements of organizational mission to guide strategic decisions
- new management techniques to install strategic plans
- sophisticated processors to realistically engage the relevant interest groups
- training and development of staff to take on new tasks and technologies
- adjustments in communication and authority networks
- skills in consensual decision making, problem solving, and conflict resolution
- motivation of people to change their behaviors
- management of the informal communication and influence networks that exist in the organization

It is noteworthy that each of Tichy's nine change levers are essentially communication skills or processes. Comprehensive communication campaigns and standards, using skills in training, public relations, advertising, persuasion, and information systems management can affect cultural change and organizational renewal.

FIGURE 7.6
(*Continued*)

April 17 Executive Conference Presentation
BRIEFING NOTES

Background

One year ago, Bank of Montreal executives were presented with a Corporate Strategic Plan developed by the senior executives over a period of several months. The plan was adopted unanimously by the senior personnel of the Bank, and each area of the Bank then began communication and implemention of the plan in their own divisions.

Following the rollout of the Plan, Matt Barrett personally visited more than 900 branches in 30 cities, and met more than 10,000 employees. The Plan focuses on excellence in customer service delivered by competent, caring employees, and this is the message Matt took to the troops.

Public Affairs' mandate from the Plan was to develop a communications program that would carry forth a unified corporate image which would support business building efforts.

Planning and Development

The first step was to examine existing multiple advertising supplier relationships, with the objective of achieving coordinated marketing communications consistent with the Bank's corporate image through one English and one French advertising agency. Vickers & Benson and Publicite Martin were chosen to carry out this objective. Public Affairs was designated as the sole means of contact with the agencies.

The next step was to take the basic premises in the Corporate Strategic Plan regarding the Bank's image and test those with consumers and employees to provide the basic framework for a unified image communications program. Insight Canada was contracted to develop and conduct a national research program.

Focus groups, telephone and in-home interviews were conducted with employees, customers and the general public to measure awareness, familiarity, attitudes and usage. The results were the planks upon which the Bank's communication strategy is based.

Creative concepts based on the research findings were developed in conjunction with Vickers & Benson and Publicite Martin, and tested at various stages with customers, and with employees. The finished creative is market driven, and responds to the key attributes that will most positively affect the Bank's image with its customers, its employees and the general public.

Creative

With a Strategic Plan in place directing the Bank's overall efforts, and with the information at hand to drive our corporate communications strategy, the Bank of Montreal is poised to take on the challenges of the 1990's.

The Bank has a vibrant new Chairman and senior management team with a more collegial approach; a strong clear strategy; a non-traditional, inspirational, "we-can-make-anything-happen" attitude; a committed sense of community; and a return to our roots in banking -- customers first and service above all. These principles combined with a recognition of market differences led to the development of distinct creative approaches for the French and English markets.

French Quebec

In French Quebec, television audiences will see Bank of Montreal with a creative approach which is:

- Pragmatic and dynamic
- Concrete and tangible
- Pertinent to today's attitudes and aspirations
- Different from anything thing the competition is doing
- Highly credible

The creative executions consist of four 30-second commercials which involve customers and employees -- real people dealing with real issues. While different from each other the commercials draw on the same communication objectives and emphasize a spirit of change, a new emphasis on customer service, and a dedication to excellence.

(*continued*)

FIGURE 7.6
(*Continued*)

The French tagline exemplifies this philosophy -- "Au dela de l'argent, il y a les gens" (Beyond the money, there are the people).

The first commercial, "Customers" demonstrates that the Bank of Montreal is different than other financial institutions. We are listening and Bank of Montreal people care about their customers.

"Efficiency-Claire Paris" shows an employee of the Bank, in a typical work day -- confident and happy to serve her clients to the best of her ability. She has decision-making power at the counter and uses it to serve her customers better.

"Restaurant-Julien Lambert" -- Every day this Bank of Montreal customer strives to serve his customers better and he wants his Bank to treat him the same way. Bank of Montreal is committed to delivering this level of service to their clients.

"Journaliste-Diane Demers" features a customer with a very demanding schedule, days that are too short and the pressure to try and get every thing done at once. There is a Bank that can help by offering flexible business hours, branches that are open on Saturdays. Bank of Montreal has all of this and more.

English Canada

The English television campaign is a recognition of the television generation that has lived with a changing medium, almost all their lives. Consumers know how to absorb multiple images, how to interpret them, register them and relate to them.

The commericals are vital and quick paced, and through the images presented recognize and establish contact with virtually all our constituents. You'll see the Canadian mosaic -- a tribute to the rural soul of the country juxtaposed against the vibrant urban society we are part of daily. You'll see the Bank's connection with youth, young families, with seniors, with business, large and small. And through it all the Bank is doing and showing things that are part of our corporate being. The commercials are a visual expression of our commitment to Canadians from coast to coast.

The English campaign tagline focuses on our dedication to customer service -- "We're paying attention".

Phase One will air on May 4th and begins with "Vision." In two versions -- 30-second and 60-second -- it reintroduces Canadians to our commitment to change for the better and our promise of bringing service back to banking.

"Employees" features real Bank of Montreal employees discussing their increased participation within the new corporate strategy, and their individual approaches to better customer service.

"Customers" shows actual customers sharing their views on what sets Bank of Montreal apart from and above other financial institutions.

Phase Two is set to air in September and, depending on what our tracking tells us, may re-use some of the spots from Phase One. It will also feature individualized examples of the commitment to excellence introduced in Phase One.

"Mrs. McIlquham" features a Bank of Montreal customer and emphasizes our respect for individual needs by showing a unique customer situation and our approach to serving her, from the point of the customer.

"Philosophy" features a Manager talking about the importance of attention to the little details, and shows real employees going about their normal daily activities. It's the little things in the banking equation that make the difference.

Supporting Materials

To reinforce the television advertising, print support materials will be provided to branches and other front-line customer service units. These include:

- Window banners
- Branch posters
- ABM posters
- Display kiosks
- "Signature" posters

- Customer handouts
- Statement stuffers

All of this material will be shipped during the last two weeks of April and instructions have been provided to prominently display them during May.

Employee Communications Program

To communicate this program internally, a video has been prepared and will be distributed to operating units in the Bank. Employees will have the opportunity to see the Phase One commercials and hear how they were developed before they go on air, so that they will be familiar with them when customers comment.

In addition to showing the commercials to all employees prior to their airing, we are seeking their feedback and suggestions as to the believability, accurateness and content of the commercials, and how they were communicated to employees. It is vital to the success of the corporate image campaign that all our employees feel a part of this process, and their input will assist in the next steps.

Television Programming

Bank of Montreal will launch the new image campaign nationally on the fourth of May. We will spend roughly equal amounts to that of our major competitors. The difference is that rather than splitting the message impact across several products and services as our competitors we will concentrate our efforts on our service image.

The commercials will be seen nationally on major networks and our reach will exceed 92%. Programming which is in keeping with the images and feelings presented by the commercials has been chosen to create even greater viewer response to our messages. With increased levels of awareness and familiarity in the minds of Canadians, the likelihood that they will consider doing business, or increased levels of business, with us will increase.

Merchandising support

Public relations

The public relations story will be a four part program of activities including;

part one	trade and specialty
part two	business and news media
part three	feature stories
part four	human stories

It is intended that this program be a proactive media relations program. The story will be that the Bank is coming back into the market with a bold new campaign. Customer service is the keynote strategy behind the campaign.

The Media profile document (attached) will serve as the full background document.

Communicators will be at the center of organizational design projects, and their interventions will be creative and expansive, as you can see from the Bank of Montreal "Image in the Making" campaign (see Figure 7.6).

Professional communicators can help to transform not only the organizations in which they work, but also society at large.

> Literature suggests that societal culture has an effect on organizational culture which affects public relations. Excellent management and excellent PR can flourish only in collaborative, participative cultures. . . . Optimistically, symmetrical PR can help to change organizational cultures. Collaborative organizations, in turn, could influence societal cultures—in large part because they are effective. But the task is difficult and long-term. PR excellence cannot come easily in organizations and societies that are not excellent. But PR could be the catalyst that begins to change them. (Grunig, 1992, p. 1)

Renaissance communicators will provide people with clearly communicated and inspirational visions. We need to articulate values, educate and train corporate citizens, reward performance, provide information, and celebrate small victories. We have the capacity to nourish our workplace communities and assist individuals and groups to achieve their dreams. Our real mission, as D'Aprix (1977) says so well, is "helping people to comprehend their work lives and to find within that critical part of their lives meaning, hope, and satisfaction" (p. 175).

The future holds many promises for those of us who wish to apply our energies and talents in professional corporate communications.

REFERENCES/ SUGGESTED READINGS

Alter, S. (1992). *Information systems: A management perspective.* Reading, MA: Addison-Wesley.

Argyris, C., & Schon, D. (1978). *Organizational learning: A theory of action perspective.* Reading, MA: Addison-Wesley.

Bassett, G.A. (1968). *The new face of communicaton.* New York: American Management Association.

Carlberg, S. (1991). *Corporate video survival.* White Plains, NY: Knowledge Industry Publications.

Carnevale, A.P. (1992). Learning: The critical technology. *Training and Development, 46*(2), pp. s1–s16.

CEO describes the chief public relations officer of the 1990s. (1991, December). *IABC Communication World,* 12.

CEOs set standards. (1992, January), *IABC Communication World,* 42.

The changing role of today's communicator. (1992, May/June). *Communication World,* 36–39.

The Conference Board. (1972). *Information technology: Some critical implications for decision makers.* New York: Author.

The corporate image. (1979, January 22). *Business Week,* 17–19.

Cleveland, H. (1990). Epilogue: The twilight of hierarchy: Speculations on the global information society. In S. Corman, S. Banks, C. Bantz, & M. Mayer (Eds.), *Foundations of organizational communication: A reader.* (pp. 327–341). White Plains, NY: Longman.

Culture is key to how organizations respond, IABC research foundation study shows. (1991, August). *IABC Communication World,* 15–16.

Daniels, T.D. & Spiker, B.K. (1991). *Perspectives on organizational communication.* Dubuque, IA: Wm. C. Brown.

D'Aprix, R. (1977). *The believable corporation.* New York: AMACOM.

Devanna, M.A. (1992). Human resource management: Competitive advantage through people. In E.G. Collins, & M.A. Devanna (Eds.), *The Portable MBA* (pp. 219–237). New York: John Wiley & Sons.

Dixon, N.M. (1992). Organizational learning: A review of the literature. *Human Resources Quarterly, 3*(1), pp. 29–50.

Farace, R., Monge, P., & Russell, H. (1977). *Communicating and organizing.* Reading, MA: Addison-Wesley.

Follett, M.P. (1990). How must business management develop in order to possess the essentials of a profession? In S. Corman, S. Banks, C. Bantz, & M. Mayer (Eds.), *Foundations of organizational communication: A reader.* (pp. 77–82). White Plains, NY: Longman.

Fulk, J., & Steinfield, C. (1990). *Organizations and communication technology.* Newbury Park, CA: Sage.

Gilbert, T. (1988). The 10 most important lessons I've learned about productivity. In G. Dixon (Ed.), *What works at work: Lessons from the masters.* (pp. 17–21). Minneapolis, MN: Lakewood Publications.

Goldhaber, G., Dennis, H., Richetto, G., & Wiio, O. (1979). *Information strategies: New pathways to corporate power.* Englewood Cliffs, NJ: Prentice Hall.

Grunig, J. (1992, April). 12 attributes of organizational excellence—and pr's role in fostering it. *Tips and Tactics.* Exeter, NH: PR Publishing Company, Inc.

Harrison, M. (1987). *Diagnosing organizations: Methods, models, and processes.* Newbury Park, CA: Sage.

Kane, K. (1991, August). What's really happening in the job market. *IABC Communication World,* 26–29.

Kanter, R.M. (1989). *When giants learn to dance.* New York: Simon & Schuster.

Landes, L. (1992, February). Down with

QUALITY program-itis. *Communication World*, 29–32.

Latest 'excellence' study results. (1992, January). *IABC Communication World*, 38–39.

Lederman, L.C. (1986). Communication in the workplace: The impact of the information age and high technology on interpersonal communication in organizations. In G. Gumpert & R. Cathcart (Eds.), *Inter/Media*, 3rd ed. New York: Oxford University Press.

Lundberg, L.B. (1965). *Management looks at public relations.* Conference presentation. Annual conference of the Public Relations Society of America, Denver, Colorado.

Marshall, K. (1992, May). Do electronic communications threaten managers' role? *CCM Communicator*, 5–6.

Meltzer, M. (1981). *Information: The ultimate management resource.* New York: AMACOM.

Nathan, E., & Gayeski, D. (1992). Theory vs. practice: An international lesson in CBT. *Performance technology 1992: Selected proceedings of the 30th NSPI conference.* (pp. 75–82). Washington, DC: National Society for Performance and Instruction.

Observations of Communication World editorial advisory board member. (1992, January). *IABC Communication World*, 49.

Pearlstein, G. (1989, August). Preston Trucking shifts to performance management. *Performance & Instruction*, 1–5.

Petrella, R. (1990). Preface: Competitiveness for what? In M. Cooley (Ed.), *European competitiveness in the 21st century: Integration of work, culture, and technology.* (pp. 1–2). Brussels, Belgium: FAST Programme.

Powers, B. (1992). Strategic alignment. In H. Stolovitch & E. Keeps (Eds.), *Handbook of human performance technology.* (pp. 247–258). San Francisco: Jossey-Bass.

Revans, R.W. (1980). *Action learning: New techniques for management.* London: Blond & Briggs.

Ruch, R.S., & Goodman, R. (1983). *Image at the top: Crisis and renaissance in corporate leadership.* New York: The Free Press.

Schramm, W. (1988). *The story of human communication: Cave painting to microchip.* New York: Harper & Row.

Shockley-Zalabak, P. (1988). *Fundamentals of organizational communication.* New York: Longman.

Thaler-Carter, R.E. (1991, August). Power moved as communicators met to discuss the state of the profession. *IABC Communication World*, 9–16.

Tichy, N. (1983). *Managing strategic change.* New York: John Wiley, & Sons.

Tichy, N., & Devanna, M.A. (1986). *The transformational manager.* New York: John Wiley & Sons.

Toffler, A. (1990). *PowerShift: Knowledge, wealth, and violence at the edge of the 21st century.* New York: Bantam Books.

Turkle, S. (1986). Computer as rorschach. In G. Grumpert & R. Cathcart (Eds.), *Inter/Media*, 3rd ed. (pp. 412–434). New York: Oxford University Press.

Welch, J.E., & Hood, E.E. (1992). To our share owners. *General Electric 1992 Annual Report.* Fairport, CT: General Electric Co.

White, E. (1986). Interpersonal bias in television and interactive media. In G. Gumpert & R. Cathcart (Eds.), *Inter/Media*, 3rd ed. New York: Oxford University Press.

Wiig, C. (1990). The company as a learning organization. *Proceedings of the 4th international conference on computer and video in corporate training*, (pp. 37–52). Lugano, Switzerland: Instituto Dalle Molle di Metodologie Interdisciplinari.

Winner, L. (1989). Mythinformation in the high-tech era. In T. Forester (Ed.), *Computers in the human context*, (pp. 82–96). Cambridge, MA: MIT Press.

Zuboff, S. (1988). *In the age of the smart machine.* New York: Basic Books.

A Organizations in Corporate Communications

Academy of Management (AM)
P.O. Box 39
Ada, OH 45810
Professors in accredited universities and colleges who teach management, and selected business executives who have made significant written contributions to the literature in the management and organization field. Publishes *Academy of Management Executive, Academy of Management Journal,* and *Academy of Management Review,* which are all quarterly journals, as well as a newsletter and conference proceedings.

American Society for Training & Development (ASTD)
1630 Duke Street
Alexandria, VA 22313
Members are professionals in corporate training and human resources/employee development. Publishes *Training & Development* and *Technical Skills Training,* which are monthly journals, and a quarterly research journal, *Human Resources Development Quarterly.* In addition, the ASTD publishes studies and monographs on training practices and economic trends.

Association for the Development of Computer-based Instructional Systems (ADCIS)
1601 West Fifth Avenue, Suite 111
Columbus, OH 43212
Primarily professors and developers of computer-based instruction. Publishes *Journal of Computer-based Instruction,* which appears quarterly.

Association for Educational Communications & Technology (AECT)
1025 Vermont Ave, N.W.
Washington, DC 20005
Members include school/college media specialists, professors of educational technology, and some instructional designers in industry. Gives awards for student media productions, publishes monthly, *Tech Trends,* and quarterly research journal, *Educational Technology: Research and Development,* as well as books and reports relating to research, practice, and legal issues.

Council of Communication Management (CCM)
Oak West Office Plaza
17W 703 E. Butterfield Road
Oakbrook Terrace, IL 60181
Members are in policy-making roles in corporate communications. Publishes monthly newsletter and promotes member networking.

International Association of Business Communicators (IABC)
One Hallidie Plaza, Suite 600
San Francisco, CA 94102
Professionals involved in employee communications and public relations within industry. Publishes *Communication World* on a monthly basis, as well as professional reports and studies.

International Communication Association (ICA)
P.O. Box 9589
Austin, TX 78766
Primarily professors and researchers interested in studying the nature of human communication and its functions in society. Publishes *Communication Theory* and

Human Communication Research, which are quarterly research journals, and *Communication Yearbook* and *Communication Abstracts,* which are annual publications.

International Communications Association (ICA)
12750 Merit Drive, Suite 710 LB-89
Dallas, TX 75251
Members are responsible for telecommunications services and facilities of major organizations. Publishes *Communique,* a bimonthly industry magazine, and a triennial *Telecommunications Professional Profile Survey,* which is a report that provides analysis of telecommunications departments, compensation at various levels and industries, and telecommunications in the corporate hierarchy.

International Communications Industries Association (ICIA)
3150 Spring Street
Fairfax, VA 22031
Dealers, manufacturers, producers, and suppliers of audiovisual, video, and microcomputer products and materials. Bestows awards for innovative use of audiovisuals, industry contributions, and advanced media use. Publishes *Communications Industries Report,* a monthly association and industry newsletter; *Management News,* a bimonthly newsletter containing information and tips on improving management practices and skills; and *Equipment Directory of Audio-Visual, Computer and Video Products,* an annual sourcebook of over 2500 audiovisual- and computer-based communications products.

International Interactive Communications Society (IICS)
P.O. Box 1862
Lake Oswego, OR 97035
Producers of multimedia software and hardware. They present awards for outstanding interactive systems and publish a newsletter.

International Television Association (ITVA)
6311 N. O'Connor Road, LB-51
Irving, TX 75039
Corporate/educational video and interactive video producers with active regional chapters. Publishes monthly newsletter as well as annual salary surveys and other professional reference tools. Maintains a library of award-winning videotapes that can be borrowed.

National Association of Desktop Publishers (NADP)
462 Old Boston Road
Topsfield, MA 01983
Members are writers and graphic artists who use desktop publishing systems. Provides a buy-direct discount service and publishes *NADP Journal* on a monthly basis.

National Society for Performance & Instruction
1300 L Street, N.W., Suite 1250
Washington, DC 20005
Members are involved in training and human performance technology. Publishes *Performance & Instruction,* which is a monthly journal, and *Performance Improvement Quarterly,* which appears four times yearly, and sponsors handbooks and reference texts.

Public Relations Society of America (PRSA)
33 Irving Pl., 3rd Floor
New York, NY 10003-2376
Professional society of some 15,000 public relations practitioners within organizations and PR agencies. Publishes monthly *PRSA News* and *Public Relations Journal,* as well as research studies.

Society for Applied Learning Technology (SALT)
50 Culpeper Street
Warrenton, VA 22186
Business, military, and educational producers of interactive media. Hold semiannual conferences and trade shows. Publishes *Instruction Delivery Systems,* a monthly journal, and *Journal of Interactive Instruction Development,* a quarterly research journal.

Society for Technical Communication (STC)
901 N. Stuart Street, Suite 304
Arlington, VA 22203
Members are professionally engaged in or interested in some phase of the field of technical communications and writing. Publishes *Intercom Newsletter,* a monthly newsletter; *Technical Communication,* a quarterly journal; and *ITCC Proceedings,* an annual compilation of conference presentations.

B Publications in Corporate Communications

Academy of Management Executive
Academy of Management Journal
Academy of Management Review
Academy of Management (AM)
P.O. Box 39
Ada, OH 45810

AVC (Audio Visual Communications)
PTN Publishing
445 Broad Hollow Road
Melville, NY 11717

AV/Video
Montage Publishing
701 Westchester Avenue
White Plains, NY 10604

CBT Directions
Weingarten Publications, Inc.
38 Chauncey Street
Boston, MA 02111

Communication Theory
International Communication Association (ICA)
P.O. Box 9589
Austin, TX 78766

Communication World
International Association of Business Communicators (IABC)
One Hallidie Plaza, Suite 600
San Francisco, CA 94102

Communications Industries Report
International Communications Industries Association (ICIA)
3150 Spring Street
Fairfax, VA 22031

Communique
International Communications Association (ICA)
12750 Merit Drive, Suite 710 LB-89
Dallas, TX 75251

Educational Technology
Educational Technology Publications, Inc.
700 Palisade Avenue
Englewood Cliffs, NJ 07632

Educational Technology: Research and Development
Association for Educational Communications & Technology (AECT)
1025 Vermont Avenue, N.W.
Washington, DC 20005

Human Communication Research
International Communication Association (ICA)
P.O. Box 9589
Austin, TX 78766

Human Resources Development Quarterly
American Society for Training & Development (ASTD)
1630 Duke Street
Alexandria, VA 22313

Instruction Delivery Systems
Society for Applied Learning Technology (SALT)
50 Culpeper Street
Warrenton, VA 22186

Journal of Computer-based Instruction
Association for the Development of Computer-based Instructional Systems (ADCIS)
1601 West Fifth Avenue
Suite 111
Columbus, OH 43212

Journal of Interactive Instruction Development
Society for Applied Learning Technology (SALT)
50 Culpeper Street
Warrenton, VA 22186

MPC World (multimedia computing)
PC World Communications
524 Second Street
San Francisco, CA 94107

Multimedia and Videodisc Monitor
Future Systems, Inc.
P.O. Box 26
Falls Church, VA 22040

Multimedia Review
Meckler Corporation
11 Ferry Lane West
Westport, CT 06880

Multimedia Solutions
IBM
4111 Northside Parkway HO4L1
Atlanta, GA 30327

National Association of Desktop Publishers Journal
Desktop Publishing Institute
462 Old Boston Street
Topsfield, MA 01983

NewMedia
P.O. Box 5948
San Mateo, CA 94402-9847

Performance & Instruction
Performance Improvement Quarterly
National Society for Performance & Instruction
1300 L Street, N.W., Suite 1250
Washington, DC 20005

Public Relations Journal
Public Relations Society of America (PRSA)
33 Irving Place, 3rd Floor
New York, NY 10003-2376

Publish
Integrated Media
501 Second Street
San Francisco, CA 94107

T.H.E. Journal (Technological Horizons in Education)
150 El Camino Real, Suite 112
Tiston, CA 92680-3670

Tech Trends
Association for Educational Communications & Technology (AECT)
1025 Vermont Avenue, N.W.
Washington, DC 20005

Technical & Skills Training
American Society for Training & Development (ASTD)
1630 Duke Street
Alexandria, VA 22313

Technical Communication
Society for Technical Communication (STC)
901 N. Stuart Street, Suite 304
Arlington, VA 22203

Training
Lakewood Publishing
50 South Ninth Street
Minneapolis, MN 55402

Training & Development
American Society for Training & Development (ASTD)
1630 Duke Street
Alexandria, VA 22313

Videography
PSN Publications
2 Park Avenue, Suite 1820
New York, NY 10016

Video Systems
Intertec Publishing Co.
9221 Quivira Road
Overland Park, KS 66215

Index

Page numbers followed by "t" and "f" indicate tables and figures, respectively.